Microscale General Chemistry Laboratory

With Selected Macroscale Experiments

Microscale General Chemistry Laboratory

With Selected Macroscale Experiments

Second Edition

Zvi Szafran

Department of Chemistry
New England College

Ronald M. Pike

Professor of Chemistry, Emeritus
Merrimack College

Judith C. Foster

Department of Chemistry
Bowdoin College

John Wiley & Sons, Inc.

Executive Editor	David Harris
Acquisitions Editor	Jennifer Yee
Production Editor	Sandra Russell
Marketing Manager	Robert Smith
Designer	Kevin Murphy
Production Management Services	Techbooks, Inc.
Cover Photo	©Charlotte Raymond/Photo Researchers

This book was typeset in 10/12 Souvenir Light by TechBooks, Inc. and printed and bound by Courier Westford. The cover was printed The Lehigh Press, Inc.

The paper in this book was manufactured by a mill whose forest management programs include sustained yield harvesting of its timberlands. Sustained yield harvesting principles ensure that the number of trees cut each year does not exceed the amount of new growth.

This book is printed on acid-free paper. ♾

Szafran, Zvi., Ronald M. Pike, Judith C. Foster

ISBN 0-471-20207-X (paper: alk. paper)

Printed in the United States of America.

10 9 8 7 6 5 4 3 2 1

Preface, Second Edition

It was almost twenty years ago that the first laboratory sections taught at the microscale level were conducted in the organic chemistry laboratories at Merrimack College and Bowdoin College in 1983. Today, microscale chemistry is commonly used in general chemistry, organic chemistry, inorganic chemistry and analytical chemistry. It is used in more than 2000 colleges and universities throughout the world, with microscale centers in place in such diverse countries as Mexico, Austria, Sweden, Finland, Australia, the People's Republic of China, and of course, the United States. Papers on microscale chemistry are routinely published in the leading journals on Chemical Education. The microscale revolution seedling has now come into full bloom.

At the microscale level, we use much smaller quantities of both chemical reagents and solvents than were previously common—often less than 1%. Typically, this means reducing quantities from 10–20 g of solid to about 100–200 mg, and from 100–250 mL of solvent to about 1 mL. This has a number of important consequences in the laboratory:

- It minimizes exposure to potentially dangerous chemicals in the lab
- It ensures that the laboratory air quality is as good as possible
- It cuts the amount of chemical waste produced by more than 99%
- It lowers the costs of operating a chemistry laboratory
- It sharply reduces the chances of a fire or explosion

We can use smaller quantities of chemicals because analytical techniques in chemistry have advanced such that only milligrams of a chemical are needed for complete analysis. The result is a laboratory which is safer and more healthful, and that minimizes harm to the environment.

We would like to thank Victor Obendrauf (University of Graz, Austria) for introducing us to his interesting work in microscale gas generation (see Experiment 12), and Dr. Mono M. Singh, Director of the National Microscale Chemistry Center (Merrimack College, North Andover, MA) for his insights, good humor and general support. We would further like to thank our editor at John Wiley and Sons, Jennifer Yee.

As is always the case, although we have repeatedly tested all the laboratory experiments, problems may still arise through use of this manual. We would appreciate hearing about them! We're glad you are part of this ongoing enterprise.

Zvi Szafran

Ronald M. Pike

Judith C. Foster

Henniker, New Hampshire
November, 2001

Table of Contents

Chapter 9: Inorganic Qualitative Analysis /383

Chapter 10: Organic Chemistry /459

Appendices /495

Glossary /511

Chapter 1 Safety in the Laboratory

1.1 GENERAL SAFETY RULES

Introduction

Safety in the laboratory is a subject of the utmost importance. Since all chemicals are harmful to some degree, the best way to ensure safety in the laboratory is by minimizing contact with all chemicals. Thus, the main way in which we promote safety in the General Chemistry Laboratory is by using the microscale technique. This technique lets us reduce the amounts of chemicals used by a factor of 100 to 1000 from the traditional multigram scale previously used in introductory chemical laboratories. There are several advantages to doing this, many of which are safety related:

- Less toxic waste is generated. This saves money on disposal costs, and helps protect both the chemist involved (you!), your instructor, and the environment.
- Chances of fire or explosion are markedly reduced.
- The air quality in the laboratory is improved. Using smaller amounts of volatile compounds cuts down sharply on the amounts of chemicals present in the air, improving both the smell and the healthfulness of the laboratory.
- Exposure to potentially dangerous chemicals is minimized.

At this point, it may seem to you that the adoption of the microscale approach eliminates all the risks involved in a chemical laboratory. It is certainly true that using microscale techniques will minimize the risks. However, even using small amounts of material, some chemicals are still highly toxic, spills or splattering of a corrosive material still can occur, or a compound may still decompose to generate a noxious gas. Furthermore, in future work, a required scale-up of a reaction may be necessary. It may be that you need more than a micro amount of a particular product. For these reasons, plus the fact that it just makes plain sense, it is prudent for each of us to be aware of several safety regulations concerning work in a chemical laboratory. It should be emphasized that, as an individual, you have an obligation to protect yourself and your fellow workers.

Before the Laboratory

Safety in the laboratory does not begin when you walk in the laboratory door. There are three initial steps that should be carried out before the experiment begins.

1. Read the directions of the experiment to be carried out with a critical eye, **in advance of the laboratory.**
2. **Think about what you are reading and visualize the experimental sequence as to the chemicals used and the arrangement of equipment.** Especially note any safety warnings such as "Use the hood" or "No flames allowed." The safety warnings in this laboratory text are there for your and your neighbors' protection. Many people find it useful to prepare a flow chart for each experiment, listing each step of the laboratory procedure in sequence.
3. You may wish to check the toxicities of the chemicals involved. Safety data, such as Material Safety Data Sheets (MSDS) or the Merck Index, discussed below, is most likely available in the laboratory or posted in the stockroom area. Your instructor may also provide information on the chemicals to be used.

Dressing for the Laboratory

To enhance safety in the laboratory, there are several things that should be considered with regard to what you wear.

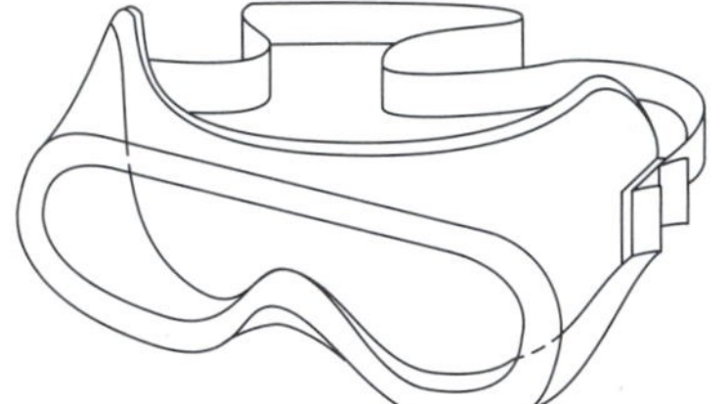

Figure 1.1 *Safety goggles.*

1. **Safety goggles** (Figure 1.1) or suitable protective glasses *are crucial.* They should be worn at all times when experimentation is taking place anywhere in the laboratory. Any visitors to the laboratory should also be required to have suitable protection. Contact lenses *are not* to be worn, as corrosive fumes or chemicals may get underneath them and prevent effective breathing and flushing of the eyes, if an accident should occur. Learn the location of the eyewash fountain (Figure 1.2) and safety shower (Figure 1.3) in your laboratory.
2. Open-toed shoes, canvas sneakers, or sandals offer no protection to the feet from chemical spills. *They should never be worn in the laboratory.*
3. Long hair should be tied back, and if ties or similar loosely hanging clothing are worn, they should be tucked in an appropriate manner.
4. Clothing that offers protection against an accidental spill is most appropriate. Suitable laboratory aprons or coats are highly recommended. Similarly, clothing that leaves the body exposed (such as a cut-off T-Shirt, short skirts, or shorts) should never be worn. Wear nonsynthetic clothing (*cotton or in the winter, wool,* if you are in a really cold climate. *Leather* is also suitable). In the case of fire, synthetic material such as nylon or spandex ignites easily, and will drip and also stick to the skin!
5. Use impermeable gloves when necessary. Do not use latex gloves. They are permeable, and some individuals are allergic to them. Wash your hands often and especially after leaving the laboratory session.

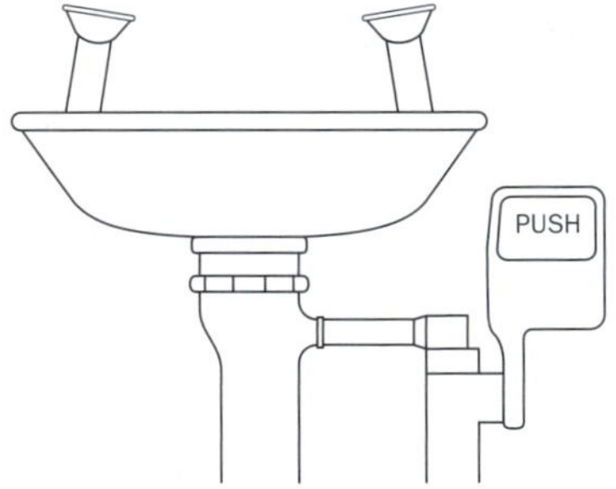

Figure 1.2 *Eyewash fountain.*

Safety Rules In the Laboratory

Specific rules and regulations have been developed based on the experience of those who have extensively studied the safety aspects of the laboratory. It is imperative that you learn these safety rules and follow them at all times. They will form a large part of your code of conduct in the laboratory.

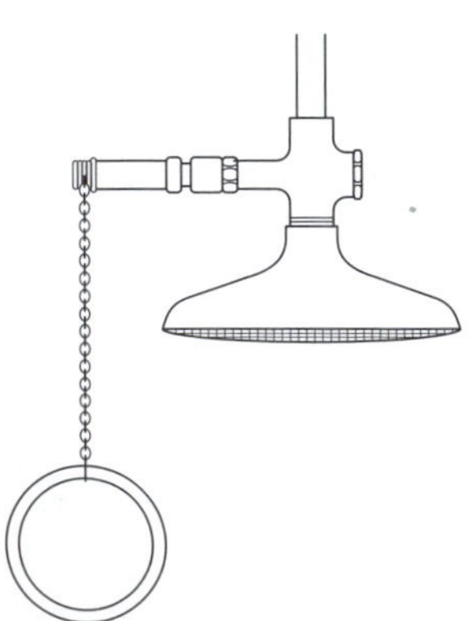

Figure 1.3 *Safety shower.*

1. **Use your common sense.** Think before you act!! *Attention to detail is critical.*

2. **Don't rush, and don't take short cuts.** If you rush your work, at best you will get poor results. At worst, you will be a danger to yourself and to those around you.
3. **Report any spill or accident *immediately*** to your instructor. Medical assistance may then be obtained, if necessary. Telephone numbers should be posted on where to call for aid. Above all—*do not panic!*
4. **Know the location and operation of safety equipment** in the laboratory from the very first meeting of the laboratory course. This equipment includes:

   - eyewash fountains (Figure 1.2)
   - safety showers (Figure 1.3)
   - fire extinguishers (Figure 1.4)
   - fire blankets (Figure 1.5)
   - first aid kits
   - fire exits

5. **Smoking is absolutely forbidden in the laboratory.** Volatile, flammable solvents can ignite easily and result in an explosion or fire. Severe burns can result from carelessness due to smoking or the use of an open flame in the vicinity of flammable solvents. The microscale laboratory markedly reduces the possibility of this aspect of potential injury but we always must be on our guard.
6. **Never work alone** in a chemical laboratory. As in swimming, apply the "buddy" system.

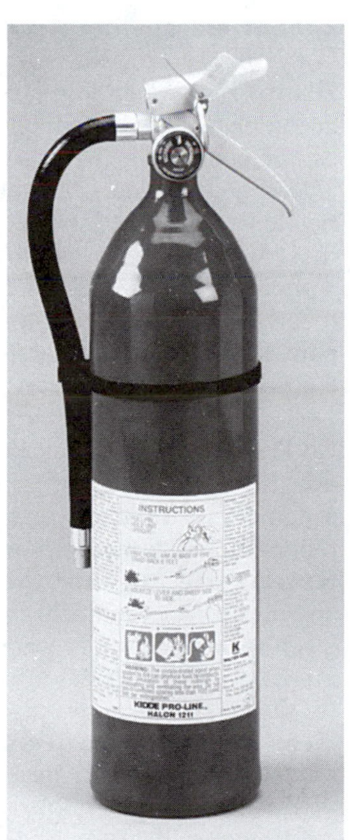

Figure 1.4 *Fire extinguisher. (Courtesy of Fisher Scientific, 711 Forbes Ave., Pittsburgh, PA 15219.)*

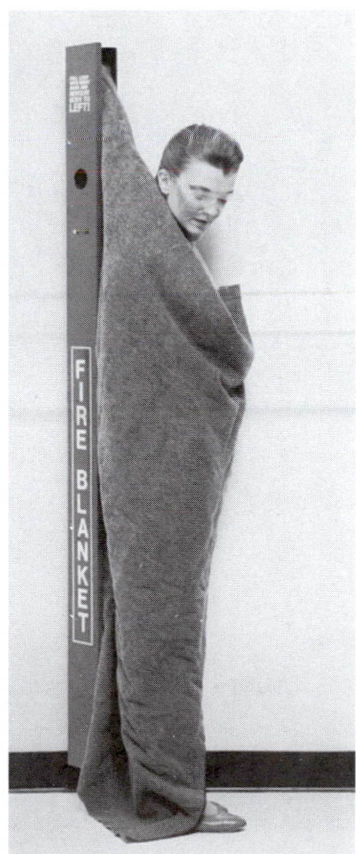

Figure 1.5 *Fire blanket. (Courtesy of Fisher Scientific, 711 Forbes Ave., Pittsburgh, PA 15219.)*

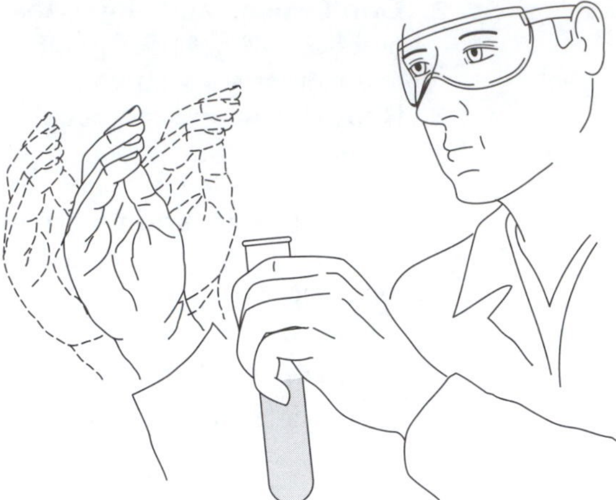

Figure 1.6 *The proper way to "smell" chemical vapors.*

7. **Minimize exposure to all chemicals,** whether they are considered toxic or not. Handle all chemicals according to any specific directions on the container, or those given to you by your instructor. Never pick up spilled solids with your bare hands. **We repeat:** Use impermeable gloves when necessary.
8. **Never directly smell a chemical.** Use your hand to waft a tiny amount of vapor toward your nose (Figure 1.6).
9. **Do not put anything in your mouth** under any circumstances while in the laboratory. This rule pertains to *food, drink, and pipets.*
10. **Dispose of chemicals properly,** in the containers provided, and according to the instructions given by the laboratory instructor. Do not simply pour waste chemicals down the sink. Recent governmental regulations have placed stringent rules on industrial and academic laboratories for the proper disposal of chemicals. Severe penalties are levied on those who do not follow proper waste disposal procedures.
11. **Special care should be taken when working with glass.** Always follow proper procedures for assembling and dismantling a particular apparatus. Your instructor will introduce you to the techniques required. **Be particularly careful** when inserting thermometers into rubber stoppers. *Be sure the stopper is lubricated with water or glycerin.* When working with glass, breakage may occur and a cut or laceration may result. Report any accident to your instructor immediately, no matter how insignificant it may seem. Quick treatment can often prevent infection or other complications from occurring.
12. **Never carry out unauthorized experiments.** At this stage in your chemical development, it is imperative that you follow the procedures given in the laboratory manual for basic safety reasons. Immediate expulsion from the laboratory and failure of the course is usually the penalty for disobeying this rule.
13. **Keep your laboratory space clean.** This also pertains to the balance area and where the chemicals are dispensed. You or your fellow students unknowingly can be burned or exposed to toxic chemicals if you do not clean up a spill.
14. **Replace caps on containers immediately after use.** An open container is an invitation for a spill. Furthermore, some reagents are very sensitive to moisture, and may decompose if left open. **Never return**

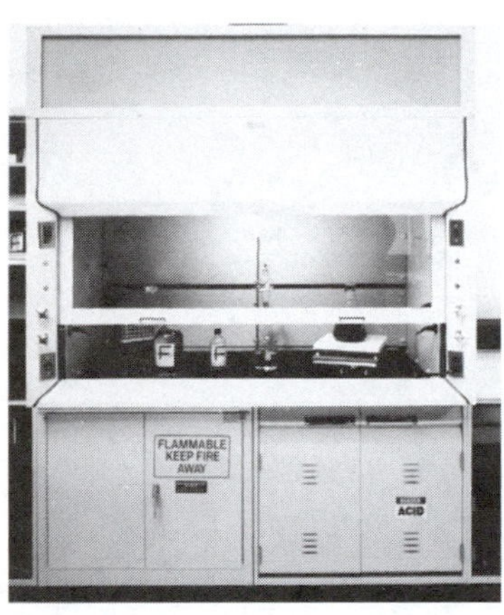

Figure 1.7 *Chemical Fume Hood. (Courtesty of Fisher Scientific, 711 Forbes Ave., Pittsburgh, PA 15219.)*

an unused portion of a reagent to its original container. Suitable disposal sites will be designated in the laboratory for chemical wastes.

15. **Never heat a closed system.** Always provide a vent to avoid an explosion. Provide a suitable trap for any toxic gases generated such as sulfur dioxide, hydrogen chloride, and chlorine. Directions will be found in the experimental procedures.
16. Use a **HOOD** (Figure 1.7) when required. The experimental procedures in the Text will indicate when a **HOOD** is to be used.
17. **Do not engage in games or horseplay** in the laboratory.
18. **Never leave a lighted burner unattended.**
19. **Never add water to a concentrated reagent when diluting the reagent.** Always add the reagent to the water. If done in reverse, local heating and density differences may cause the solution to splash back on you.
20. **Never remove any chemical substance from the laboratory.** This is grounds for expulsion or other severe disciplinary action.
21. **Know where to go** in case of injury or to evacuate the building.

1.2 PLANNING FOR CHEMICAL SAFETY

Use of MSD Sheets

Many chemicals, as pointed out above, have dangers associated with their use. One way of combating these dangers is by the use of Material Safety Data Sheets (MSDS), which are provided by each manufacturer or vendor *as required by law* for the chemicals purchased and used in your laboratory. The information given relates to the risks involved when using a specific chemical. These sheets are available to you as a laboratory worker.

A typical MSD sheet is shown in Figure 1.8 for sodium chloride. The sheet is divided into several sections: Identification, Toxicity Hazards, Health Hazard Data, Physical Data, Fire and Explosion Data, Reactivity Data, Spill or Leak Procedures, and Additional Precautions and Comments. The identification section provides additional names by which the compound is known (salt, for example), the CAS (Chemical Abstract Service) number, and the Sigma or Aldrich catalog product number. The CAS number is especially useful, as one can access several databases using this number to obtain listings of papers and books in which this compound is discussed and used.

The **Toxicity Hazards** section contains results of studies detailing the toxicity of the compound in various animal and inhalation tests. Sodium chloride is a well-studied compound, so many such tests have been performed. There are several common abbreviations used:

HMN	Human
IVN	Intravenous (in the blood stream)
LD50	The dose with which 50% of the test subjects will die
LDLo	Lowest lethal dose
MUS	Mouse
ORL	Oral dose
SKN	Skin

Thus, the listing ORL-RAT LD50: 3000 mg/kg indicates that when sodium chloride was given via oral dose to a test sampling of rats, the dose that would kill 50% of the rats was 3000 mg per kg of the rat's weight. If one could extrapolate directly from a rat to a human being, it would require an oral dose of 240 g of sodium chloride to kill half of a random group of 80 kg humans.

MATERIAL SAFETY DATA SHEET
Sigma-Aldrich Corporation
1001 West Saint Paul Ave, Milwaukee, WI 53233 USA

------------------ IDENTIFICATION ------------------

PRODUCT #: S9888 NAME: SODIUM CHLORIDE ACS REAGENT
CAS#: 7647-14.5
MF: CL1NA1

SYNONYMS
COMMON SALT * DENDRITIS * EXTRA FINE 200 SALT * EXTRA FINE 325 SALT* HALITE * H.G. BLENDING * NATRIUMCHLORID (GERMAN) * PUREX * ROCK SALT * SALINE * SALT * SEA SALT * STERLING * TABLE SALT * TOP FLAKE * USP SODIUM CHLORIDE * WHITE CRYSTAL *

------------------ TOXICITY HAZARDS ------------------

RTECS NO: VZ4725000
SODIUM CHLORIDE

IRRITATION DATA

SKN-RBT 50 MG/24H MLD	BIOFX* 20-3/71
SKN-RBT 500 MG/24H MLD	28ZPAK -,7,72
EYE-RBT 100 MG MLD	BIOFX* 20-3/71
EYE-RBT 100 MG/24H MOD	28ZPAK -,7,72
EYE-RBT 10 MG MOD	TXAPA9 55,501,80

TOXICITY DATA

ORL-RAT LD50:3000 MG/KG	TXAPA9 20,57,71
ORL-MUS LD50:4000 MG/KG	FRPPA0 27,19,72
IPR-MUS LD50:6614 MG/KG	COREAF 256,1043,63
SCU-MUS LD50:3 GM/KG	ARZNAD 7, 445,57
IVN-MUS LD50:645 MG/KG	ARZNAD 7,445,57
ICV-MUS LD50:131 MG/KG	TYKNAQ 27,131,80

REVIEWS, STANDARDS, AND REGULATIONS
EPA GENETOX PROGRAM 1988, NEGATIVE: IN VITRO CYTOGENETICS-NONHUMAN; SPERM MORPHOLOGY-MOUSE
EPA GENETOX PROGRAM 1988, INCONCLUSIVE: MAMMALIAN MICRONUCLEUS
EPA TSCA CHEMICAL INVENTORY, 1986
EPA TSCA TEST SUBMISSION (TSCATS) DATA BASE, JANUARY 1989
MEETS CRITERIA FOR PROPOSED OSHA MEDICAL RECORDS RULE FEREAC 47,30420, 82

GET ORGAN DATA
MATERNAL EFFECTS (OVARIES, FALLOPIAN TUBES)
EFFECTS ON FERTILITY (PRE-IMPLANTATION MORTALITY)
EFFECTS ON FERTILITY (POST-IMPLANTATION MORTALITY)
EFFECTS ON FERTILITY (ABORTION)
EFFECTS ON EMBRYO OR FETUS (FETOTOXICITY)
EFFECTS ON EMBRYO OR FETUS (FETOTOXICITY)
SPECIFIC DEVELOPMENTAL ABNORMALITIES (MUSCULOSKELETAL SYSTEM)

------------------ HEALTH HAZARD DATA ------------------

ACUTE EFFECTS
MAY BE HARMFUL BY INHALATION, INGESTION, OR SKIN ABSORPTION.
CAUSES EYE IRRITATION.
CAUSES SKIN IRRITATION.
MATERIAL IS IRRITATING TO MUCOUS MEMBRANES AND UPPER RESPIRATORY TRACT.

FIRST AID
IN CASE OF CONTACT, IMMEDIATELY FLUSH EYES WITH COPIOUS AMOUNTS OF WATER FOR AT LEAST 15 MINUTES.
IN CASE OF CONTACT, IMMEDIATELY WASH SKIN WITH SOAP AND COPIOUS AMOUNTS OF WATER.
IF INHALED, REMOVE TO FRESH AIR. IF NOT BREATING GIVE ARTIFICIAL RESPIRATION. IF BREATING IS DIFFICULT, GIVE OXYGEN. CALL A PHYSICIAN.

------------------ PHYSICAL DATA ------------------

MELTING PT: 801 C
SPECIFIC GRAVITY: 2.165

APPEARANCE AND ODOR
WHITE CRYSTALLINE POWDER

------------------ FIRE AND EXPLOSION HAZARD DATA ------------------

EXTINGUISHING MEDIA
NON-COMBUSTIBLE.
USE EXTINGUISHING MEDA APPROPRIATE TO SURROUNDING FIRE CONDITIONS.

SPECIAL FIREFIGHTING PROCEDURES
WEAR SELF-CONTAINED BREATHING APPARATUS AND PROTECTIVE CLOTHING TO PREVENT CONTACT WITH SKIN AND EYES.

------------------ REACTIVITY DATA ------------------

INCOMPATIBILITIES
STRONG OXIDIZING AGENTS
STRONG ACIDS

HAZARDOUS COMBUSTION OR DECOMPOSITION PRODUCTS
NATURE OF DECOMPOSITION PRODUCTS NOT KNOWN

------------------ SPILL OR LEAK PROCEDURES ------------------

STEPS TO BE TAKEN IF MATERIAL IS RELEASED OR SPILLED
WEAR RESPIRATOR, CHEMICAL SAFETY GOGGLES, RUBBER BOOTS AND HEAVY RUBBER GLOVES.
SWEEP UP, PLACE IN A BAG AND HOLD FOR WASTE DISPOSAL.
AVOID RAISING DUST.
VENTILATE AREA AND WASH SPILL SITE AFTER MATERIAL PICKUP IS COMPLETE.

WASTE DISPOSAL METHOD
FOR SMALL QUANTITIES: CAUTIOUSLY ADD TO A LARGE STIRRED EXCESS OF WATER. ADJUST THE PH TO NEUTRAL, SEPARATE ANY INSOLUBLE SOLIDS OR LIQUIDS AND PACKAGE THEM FOR HAZARDOUS-WASTE DISPOSAL. FLUSH THE AQUEOUS SOLUTION DOWN THE DRAIN WITH PLENTY OF WATER. THE HYDROLYSIS AND NEUTRALIZATION REACTIONS MAY GENERAL HEAT AND FUMES WHICH CAN BE CONTROLLED BY THE RATE OF ADDITION.
OBSERVCE ALL FEDERAL, STATE, AND LOCAL LAWS.

— PRECAUTIONS TO BE TAKEN IN HANDLING AND STORAGE —

CHEMICAL SAFETY GOGGLES.
USE PROTECTIVE CLOTHING, GLOVES AND MASK.
SAFETY SHOWER AND EYE BATH.
MECHANICAL EXHAUST REQUIRED.
DO NOT BREATHE DUST.
DO NOT GET IN EYES, ON SKIN, ON CLOTHING.
WASH THOROUGHLY AFTER HANDLING.
IRRITANT.
KEEP TIGHTLY CLOSED.
HYGROSCOPIC
STORE IN A COOL DRY PLACE.

------ ADDITIONAL PRECAUTIONS AND COMMENTS ------

SECTION 9 FOOTNOTES
REACTS VIOLENTLY WITH BROMINE TRIFLUORIDE AND LITHIUM.
THE ABOVE INFORMATION IS BELIEVED TO BE CORRECT BUT DOES NOT PURPORT TO BE ALL INCLUSIVE AND SHALL BE USED ONLY AS A GUIDE. SIGMA-ALDRICH SHALL NOT BE HELD LIABLE FOR ANY DAMAGE RESULTING FROM HANDLING OR FROM CONTACT WITH THE ABOVE PRODUCT. SEE REVERSE SIDE OF INVOICE OR PACKING SLIP FOR ADDITIONAL TERMS AND CONDITIONS OF SALE

Figure 1.8 *Material safety data sheet: sodium chloride. (Reprinted with permission of Aldrich Chemical Co., Inc., Milwaukee, WI.)*

Needless to say, this is well above the amount of sodium chloride one would expect to ingest by accident in a laboratory. We can conclude that sodium chloride is not very risky in this regard.

The **Health Hazard Data** section indicates that inhalation, ingestion, or skin absorption may be harmful, and that the material is irritating to mucous membranes and the upper respiratory tract. This may be surprising for as "innocent" a material as salt, but it is certainly well enough known that salt will sting the eyes, and that prolonged exposure of the skin to salt water can be harmful. While spilling a small amount of sodium chloride on the skin would not be harmful, this warning illustrates the general principle of trying to minimize contact with any chemical. The section also gives the treatment for having contact with salt in the eyes: flushing with water for at least 15 minutes.

The **Fire and Explosion Hazard Data** and **Reactivity Data** sections provide information about chemical incompatibilities and other chemical reaction dangers. We are told that sodium chloride does not combust, and that it may react with strong oxidizing agents or strong acids.

The **Spill or Leak Procedures** sections give the steps to be taken if material is released or spilled, and generally refer to large, industrial amounts. Specific information will be provided in the experimental procedures for materials with unusual handling characteristics. Waste disposal methods are also given. Again, they generally refer to quantities much larger than those used in these laboratory experiments. Also, the methods are designed for materials that are less than pure. Pure sodium chloride can be added to water, and the solution will be neutral. This is not necessarily true of industrial grades of sodium chloride, so that care should be indicated.

The **Handling and Storage** section gives some practical advice in how to deal with the compound, as well as recommendations about safety equipment that should be on hand (shower, eye bath).

Finally, the **Additional Precautions and Comments** section details specific dangers associated with this compound. Sodium chloride is known to react violently with lithium or bromine trifluoride under certain conditions. These materials must never be used in the same reaction step.

It may seem to you that the MSD sheet is too detailed. This is certainly true for the microscale usage of sodium chloride, but keep in mind that these sheets are designed for many different kinds of use. A judicious reading of the sheets will provide the chemist with much useful information, and you will quickly learn what aspects of safety to "focus in" on. It is much better to have the detailed information and not to need it than to be in the opposite predicament.

The Merck Index

Similar information to a MSD sheet in a more compact form can be found in the Merck Index (Merck and Co., Rahway, NJ). This basic reference work gives the "bottom line" on the toxicity of chemicals, and their incompatibilities. In the case of sodium chloride (see Figure 1.9), the index lists under Human Toxicity: "Not generally considered poisonous. Accidental substitution of NaCl for lactose in baby formulas has caused fatal poisoning." While the information in the Merck Index is not as complete as on the MSD sheets, it is generally sufficient for our purposes at the microscale laboratory level. The Index also supplies some interesting information about the common usages of the chemicals listed, with a special emphasis on medical usages. References to the chemical literature are also provided.

CRC Handbook of Chemistry and Physics

This reference work, which is updated each year, contains a wide range of data (in table form) in the area of health, safety, and environmental protection.

8430. Sodium Chloride, Salt; common salt. NaCl; mol wt 58.45. Cl 60.66%, Na 39.34%. NaCl. The article of commerce is also known as *table salt, rock salt* or *sea salt.* Occurs in nature as the mineral halite. Produced by mining (rock salt), by evaporation of brine from underground salt deposits and from sea water by solar evaporation: Faith, Keyes & Clark's *Industrial Chemicals*, F. A. Lowenheim, M. K. Moran, Eds. (Wiley-Interscience, New York, 4th ed., 1975) pp 733–730. Comprehensive monograph: D. W. Kaufmann, *Sodium Chloride*, ACS monograph Series no. 145 (Reinhold, New York, 1960) 743 pp.

Cubic, white crystals, granules, or powder; colorless and transparent or translucent when in large crystals. d 2.17. The salt of commerce usually contains some calcium and magnesium chlorides which absorb moisture and make it cake. mp 804° and begins to volatilize at a little above this temp. One gram dissolves in 2.8 ml water at 25°, in 2.6 ml boiling water, in 10 ml glycerol; very slightly sol in alcohol. Its soly in water is decreased by HCl and it is almost insol in concd HCl. Its aq soln is neutral. pH: 6.7–7.3 d of said aw soln at 25° is 1.202. A 23% aq soln of sodium chloride freezes at −20.5°C (5°F). LD_{50} orally in rates: 3.75 g/kg, Boyd, Shanas, *Arch. Int. Pharmacodyn. Ther.* **144,** 86 (1963).

Note: Blusalt, a brand of soldium chloride contg trace amounts of cobalt, iodine, iron, copper, manganese, zinc is used in farm animals.

Human Toxicity: Not generally considered poisonous. Accidental substitution of NaCl for lactose in baby formulas has caused fatal poisoning.

USE: Natural salt is the source of chlorine and of sodium as well as of all, or practically all, their compds, e.g., hydrochloric acid, chlorates, sodium carbonate, hydroxide, etc; for preserving foods, manuf soap, dyes—to salt them out; in freezing mixtures; for dyeing and pritning fabrics, glazing pottery, curing hides; metallurgy of tin and other metals.

THERAP CAT: Electrolyte replenisher, emetic; topical antiinflammatory.

THERAP CAT (VET): Essential nutrient factor. may be given orally as emetic, stomachic, laxative or to stimulate thirst (prevention of calculi). Intravenously as isotonic solution to raise blood volume, to combat dehydration. Locally as wound irrigant, rectal douche.

Figure 1.9 Merck Index *listing: sodium chloride. [Reprinted by permission [10th ed., 1983] Merck & Co., Inc., Rahway, NJ.]*

It also includes directions for the handling and disposal of laboratory chemicals. Copies of this handbook are generally found in most laboratories.

Labeled Containers

Most chemical supply houses now label their containers with data not only showing the package size, physical properties, and chemical formula, but also **pictures** or **codes** that illustrate hazardous information. **Check each container for such information.**

1.3 FIRE SAFETY

Many solvents commonly used in the general chemistry laboratory are nonflammable. However, solvents such as ethanol, methanol, diethyl ether, and acetone are flammable. (Appendix A gives safety data for common organic solvents, including fire safety data.) Chemical fires can also occur. There is also a small risk (due to the use of electronic equipment) of an electrical fire due to a short circuit, frayed electrical cord or power surge.

If a fire should occur, the most important thing to remember is to keep calm. **Call your instructor.** Several types of equipment for dealing with small fires are contained in most laboratories. The most obvious is the fire extinguisher. Laboratory extinguishers (Figure 1.4) should weigh no more than 10 lbs., so as to be of convenient size to lift and employ rapidly. Ideally, there should be at least one fire extinguisher for every laboratory bench. Several types are available, the most common being of the dry chemical (bicarbonate powder under pressure) or compressed carbon dioxide type.

Most small fire extinguishers are activated by pointing the nozzle toward the base of the fire and squeezing the handle. A jet of compressed powder or foam will then discharge from the nozzle, smothering the fire. In some cases, it may be necessary to pull a pin from the handle before it can be squeezed. Some fire extinguishers only operate when turned upside-down. It is imperative that you become familiar with the proper use of the fire extinguishers located in the laboratory. In most cases of fire, you should leave the laboratory, and allow the instructor to take the necessary steps.

Alternate ways also exist for putting out small fires. Fires in small vessels can be extinguished by inverting a beaker or other such item over the burning vessel, thereby excluding oxygen. A second way of putting out such fires is by covering the vessel with soaking wet towels. Never use dry towels for this purpose. If a fire should occur, it is important to immediately remove any flammable material from the vicinity, especially bottles of flammable solvents and gas cylinders. If fire comes in contact with these items, an explosion can occur.

Whenever a fire occurs, there is also an associated danger of inhalation of smoke or toxic fumes. This is potentially more dangerous than the fire itself. Thus, if the air is not fit to breathe, the fire should be abandoned, and the fire department called. Any persons overcome by fumes should be removed to a well-ventilated area, and health professionals should be called immediately. If the fire is too large to contain, the area should be evacuated, and the fire department should be called immediately. The fire department should be apprised of the specific nature of the laboratory fire, so that the proper equipment can be brought.

In the event that your clothing should catch on fire, all laboratories should be equipped with safety showers (Figure 1.3) and fire blankets (Figure 1.5). To activate a safety shower, stand beneath it and pull down on the lever or chain. *Remain under the shower until you are thoroughly soaked*. To use a fire blanket, grasp the rope or material at the end of the blanket, and turn so that the blanket surrounds you tightly. This will smother the fire.

Name: ______________________________

Date: __________ Section: ______________________________

Questions

1. Draw a diagram of your laboratory and attach it to your report. Show your drawer location. Also locate the position of the following safety items:

eyewash stations	fire extinguishers	fire alarms	fire blankets
safety showers	first aid kits	safety exits	fume hood

2. Describe how to use the fire extinguisher found in your laboratory.

3. Is it wise to eat lunch in the laboratory? Why or why not?

4. Explain the danger of wearing open-toed sandals in the laboratory.

5. Obtain an MSD sheet for a chemical of your choice. On the sheet, underline the CAS #, the solubility data, fire and explosion data, reactivity data, and protective equipment required when using the chemical. Does your laboratory meet the safety requirements for using this chemical? Why or why not?

Name: ____________________

Date: ________ Section: ____________________

Safety Quiz

Complete the following questions and turn them in to your Instructor at the beginning of your laboratory session.

1. List three items that are essential for the laboratory in terms of clothing and/or protective wear.

2. If a spill of a corrosive chemical occurs, what is the first thing you should do?

3. Is it wise to work in the laboratory alone? Why or why not?

4. Where can you find the toxicity data of a chemical you are using?

5. Why is it prudent to keep the balance area clean in the laboratory?

6. Why is it recommended not to wear contact lenses while working in the laboratory?

7. Where is the best place to dispose of broken glass?

8. Should you return any unused portion of a reagent to its original container? Why or why not?

9. What is the purpose of a fire blanket?

GENERAL REFERENCES

1. American Chemical Society, *Safety in Academic Laboratories,* 6th ed., ACS Books: Washington, DC, 1995.
2. Lewis, R. J., Sr., *Hazardous Chemicals Desk Reference,* 4th ed., Van Nostrand Reinhold: New York, 1997.
3. American Chemical Society, *Less is Better (Laboratory Chemical Management for Waste Reduction)*, American Chemical Society: Washington, DC, 1993.
4. *Merck Index,* 12th ed., Merck and Co.: Rahway, NJ, 1996. See also *The Merk Index: CD-ROM Windows Version 12.3,* 12th ed., CRC Press: Boca Raton, FL, 2000.
5. *Handbook of Chemical Safety,* 4th ed., Furr, A. K. Jr., Ed.; CRC Press: Boca Raton, FL, 1995.
6. Lund, G.; Sansone, E. B., "Safe Disposal of Highly Reactive Chemicals," *J. Chem. Educ.* **1994,** 71, 972.
7. Amour M. A., *Hazardous Laboratory Chemicals Disposal Guide,* 2nd, ed., CRC Press: Boca Raton, FL, 1996.
8. Lide, D. R., Ed., *Handbook of Chemistry and Physics,* 81st ed., CRC Press: Boca Raton, FL, 2000. See also *CD-ROM Version:* Chapman & Hall, 2000.
9. Young, J. A., *Improving Safety in the Chemical Laboratory: A Practical Guide,* 2nd ed., Wiley: New York, 1991.
10. Mollinelli, R. P.; Reale, M. J.; Freudenthal, R. I., *Material Data Safety Sheets,* Hill & Gernett: Boca Raton, FL, 1992.
11. Gorman, C. E., Ed., *Working Safely With Chemicals in the Laboratory,* 2nd ed., Genuum Press: Schenectady, NY, 1995.
12. Hall, S. K., Ed., *Chemical Safety in the Laboratory,* CRC Press: Boca Raton, FL, 1994.
13. Committee on Prudent Practices for Handling, Storage, and Disposal of Chemicals in Laboratories, National Research Council, *Prudent Practices in the Laboratory: Handling and Disposal of Chemicals,* National Academy Press: Washington, DC 1995.
14. Alaimo, Robert J., Ed., *Handbook of Chemical Health and Safety,* Oxford Press, 2001. Chapter 34 is entitled "Health and Safety in the Microscale Chemistry Laboratory" by Zvi Szafran, Mono M. Singh, and Ronald M. Pike.
15. It is also recommended that one refer to the numerous articles on safety that appear regularly in the *Journal of Chemical Education* and *The Chemical Educator*.

Chapter 2
Numbers and Calculations

2.1 INTRODUCTION

Chemistry is a laboratory science. By this, we mean that most chemical discoveries and investigations are carried out in a laboratory through a process called the **scientific method,** whose steps are listed below:

1. A problem worth investigating is identified, and an experiment is designed to investigate all aspects of that problem.
2. Data (experimental measurements) are then gathered.
3. The data are then organized, analyzed and used to make generalizations called **laws.**
4. We seek tentative explanations as to why the laws work, which are called **hypotheses.**
5. The hypotheses are continuously tested. If they hold up, the hypotheses become **theories.**

It is obvious that the strength of the laws, hypotheses, and theories rely directly on the quality and quantity of data that we gather. This leads to some interesting questions:

- What is the best way of expressing the data so that it can be easily understood?
- How accurate are the experimental measurements?
- How reliable and reproducible are the methods by which the data is gathered?
- How much data needs to be gathered before we can formulate a law or draw a conclusion?
- What is the best way of displaying the data so that trends and relationships can be seen?

Each of these questions must be looked at before we can proceed with any experimental work.

2.2 SCIENTIFIC NOTATION

Numbers obtained in chemical measurements can span a very wide range of values. The mass of an oxygen molecule is 0.000000000000000000000053 g. The number of oxygen molecules in a mole is 602,000,000,000,000,000,000,000. Neither of these numbers is particularly convenient to work with, as they contain too many zeroes—it is too easy to write one zero too many or one too few. Therefore, we can easily see that in chemistry, it is not always convenient to express measurements using "plain" numbers.

Chemists often use a different way to write very large or very small numbers, called **scientific notation.** In scientific notation, numbers are written as powers of 10 in the form $a.aaa \times 10^z$. The number $a.aaa$ is called the **coefficient,** and z is called the **exponent.**

Consider the number 1000. This number is the product of $10 \times 10 \times 10$, and can therefore be written as 1×10^3. Note that this can be done most easily (and without factoring the number!) by moving the decimal point at the end of the number to the left until a number between 1 and 10 is obtained. The number of places to the left that the decimal point is moved is the value of the exponent (three in this case). Thus, we obtain the following representations in scientific notation for the numbers below:

$$145 = 1.45 \times 10^2 \quad 95{,}134 = 9.5134 \times 10^4 \quad 3{,}567{,}000 = 3.567 \times 10^6$$

If the number is less than one, the decimal point needs to be moved to the right until a number between 1 and 10 is obtained. The number of places to the right that the decimal point is moved is the *negative* of the value of the exponent. Consider the number 0.0015. We need to move the decimal point three places to the right. Thus, in scientific notation, the number 0.0015 becomes 1.5×10^{-3}. The general rule is: **Whenever the decimal point is moved to the left, the exponent goes up by one for each place moved. Whenever the decimal point is moved to the right, the exponent goes down by one.**

When adding or subtracting numbers in scientific notation, the numbers must all have their exponents to the same power. Suppose that we wish to add the numbers 3.51×10^3 and 1.2×10^2. We must write both as having the same power, so we must either change the 3.51×10^3 to 35.1×10^2 by moving the decimal point one place to the right, or alternatively, write 1.2×10^2 as 0.12×10^3 by moving the decimal point one place to the left.

Choosing the second option:

$$\begin{array}{r} 3.51 \times 10^3 \\ +\ \underline{0.12 \times 10^3} \\ 3.63 \times 10^3 \end{array}$$

Note that only the number is added or subtracted—the exponent remains the same.

When multiplying numbers in scientific notation, the coefficients are multiplied and the exponents are added. Thus, when 2×10^5 is multiplied by 3×10^3, the result is 6×10^8. Similarly, when dividing numbers in scientific notation, the coefficients are divided and the exponents are subtracted. Thus, when 9×10^6 is divided by 3×10^2, the result is 3×10^4.

2.3 ACCURACY, PRECISION, AND UNCERTAINTY

Accuracy is defined as how close an experimentally determined value is to the true value. Suppose that you were measuring the mass of a 1.00 g object, and obtained values of 0.98, 0.99, 0.98, 1.01, and 1.00 g. Your values each would be reasonably accurate, as they are all close to the true value of 1.00 g.

There is an uncertainty associated with any measurement, due to the accuracy with which the measuring device is calibrated and the care with which it is read. Typically, the uncertainty is ±0.5 in the last calibrated digit. Some glassware readings can be estimated with somewhat smaller uncertainty.

Measuring Device	*Uncertainty*
triple beam balance	±0.05 g
analytical balance	±0.00005 g
100 mL graduated cylinder	±0.5 mL
10 mL graduated cylinder	±0.05 mL
25 mL buret	±0.02 mL
25 mL volumetric flask	±0.02 mL
1 mL graduated pipet (.01 grad)	±0.002 mL

Precision refers to how reproducible your measurements are. Suppose that you had very carefully measured the length of a particular object using a ruler ten times, each time obtaining the result of 3.45 cm. The value 3.45 cm would be known with great precision.

Consider the following data, obtained by students A, B, and C, who each weighed the same object twelve times:

A	*B*	*C*
2.03	1.95	2.01
2.01	1.99	2.00
2.04	2.08	2.02
2.02	2.11	2.00
2.03	2.05	2.00
2.04	2.03	1.99
2.02	2.01	2.01
2.03	2.00	2.00
2.03	2.03	1.99
2.05	2.06	2.01
2.03	2.01	1.98
2.02	2.05	2.00

A simple approach to determine the precision of these measurements is to plot how many times each value of weight occurs (y axis) vs the weight (x axis), and see how tightly bunched (precise) the readings are. This plot is shown in Figure 2.1. We can see that A and C are more precise (more tightly bunched) than B.

We now also can answer the question, "How much data needs to be gathered before a law can be formulated or a conclusion can be drawn?" In cases A and C above, enough data points seem to have been gathered, as the data is reasonably centered about a mean value. In case B, more data is needed, as there are an insufficient number of data points to establish convergence.

Note that accuracy and precision do not mean the same thing—they are not related quantities. The ruler used in the previous paragraph could have

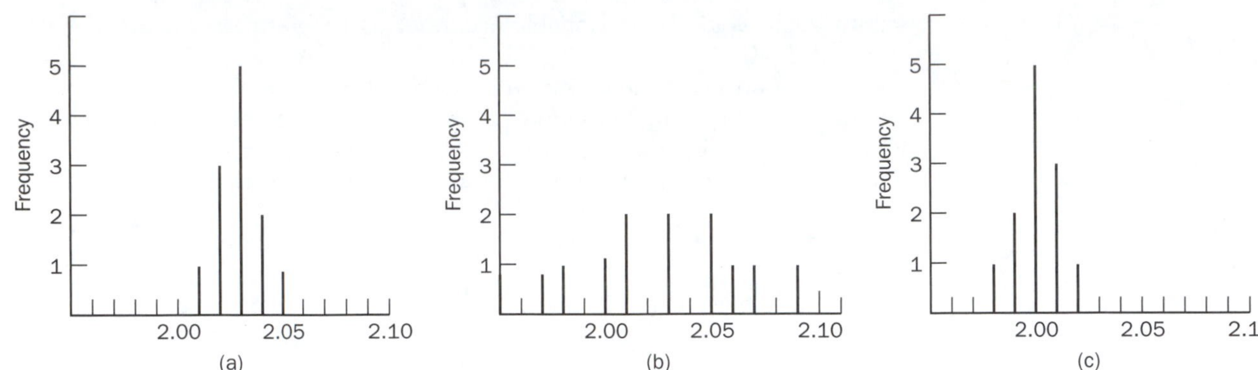

Figure 2.1 *Frequency plot for tabulated data.*

been warped, giving readings far from the true or accurate value. The measurements would still be reproducible, or precise.

2.4 STANDARD DEVIATION

The most common method of determining the precision of a series of measurements is called the **standard deviation, *s*.** The formal equation for standard deviation is

$$s = \sqrt{\frac{\Sigma (x - x_{av})^2}{n - 1}}$$

Here, the average value of the measurement (x_{av}) is obtained, and subtracted from each given measurement (x). The difference is then squared, and the squared differences are then summed up for each measurement. This is divided by $n - 1$, where n is the number of measurements. Finally, the square root is taken.

In practice, this method of taking the standard deviation is quite tedious, and a simpler method is available which is equivalent. In this approach, the equation is

$$s = \sqrt{\frac{\Sigma x^2 - \frac{(\Sigma x)^2}{n}}{n - 1}}$$

As an illustration, suppose that we weigh a number of soil samples, and obtain weights of 150, 162, 160, and 152 g. What is the standard deviation of our sample? Using the above equation, we first must add up the measurements, as well as their squares:

x	x^2
150	22,500
162	26,244
160	25,600
152	23,104
$\Sigma x = 624$	$\Sigma x^2 = 97{,}448$

Note that $n = 4$, since there have been four measurements.

Substituting into the above equation:

$$s = \sqrt{\frac{97{,}448 - \frac{(624)^2}{4}}{3}}$$

$$s = \sqrt{34.67} = 5.88 \text{ g}$$

Standard deviation is related to probability—there is a 68% chance of obtaining a value within one standard deviation of the average value, a 95% chance of obtaining a value within two standard deviations, and a 99.7% chance of obtaining a value within three standard deviations. Thus, of a group of 100 chemists, 68 of them would obtain a reading of 156 ± 5.88 g, 95 of them would obtain readings of 156 ± 11.8 g, etc.

In scientific measurements, we obviously strive for data that are both accurate and precise. How can we ensure that this sort of data is obtained? First, when a phenomenon is investigated, several duplicate trials will be obtained (helping our accuracy). Many times, the phenomenon will be investigated in more than one way, in case one type of measurement does not give an accurate value. There have been many cases where results obtained in one manner have been invalidated due to some external influence that the investigator did not take into consideration.

2.5 SIGNIFICANT FIGURES

The uncertainty in a measurement is usually indicated by the number of significant figures in the value. The more significant figures there are in a given number, the less uncertainty there is in that number. How, then, does one determine the number of significant figures in a given measurement?

In a laboratory measurement, the number of significant figures is the number of figures in the measurement that you are sure of. Suppose that you are measuring the volume of solution held in a graduated cylinder, which is calibrated to the tenth of a milliliter. You might take a reading of 62.54 mL, where you would estimate the final digit between the two hash marks on the cylinder. How many of these figures are truly significant? Only three of them have real certainty (62.5), whereas there is some uncertainty in the last digit. Properly, then, the number of significant figures in this measurement is three. It is common practice to include the first insignificant digit in calculations to overcome uncertainties with rounding off. When the calculation is completed (see below), however, the uncertain digits should be dropped.

Many times, a chemist is called upon to analyze other peoples' data. How do we know how many significant figures a number has in this case? For numbers with decimal points, the rule of thumb is that **all digits are significant except for zeroes to the left of the leftmost nonzero digit** (i.e., 0.0015 has two significant figures, not 4 or 5).

Numbers without decimal points are a little trickier. If they end in zero(es) (e.g., 12,500), are the zeroes significant or not? Normally, the assumption is that zeroes of this type are not significant, unless indicated in some manner. The most common ways to indicate a significant ending zero is to write a bar above the zero, or write the zero in boldface type (e.g., 12,5**00**). An easier way to indicate significance is to use scientific notation. If the 10's place zero were significant in 12,500, but not the 1's place zero, this number would be written as 1.250×10^4, making it clear that there are four significant figures. When reporting an average for several direct measurements, the number of significant figures in the average should not exceed the number of significant figures in the individual measurements.

In calculations, there are two major rules for significant figures, one for adding and subtracting, the other for multiplying and dividing. **When adding and subtracting numbers, retain all significant figures while calculating, then round off to the precision of the least precise value.**

Suppose you have recorded the relative amount of light absorbed by a series of solutions of known concentration. In performing calculations on these data, the concentrations and absorbances must first be added up:

Concentration	1.24×10^{-4}	3.00×10^{-4}	6.00×10^{-4}	1.10×10^{-3}
Absorbance	0.010	0.026	0.53	1.1

Converting the 1.10×10^{-3} to 11.0×10^{-4}, we add the concentrations, getting a result of 21.24×10^{-4}. The least precise of the values, however, is 11.0×10^{-4} (one digit to the right of the decimal point), so the sum is rounded to match, resulting in 21.2×10^{-4} or 2.12×10^{-3}. Likewise the sum of the absorbancies, properly rounded, is 1.7 (one digit to the right of the decimal point), not 1.666.

When subtracting, the answer may have fewer significant figures than the initial numbers. Let's say the mass of an empty filter crucible was 25.8367 g. It is used to collect a product, then dried and reweighed, and the final mass is 26.2391 g. These masses each have six significant figures, and each is precise to 0.0001. The difference is 0.4024 g, which has the same precision (four digits to the right of the decimal point), but only has four significant figures.

When multiplying and dividing, the number of significant figures in the answer is the same as the factor with the least number of significant figures. Assume that the product collected above has a molecular weight of 392.18 g mol^{-1}. To calculate the number of moles, the mass (0.4024 g) must be divided by the molecular weight, and the result is obtained as $1.026059462 \times 10^{-3}$. How many significant figures should be reported? Of the two original numbers, the mass has four significant figures, and the molecular weight has five, so the answer is properly reported as 1.026×10^{-3} (four significant figures).

When you have a combination of adding/subtracting and multiplying/dividing, both rules must be used. If we were to combine the two previous calculations into one equation, we would have

$$\frac{26.2391 - 25.8367 \text{ g}}{392.18 \text{ g mol}^{-1}}$$

Note that the numerator has six significant figures and the denominator has five, so it might seem that the answer should have five. However, note that the numerator has a subtraction step in it, and the result of the subtraction has only four significant figures. Thus, the answer should have only four significant figures.

2.6 GRAPHING OF DATA

A graph is often the preferred way to display analyzed chemical data, since it gives a visual representation of the material. It is often difficult to recognize relationships from the written data, whereas relationships are often easier to see in a graph.

A graph illustrates the way in which one property of a substance **(the dependent variable, *y*)** changes when some other property **(the independent variable, *x*)** undergoes a controlled change. Some examples encountered in experiments in this laboratory manual include a graph of how the volume (*y*) of a fixed amount of gas at a constant pressure changes as a function of the temperature (*x*) [Experiment 9: Part B], or a graph showing how absorbance (y) changes with concentration (x) at a fixed wavelength [Experiments 13–15].

There are several elements that make up a graph. Generally, graphs of chemical data use an x–y axis system, with the origin often being the point (0,0). The origin is not always included on graphs. If constructed by hand, graphs should be drawn on graph paper with a grid of at least 10 per cm. Various computer graphing programs are also available. Some instruments provide data directly as graphs.

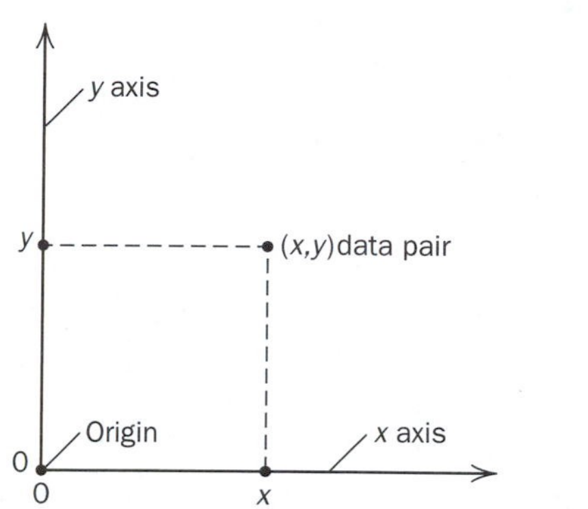

Figure 2.2 *Basic elements of a graph.*

In preparing a graph of some data, one must first lay out the graph coordinates. First, look at the range of x values and y values, and set the scale for each axis accordingly. The best graphs usually range from just below the smallest x- and y-values to just above the largest x and y values. It is often helpful to express all the values in the same power of ten for an axis, to avoid scale errors. Make sure that all your data will fit on the graph, and that the data points are distributed over the entire graph. Try to avoid bunching data so that it gives a vertical or horizontal line (unless you are trying to show that one of the variables is a constant, such as an equilibrium constant). The basic elements of a graph are shown in Figure 2.2.

Label the graph axes according to the property they represent, *including any units* (mol L^{-1}, °C, etc.). When dealing with scales containing powers of 10, it can be very confusing to write all the exponents on the scale. A more convenient method is to multiply all the data for that variable by some power of 10, then include that notation in the axis label. For example, suppose the x axis of a graph is "concentration of Fe^{2+}," with values of 2×10^{-4}, 4×10^{-4}, 6×10^{-4}, and 8×10^{-4} mol L^{-1}. It is more convenient to multiply all the values by 10^4. The scale would now have just the values 2, 4, 6, and 8 and the axis label would be "concentration of Fe^{2+}, mol $L^{-1} \times 10^4$" (meaning that the values on the axis are 10^4 times the actual values).

A data point is an (x, y) data pair, plotted on your graph. After plotting all data, a best-fit straight line or a smooth curve should be drawn (not a jagged point-to-point line). The line or curve does not necessarily have to touch all (or any) of the data points—it should go through a visual average of the points.

Straight-line graphs are the most useful in analyzing chemical data, as they provide for easy interpolation. Raw data often do not give a straight line when the two variables are graphed. The data can sometimes be linearized, or transformed, to give a linear relationship. Transformation is a mathematical manipulation of a variable (x), such as squaring it (x^2), taking its inverse ($1/x$), its logarithm ($\ln x$) or some other mathematical function. Sometimes, both variables need to be transformed to give a linear relationship. If "y" is the quantity being plotted on the y axis, and "x" is the quantity plotted on the x axis, the equation of the resulting line is

$$y = mx + b$$

where m is the slope of the line ($\Delta y/\Delta x$), and b is where the line will cross the y axis (the y intercept.)

Some examples are given here, taken from various experiments:

1. [Experiment 9: Part A] The pressure (P) and volume (V) are recorded for 0.100 mole of a gas at room temperature. When P (y axis) is plotted against V (x axis), a straight line is not obtained—a hyperbola is. This graph can be made linear by transforming P by taking its inverse, $1/P$, and plotting this against V. The equation of the resulting line is therefore

$$1/P = mV + b$$

From the graph below, it is easily seen that the y intercept is 0.

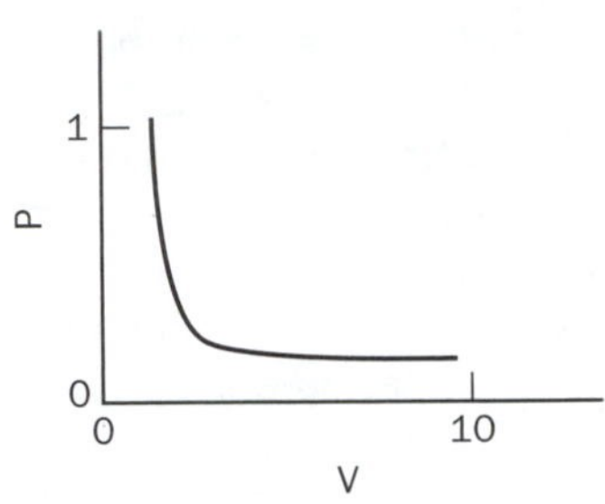

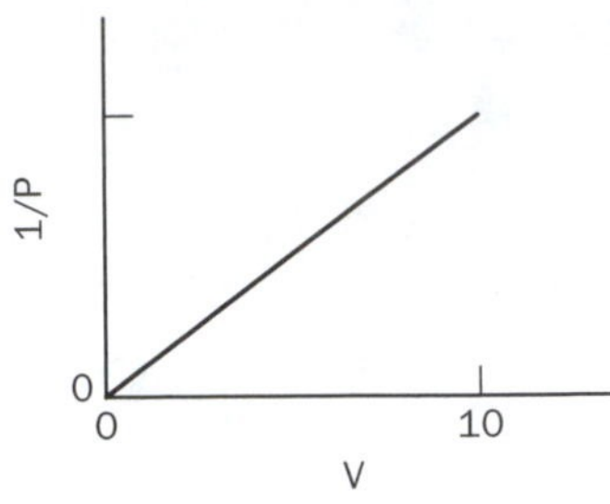

2. [Experiments 13–15] The percentage transmittance (%T) is recorded for several different concentrations of a solution. A plot of %T (y axis) versus concentration (x axis) is not linear. The graph can be linearized by converting %T to absorbance (abs = 2 − log %T) and plotting the absorbance (y axis) against concentration (x axis).

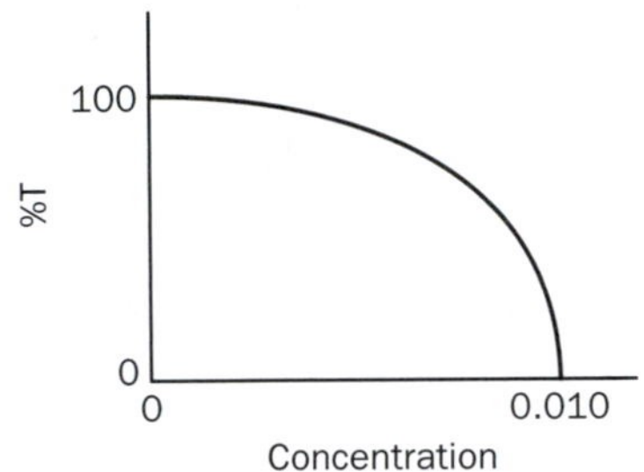

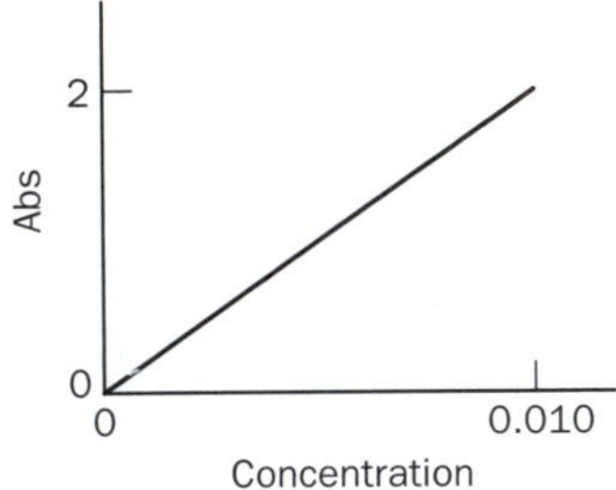

The resulting equation is

$$Abs = m\ (\text{concentration}) + b$$

3. [Experiment 26] The rate constant (k) is determined at several different temperatures in a kinetics experiment. The plot is made linear by transforming both variables and plotting log k (y axis) as a function of $1/T$.

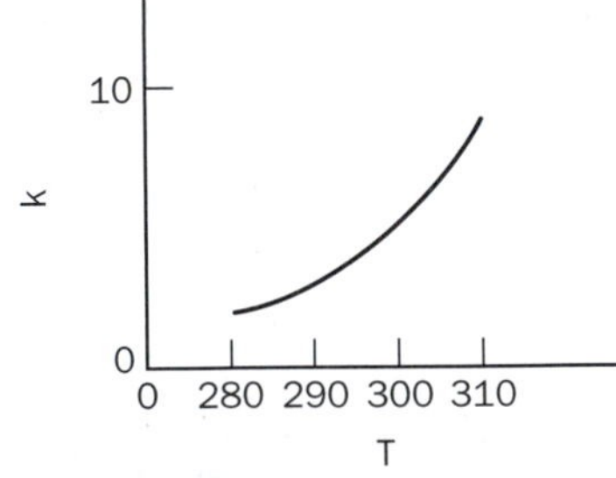

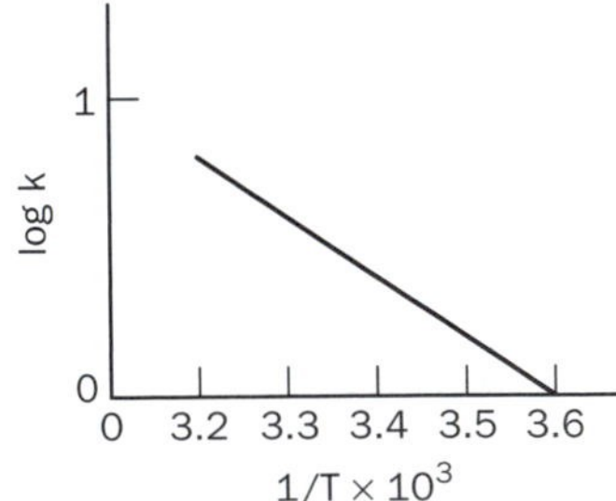

Some graphs cannot be made linear, as there is no transformation that will linearize the data. Any extrapolation must be done on the nonlinear graph. Two common examples are given below:

1. [Experiments 23 and 24] For calculating molar enthalpies, temperature (T) is plotted versus elapsed time after mixing the reagents. A straight line is drawn through the relatively linear part of the data after the temperature has peaked and begins to return to room temperature. Generally there are a few points above or below the postmixing line that should be ignored in drawing the line, as they represent incomplete reaction or incomplete mixing. The line is extended to the time of mixing (0 minutes) to extrapolate the temperature that would have been observed had an instantaneous reaction taken place. The point at which the straight line intersects the y axis (time of mixing) is T_{mix}. This is plotted below:

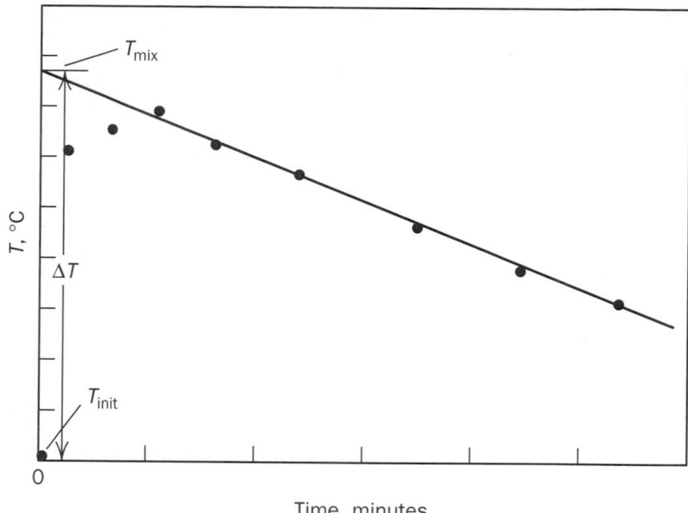

ΔT for the reaction is the extrapolated temperature, T_{mix}, minus the initial temperature, T_{init}, as shown. ΔT may be either positive or negative, depending on whether the mixture heats up or cools down.

2. [Experiment 18] A weak acid is titrated with sodium hydroxide, and the titration is monitored with a pH meter. The data collected consists of pH as a function of volume of NaOH added. A typical graph would look like the one shown below:

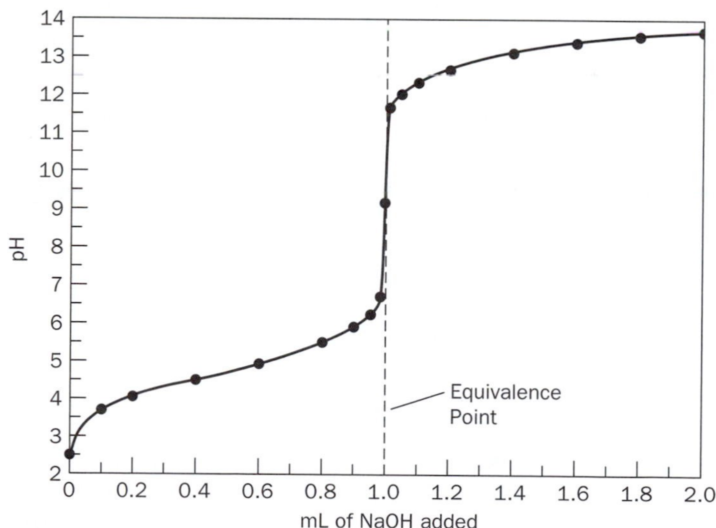

A smooth curve is drawn through the data points. The important area of the curve is where there is a rapid rise in the pH over a small volume of NaOH added. This is called the "equivalence point." The exact equivalence point is taken to be the midpoint of the steep rise. From that point, drop a vertical line to the x axis to read the volume of NaOH (1.0 mL in this example). This graph is not linear, and no transformation will make it linear.

Name: ______________________________

Date: __________ Section: ______________________

Questions

1. Write the following numbers in scientific notation.

a. 0.0045

b. 1315.88

c. 132,000

2. Determine the number of significant figures in each of the following numbers:

a. 321.55

b. 320

c. 0.0054

d. 320.040

3. The following physical data is obtained for the vapor pressure of CCl_4:

Pressure, bar	0.044	0.075	0.122	0.190	0.288	0.422	0.601	0.829	1.124
Temperature, K	273	283	293	303	313	323	333	343	353

Plot a graph of ln P (y axis) versus T^{-1} (x axis), and determine the slope and intercept.

Chapter 3
Laboratory Equipment and Experimental Techniques

3.1 GLASSWARE

The majority of experiments in the microscale laboratory involve the use of glassware, much of which is expensive. Your responsibility is to keep this glassware clean and to give it proper care.

Figure 3.1 illustrates the glassware you will use in the majority of experiments presented in this text. Figure 3.1 presents the more common glassware found in your laboratory locker. This includes, for example, beakers and Erlenmeyer flasks (10, 25, and 50 mL), a Hirsch funnel (plastic or porcelain), graduated cylinders (10 and 25 mL), suction filter flasks (25 and 50 mL), a separatory funnel, graduated glass or plastic pipets, and glass funnels and test tubes of various sizes.

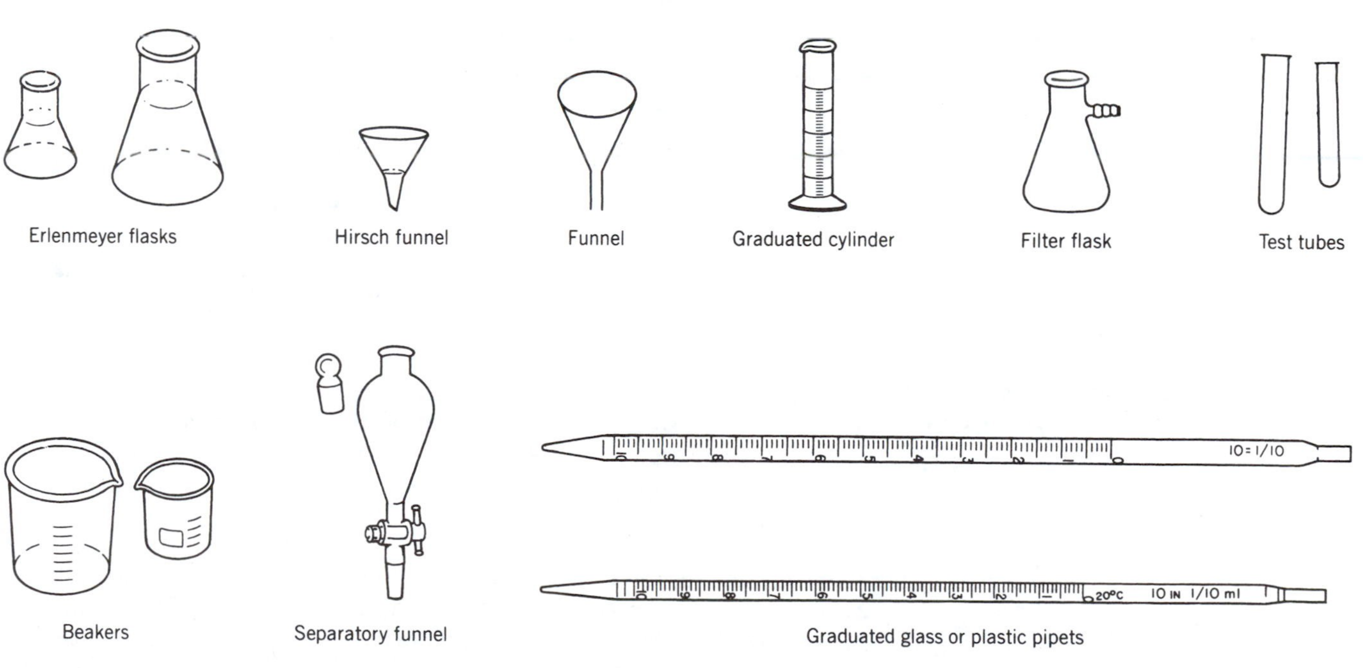

Figure 3.1 *Common glassware.*

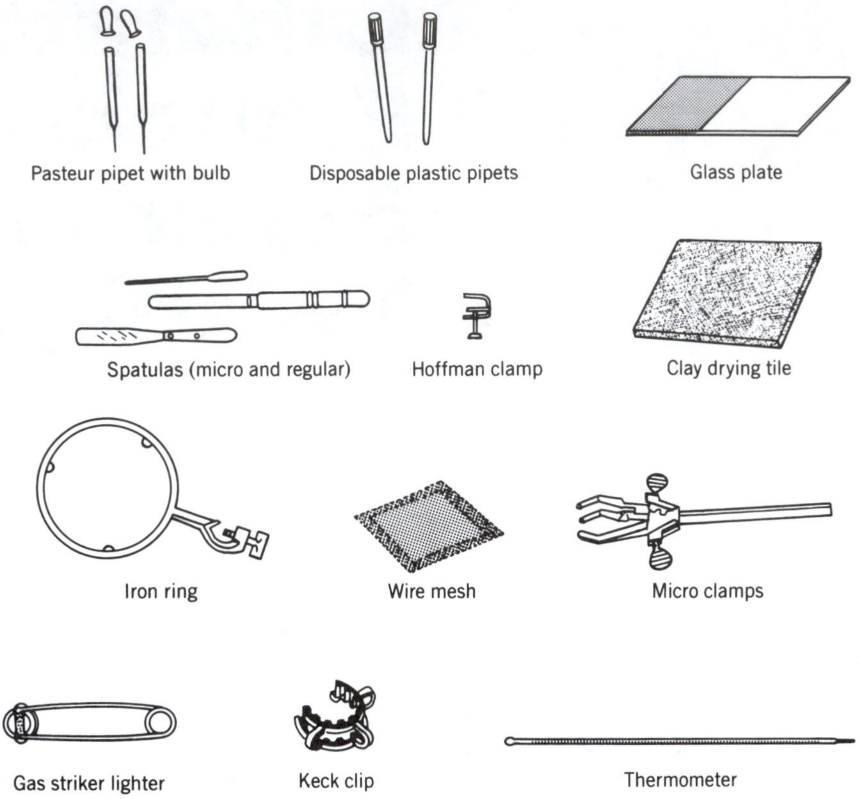

Figure 3.2 *Standard locker equipment.*

Other Locker Equipment

Your locker contains other pieces of equipment that are essential in carrying out the basic manipulations in a microscale laboratory. This may include items such as Pasteur pipets with bulbs, plastic disposable pipets, spatulas (micro and regular), a Hoffman screw clamp, a clay drying tile, assorted micro clamps, an iron ring, a wire mesh, a gas striker, a thermometer, and an approximately 12″ × 12″ glass plate with a half-white/half-black background. A collection of these items is shown in Figure 3.2.

Special mention is made of the glass plate indicated above. This item is strongly recommended for use when working at the microscale level. Since a large portion of the equipment is small and various stoppers, vials, pipets, etc., are used, placement of these items on the glass plate will assist you in keeping a neat and orderly approach to the operations at hand. The dark and white background gives a good contrast to various objects.

3.2 TRANSFERRING OF CHEMICALS

Solids

A number of steps are involved when transferring solids from a reagent bottle to a reaction flask or weighing paper. If you are using weighing paper, the paper should be sharply creased down the middle, so that the chemical can pour off more easily.

1. Read the label and make sure that you have the proper reagent bottle.
2. Tilt the sealed bottle until some chemical gets near the bottle's mouth.
3. Remove the bottle's cap. It is generally a bad idea to lay the cap on a benchtop, due to the possibility of contamination. Hold the cap in the same hand as the bottle. If the bottle's cap is flat, it can be safely set down **upside-down.**

4. Center the bottle over the reaction flask or weighing paper. Gently roll the bottle, until the desired amount of solid has been transferred (Figure 3.3a). Alternatively, the chemical can be scooped from the bottle using a spatula or scoopula (Figure 3.3b). The spatula is then tapped until the desired amount of solid has been transferred (Figure 3.3c). **Never use the same spatula to obtain different chemicals, unless the spatula has been carefully cleaned and dried in between.**

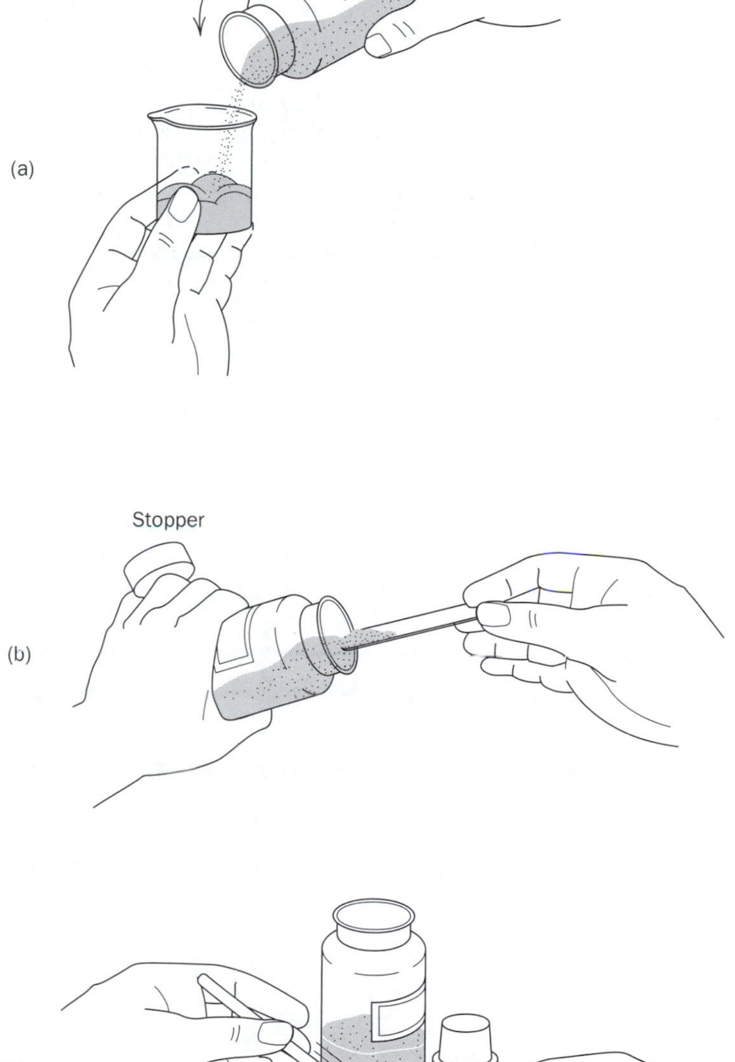

Figure 3.3 *Proper method of transferring solids: (a) Rolling the bottle to transfer a solid. (b) Use of a spatula to remove a solid from a bottle. (c) Tapping the spatula to deliver the solid.*

Liquids

Decanting, or pouring of liquids, has two major functions in the laboratory. The first is to simply measure out a quantity of liquid, usually from a reagent bottle. The second is pouring off a supernatant liquid, to separate it from a precipitated solid.

Small quantities of liquids are most efficiently transferred using an automatic delivery pipet. Several types are commercially available, three of which are shown in Figure 3.4. All have a control to adjust the pipet to deliver the desired volume. They are also designed so that the liquid being transferred comes in contact only with the plastic tip—thus, no rinsing is necessary. This tip can be ejected from the pipet, after delivery of the liquid is complete.

A few tips on the use of automatic delivery pipets:

a. They are relatively expensive, so treat them with respect.
b. Always store, hold, and use the pipet in a vertical position. Liquid running back up into the pipet mechanism can cause damage to the controls.
c. Never immerse the pipet in a liquid—only about 5 mm of the plastic tip should come in contact with the liquid to be transferred.
d. Develop a smooth depression and release technique of the control mechanism. Consistent results come with care and practice.

It is often necessary to convert volume measurements into weight measurements, and vice versa. For most calculations (unless extremely accurate data is needed), the familiar relationship,

$$\text{Density (g mL}^{-1}\text{)} = (\text{mass, g})/(\text{volume, mL})$$

can be used.

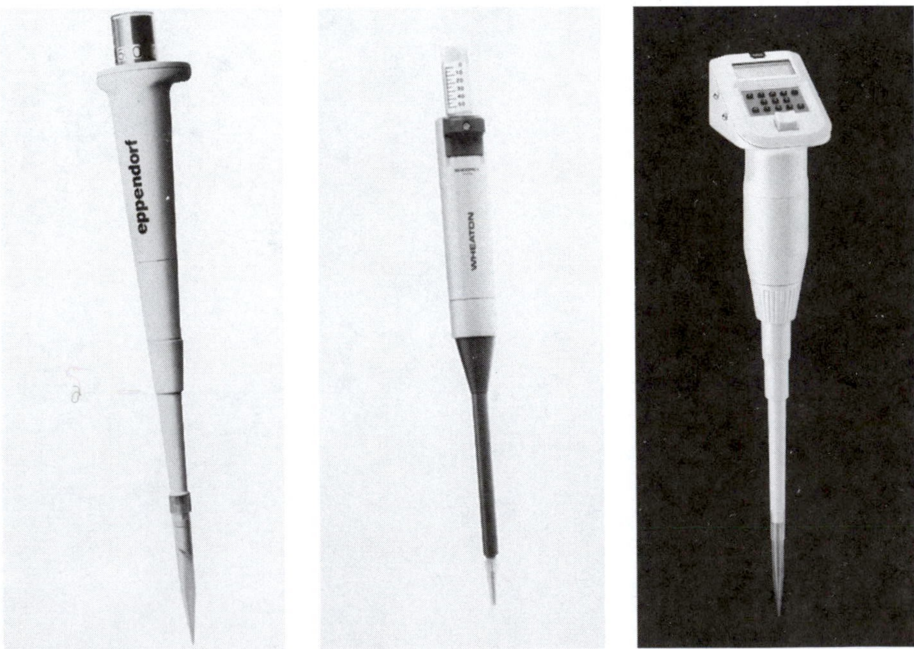

Figure 3.4 *Automatic delivery pipets. (Courtesy of Sargent-Welch Scientific Co., a VWR Company, Skokie, IL; Ace Glass, Inc., Vineland. NH: Rainin Instrument Co., Woburn, MA.)*

For larger quantities of liquid, the procedure is slightly different than for solids:

1. Read the label and make sure that you have the proper reagent bottle.
2. Remove the bottle's stopper. It is generally a bad idea to lay the stopper on a benchtop, due to the possibility of contamination. Hold the stopper in the same hand as the bottle.
3. Pour the liquid into the desired reaction flask, preferably down a glass rod (see Figure 3.5).
4. **Do not pour from a wide-mouthed bottle into a narrow-mouthed container.** This is a major cause of spills.

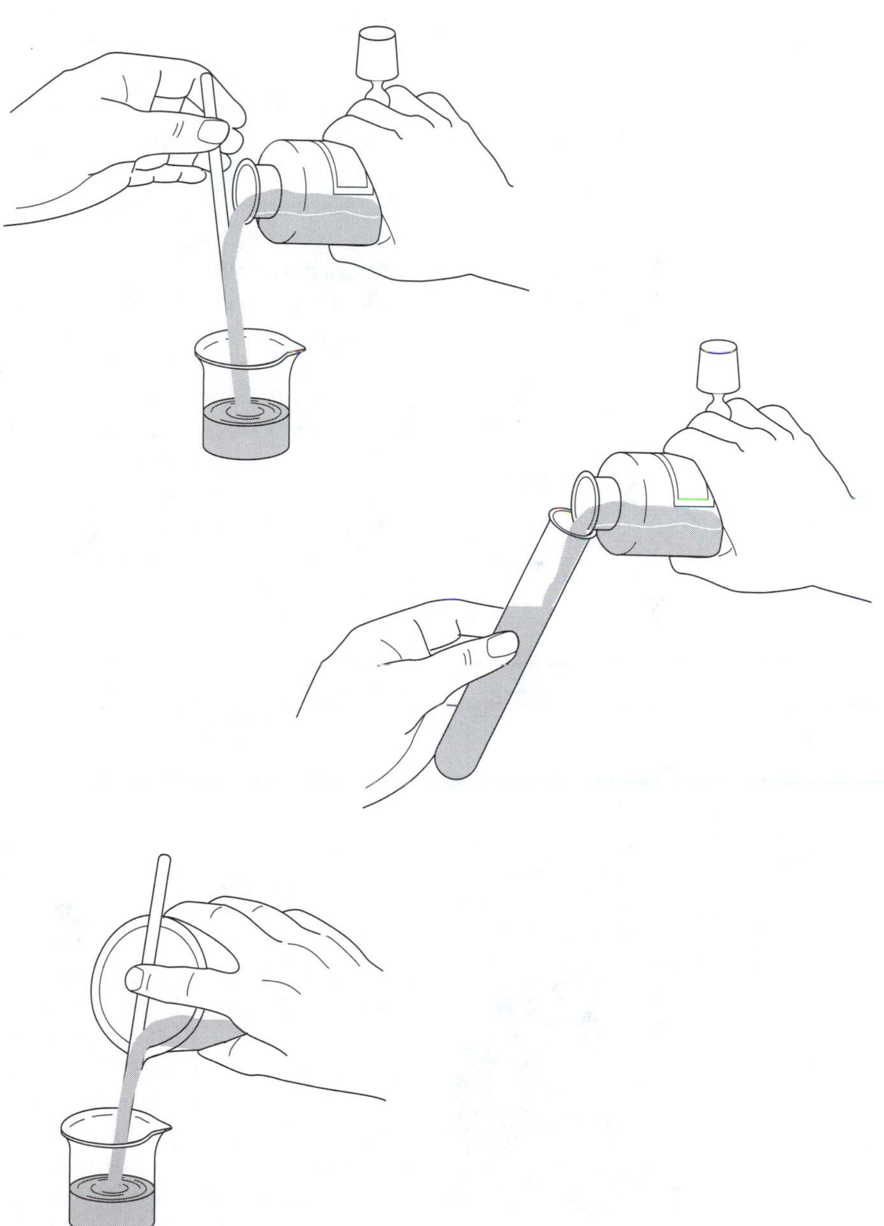

Figure 3.5 *Proper method for pouring liquids.*

5. When decanting (pouring off) a liquid from a solid, it is best to let the solid settle to one side of the container. This is most easily accomplished by placing the container at a sharp angle (in a secure location, such as a cork ring) while the solid is settling.

3.3 WEIGHING

Weighing is one of the most critical measurements made at the microscale level. The use of single-pan electronic balances with automatic taring (zeroing) and digital readout has made the move to the microscale laboratory a reality. Some laboratories will use the older, but accurate top-loading balances. It is essential that each worker take the responsibility of keeping the balance(s) clean, as well as around the balance area. These are expensive and delicate instruments. Never move a balance without the laboratory instructor's approval. Some types of balances are easily thrown out of adjustment if they are improperly moved, necessitating an expensive repair.

Chemicals should never be placed directly on the balance pan. If any chemical is spilled on or near a balance, clean it up right away. Most chemicals are corrosive, and can cause a balance to malfunction. Service calls are expensive, and new balances are even more so. Solid chemicals should be weighed using aluminum or plastic boats, or various glass containers. Weighing paper or filter paper may also be used in certain situations. Never tip a bottle of solid to pour solid into a container on the balance. Either use a scoopula to introduce small amounts of solid into the weighing container on the balance, or remove the weighed container from the balance, add solid, then reweigh. Liquid chemicals are generally weighed directly in the reaction flask being used or in the product flask when isolated as a liquid. In any event, a glass container is usually preferred. Again, do not pour liquids into a container sitting on a balance.

Objects to be weighed should be dry and at room temperature. This is particularly important on analytical balances (sensitivities of ± 0.0001 or ± 0.001 g). If the sample is not dry, its mass will decrease with time due to evaporation of water. If it is not at room temperature, it will generate air currents as it heats or cools the air around it, and this will cause inaccurate or unsteady readings.

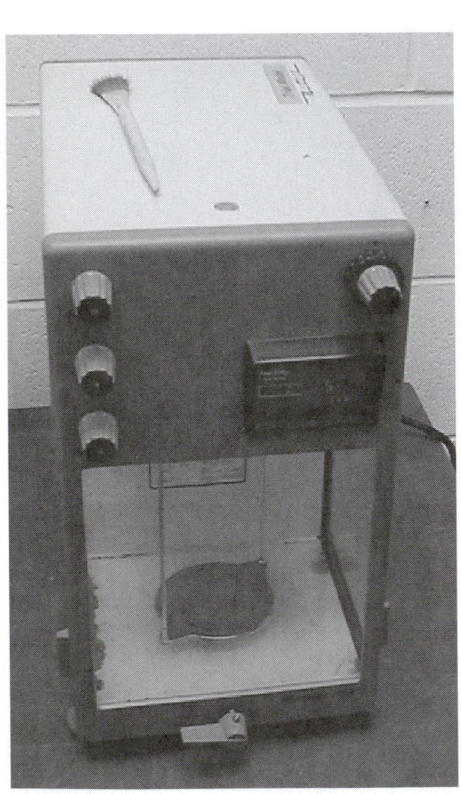

Figure 3.6 *Fisher AB-120 mechanical single pan analytical balance. (Courtesy of Fisher Scientific, 711 Forbes Ave., Pittsburgh, PA 15219.)*

Mechanical Single Pan Balance

The mechanical single pan balance (Figure 3.6) is an analytical balance that has dials to control the setting of weights. The balance often has an enclosed weighing chamber to eliminate air currents, which can cause unsteady readings.

In order to be used, the balance must first be leveled. The leveling bulb is near the front of the instrument. If the bubble is not in the center of the bulb, inform the lab instructor (the adjusting screws at the feet of the balance are adjusted until the bubble is centered). With an empty pan and all sliding doors closed, turn the **on/off knob** to the full-on position. A **moving scale** will appear on the front of the balance, and should swing freely (without jerking). Turn the **zero knob** (usually toward the rear of the balance) until the scale indicates "0."

Turn the balance off, and gently place the object to be weighed on the pan. Turn the on/off knob to the partial-on position (usually the opposite direction from full-on). The moving scale will appear, but will be jerky in motion. Start with the largest weights, and turn the dial one step at a time until the lighted scale jumps up and stays up, and then turn it back one step. Repeat this with the dials for lesser weights. Now, turn the on/off knob to the full-on position. The moving scale should swing freely, and the numbers should be visible.

Reading the weight depends on the type of balance. Most single pan balances have dials that control numbers (called a **vernier**) next to the moving scale.

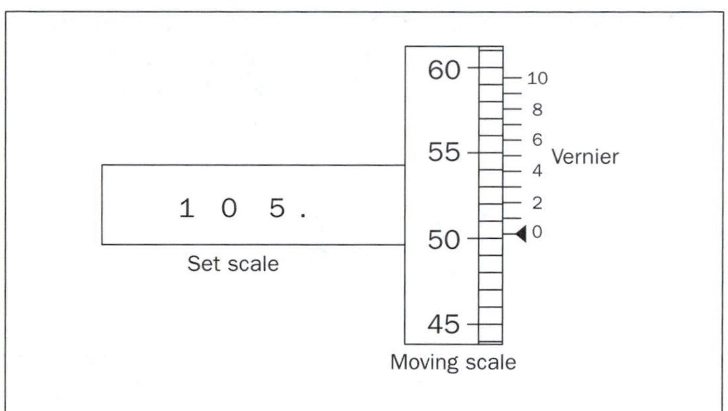

Note the location of the pointer. Read the first several numbers from the set scale and then from the moving scale. From the location of the vernier arrowhead, we can see that the balance is reading between 105.50 and 105.51 g. The last decimal place can be obtained by looking at the vernier lines, and noting where they best align with the moving scale lines. In the figure above, this is at the vernier reading of "3." The balance reading would therefore be 105.503. Other balances are slightly different—an adjustment knob is turned until the vernier arrowhead is aligned with a line on a separate vernier scale. This in turn sets additional digits on the adjustable scale (located to the right of the moving scale). This is shown below:

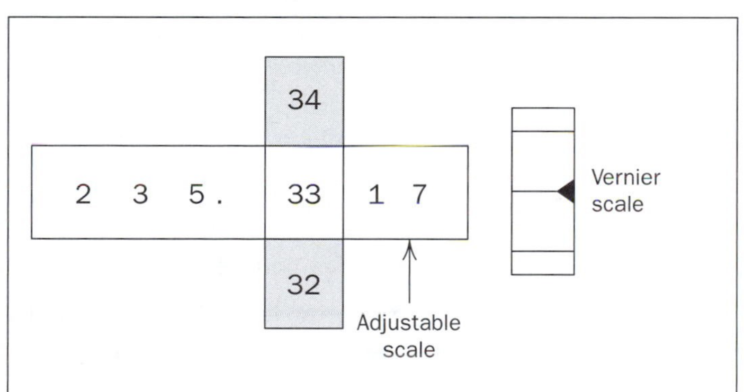

The balance pictured above reads 235.3317 g.

Digital Balances

Digital balances (Figure 3.7) come in several varieties, with sensitivities ranging from ±0.01 to ±0.0001 g. The balances should be leveled in the same manner as indicated above. Turning on the balance also zeroes the balance. If the balance should drift away from a zero reading, merely push the tare bar (or spot) on the front of the balance. Place the object to be weighed on the balance pan. When the reading comes to a constant value (often, the letter "g" will appear at this point), the weight of the object can be read directly.

One great advantage of digital balances is that the weighing container's weight can be tared out. Suppose an experiment calls for 100 mg of a particular chemical. Place a folded piece of weighing paper (or a weighing boat, or the experimental piece of equipment) on the weighing pan, and press the tare bar to remove its weight. The reading should now be 0.000 g. Using a spatula, place the substance to be weighed in the weighing container. The reading on the digital readout will be the sample's weight.

Figure 3.7 *Mettler AJ-100 digital balance. (Courtesy of Fisher Scientific, 711 Forbes Ave., Pittsburgh, PA 15219.)*

3.4 LIQUID VOLUMES

Various techniques are used to measure small volumes of liquids at the microscale level. When 1–2 mL of a reagent, solvent or solution is required to be accurately measured, it is convenient to use one of the measuring devices listed below (see Figure 3.8):

a. A glass graduated pipet with a pipet pump or bulb
b. A graduated 1 or 2 mL glass or plastic syringe
c. A calibrated Pasteur pipet with pipet pump or bulb

Several types of pipet pumps are commercially available, as are glass, polypropylene, and disposable syringes. Disposable syringes or pipets are cheap and often convenient to use, especially when aqueous solutions are dispensed.

The standard calibrated glass pipets (1, 2, 5 mL) come in various designs. Pipets are calibrated in one of two ways: to deliver a given amount (marked TD) when allowed to drain under gravity, or to contain a given amount between the limiting calibration marks (marked TC).

The use of a glass pipet is shown in Figure 3.9. It is important for you to recognize which pipet type you have. For TD glassware, the volume released is exactly the volume marked on the pipet. Some liquid will remain in the pipet when delivery is complete. Do not try to shake this liquid out—it will make your delivered volume inaccurate. **NEVER use mouth suction to draw *any liquid* into the pipet!** Serious injury may result if this advice is not followed.

The curved surface of a liquid is called the **meniscus.** When making a reading (Figure 3.10), be sure to have your eye at the same level as the meniscus in order to avoid problems due to parallax. The meniscus is always read at the curve's center, regardless whether it curves upward or downward.

Take care when reading the volume. Some scales read from 0 at the top to a maximum at the bottom (graduated pipets and burets), whereas graduated cylinders go from 0 at the bottom to the maximum at the top. In Figure 3.10 the proper reading would be 1.178 mL.

A glass Pasteur pipet may be easily calibrated by drawing a known amount of liquid from a filled 10 mL graduated cylinder into the pipet using a pump or bulb, and marking the level of the liquid. It is recommended that you calibrate several pipets showing volumes of 0.5, 1.0, 1.5, and 2.0 mL. For permanent use, a light file mark may be scored on the pipet. Plastic pipets (also called Beral pipets) may be calibrated in a similar manner.

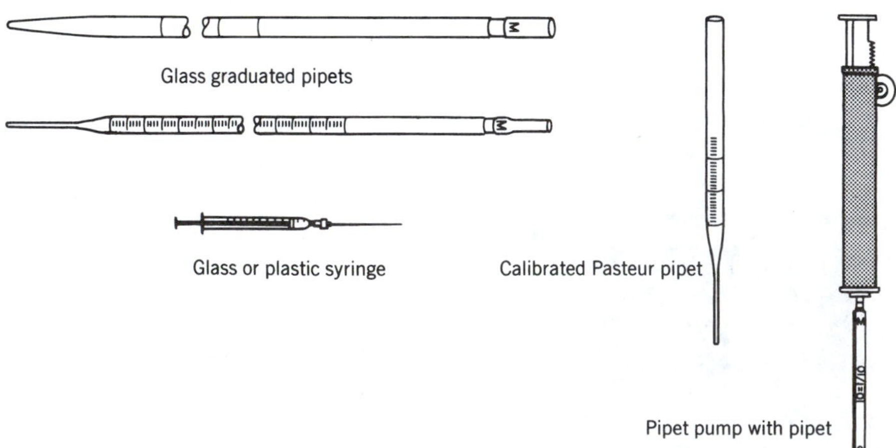

Figure 3.8 *Equipment to measure volume. (Courtesy of Ace Glass, Inc., Vineland, NJ; Thomas Scientific, Swedesboro, NJ.)*

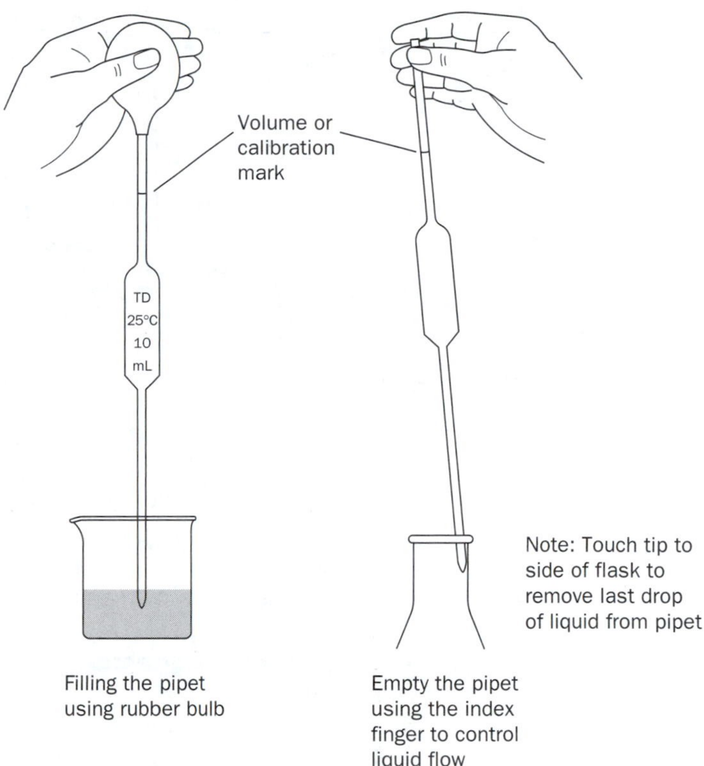

Figure 3.9 *Proper use of a pipet.*

In many experiments, it is recommended that a Pasteur *filter* pipet be used. To make one, a small cotton plug is inserted into the tip of a Pasteur pipet by use of a wire, as shown in Figure 3.11. It is important to use the correct amount of cotton, so that the plug is not too tight to prevent easy flow of the liquid or too loose to come out when a liquid is dispensed. The old saying holds: Practice makes perfect!

The Pasteur filter pipet offers two distinct advantages over the regular Pasteur pipet. First, the solution is automatically filtered each time it is taken into the pipet. This is necessary, since it is important to remove dust or other suspended material in the solution at the micro level, where purity of reagents

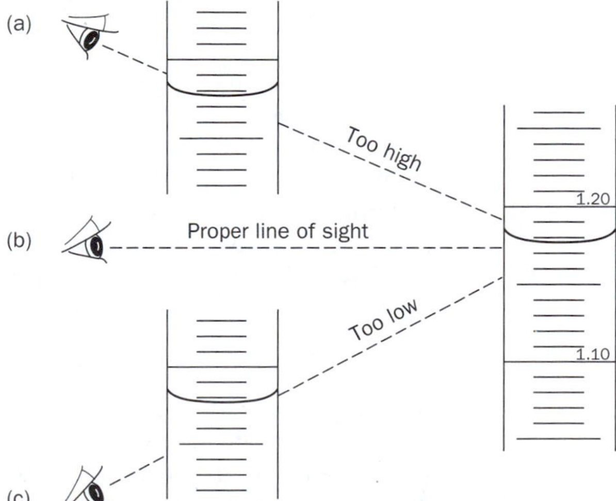

Figure 3.10 *Reading a meniscus.*

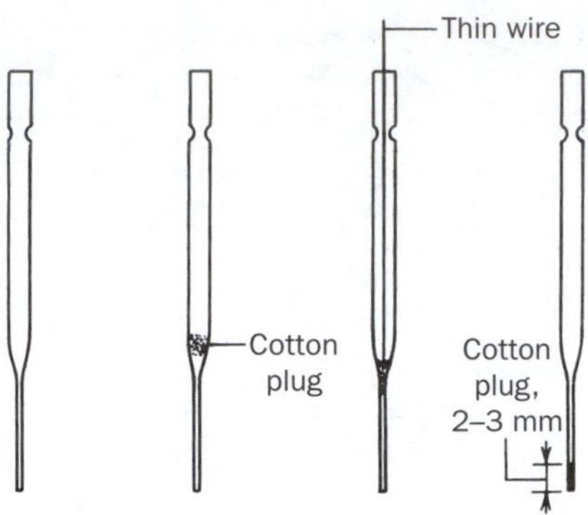

Figure 3.11 *Preparation of Pasteur filter pipet.*

and products is critical. Second, when a volatile solvent such as ether is used, a back pressure rapidly builds up in the bulb that causes the solvent or solution to drip from the tip of the pipet. The small plug allows the operator sufficient time to make a transfer rapidly and completely from one container to another without loss of material.

An Important Suggestion: *When transferring a liquid material by pipet from one container to another, hold the container containing the material to be transferred as close to the receiving container as possible, preferably holding both in one hand. This arrangement allows for a smooth, quick transfer with minimal loss.*

If the pipet is being used to measure an exact volume of a solution with an exact concentration, the pipet should first be rinsed with the solution. Draw up a small amount of the solution, then use a pipet pump to draw the rinse up and down the pipet, then expel the rinse into a waste container. Alternatively, larger pipets may be tipped on the side and rotated after removing the bulb or pump, so the entire inner surface is rinsed.

Volumetric Flasks

Glassware such as Erlenmeyer flasks, beakers, and test tubes *are not* calibrated for use as accurate measuring devices! Graduated cylinders are somewhat more accurate, burets (see below) if used properly are quite accurate, but we recommend the use of volumetric flasks or of the calibrated types of pipets discussed above for accurate measurement of small amounts of liquids at the micro level.

Volumetric flasks (Figure 3.12) should be used as the container when the concentration of the solution being prepared must be very accurate (3–4 significant figures). An example would be in the preparation of a standard solution (see Experiment 17) for a titration. Volumetric flasks are usually calibrated TC at room temperature, and must be used at the indicated temperature. There is only one graduation, which is etched on the neck of the flask. This is frequently referred to as "the mark." When the bottom of the meniscus is exactly on the mark, the flask contains exactly the calibrated volume.

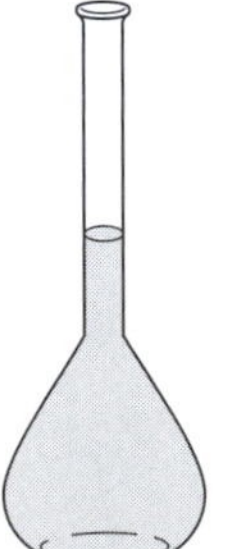

Figure 3.12 *Volumetric flask.*

The flask should be rinsed with solvent before use. The usual procedure for preparing a solution is to partially fill the volumetric flask with solvent, add the appropriate amount of solute, swirl to dissolve and mix, and then fill to a little below the mark with additional solvent. The last few drops of solvent, to bring the meniscus to the mark, should be added with a Pasteur pipet (or medicine dropper). If the solvent is added too quickly, the meniscus may

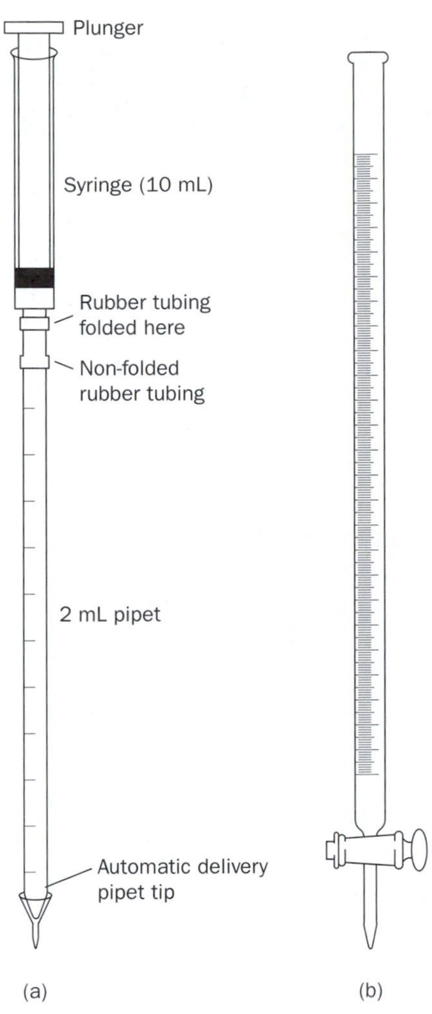

Figure 3.13 *(a) 2 mL microburet. (b) Standard 50 mL buret.*

wind up above the mark. The entire solution must then be discarded, as no liquid should ever be removed from a volumetric flask before solution preparation is complete.

When mixing the contents of the volumetric flask, be certain to hold both the flask and the stopper. Filled flasks should not be held by the neck, as they are too heavy for the glass in the flask neck to support. Use two hands. Broken flasks, severe cuts, and exposure to chemicals can be the result of improper shaking.

Microburets

The microburet shown in Figure 3.13a can be constructed from a 2-mL graduated pipet that has graduations every 0.01 mL, a disposable yellow tip (10 – 100 μL automatic delivery pipet size), a 5- or 10-mL plastic syringe (without needle), and a 2-inch piece of soft Latex rubber tubing (o.d. ~ 8 mm). Soft Tygon tubing may also be used.

1. Lubricate the rubber tubing with water and slide it (3/4 of the way) over the top of the 2-mL pipet. Let it stand for a period of at least 1 hour and then peel the tubing off so that it folds over itself. It now has a double layer.
2. Fit the folded end of the tubing to the bottom of the syringe, as shown in the figure. *Make sure it fits snugly. A loose fitting at this juncture can later lead to leakage of solution from the buret.* The plunger of the syringe should be held about 1 cm above the bottom of the syringe during this operation. This will allow extra room for manipulating the plunger so as to remove the last drop of liquid from the buret.
3. Lubricate the top end of the pipet with water and attach the other end of the tubing-syringe assembly to the top of the pipet.
4. Tightly place the yellow plastic tip on the bottom of the graduated pipet. The apparatus should now resemble the Figure in 3.13a.

The microburet is now complete. If treated properly, these microburets can be used repeatedly over long periods of time without having to dis-assemble the arrangement. The use of the microburet is described in Section 3.9 (Titration), p. 46. Larger quantities of liquids can be dispensed using a standard 50 mL buret (Figure 3.13b). The amount of material delivered from the buret is controlled via a stopcock.

3.5 HEATING METHODS

The Microburner

If certain precautions are observed (mainly no flammable solvents in use, and the gas being turned off immediately after use of the burner), the microburner or Bunsen burner (Figure 3.14) has a useful role to play in the laboratory. The microburner operates by combusting natural gas. The burner is connected to a gas supply with a short length of rubber tubing.

Matches or strikers may be used to light the burner. Strikers are generally safer to use. When using a match, strike the match first, hold it near the top of the burner (but not directly above it), and then turn the gas jet on full. The gas should light in a few seconds. Holding the match directly above the burner can cause it to blow out the match. If a striker is used, turn on the gas jet first, and then squeeze the striker, sliding the flint across the friction bar so that a spark forms. Hold the striker in the fully squeezed position above the burner, and adjust the burner flame if necessary. **If the flame goes out, turn the gas off and wait at least 30 sec before relighting.** This gives any uncombusted gas time to disperse before another flame is lit.

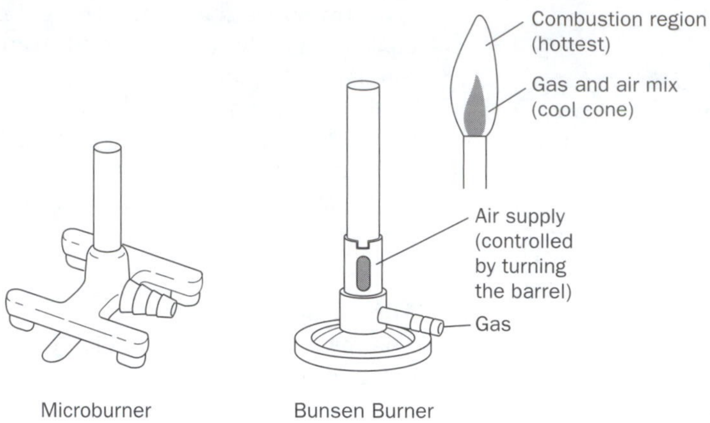

Figure 3.14 *Microburner and bunsen burner.*

While natural gas has no odor, β-mercaptoethanol is usually mixed in for safety's sake. If the distinctive odor of this sulfur-containing compound is detected, make certain that all gas jets are tightly closed. **If the odor is still apparent, the laboratory should be evacuated until the gas leak is found.**

Sand Bath or Aluminum Block with Magnetic-Stirring Hot Plate

A sand bath placed on a magnetic-stirring hot plate (Figure 3.15a) makes a very efficient heating device. This arrangement allows for stirring (see below) and heating to be performed simultaneously. This approach is very inexpensive, provides a nonflammable source of heat, and does not decompose. The sand is usually contained in a glass crystallizing dish or a metal container.

Sand, being a poor conductor of heat, in effect acts as an insulator. Thus, the temperature of a container placed in the sand will vary depending upon the depth to which it is immersed. It is therefore recommended that each individual be assigned their own equipment (sand bath and hot plate), and that a calibration of this heating source be made at the first session of the laboratory. This approach is also quite effective at temperatures below 100°C, and because of this has replaced the steam bath in many laboratories.

An alternative to the sand bath is the aluminum block (Figure 3.15b) placed on the magnetic-stirring hot plate. These blocks are now available from several supply houses (or can be easily manufactured in a metal shop). Users of this device cite two major advantages over the sand bath method: better heat transfer and no problem with breakage of a crystallization dish or spilled sand. In addition, they are relatively cheap and store easily.

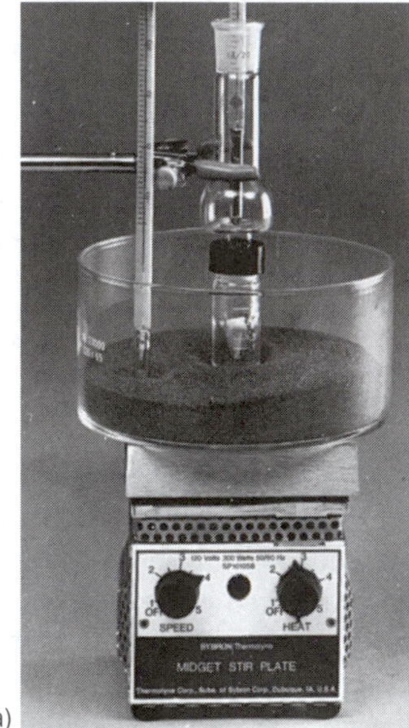

Figure 3.15 *(a) Sand bath on magnetic stirring hot plate. (b) Aluminum heating blocks. (Courtesy of Ace Glass, Inc., Vineland, NJ.)*

Heating Safety

1. When heating glass containers over a burner flame, use a wire gauze square with a nonflammable ceramic center as a support. This is placed on an iron ring clamped to a ring stand.
2. Never heat thick-walled vessels, such as filter flasks or volumetric flasks.
3. When heating liquids in small test tubes, always heat from the top down to avoid explosive boiling of the liquid (Figure 3.16). Never point the top of the tube at any other person, or at yourself.
4. Never heat a flammable object over an open flame.
5. Be aware that most solvent vapors are denser than air, and will sink and flow along counter tops and into sinks. They may even be ignited by a burner some distance away.

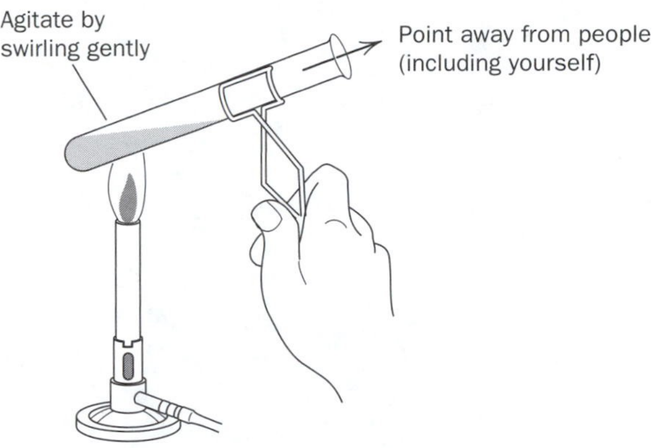

Figure 3.16 *Proper way of heating a test tube.*

6. When heating a solution containing solids, especially finely divided solids, be sure to stir or swirl continuously to avoid "bumping."
7. When boiling a liquid, it is often wise to use boiling stones (small chips of silicon carbide). These sit at the bottom of the container, promote local boiling, and prevent "bumping." Always use a fresh boiling stone.
8. **Never heat a closed container.**

3.6 STIRRING

Stirring is often required in microscale reactions, and is generally carried out using a stirring rod, microspatula, or magnetic stir bars or vanes. Heating and stirring are conveniently carried out simultaneously by using a magnetic-stirring hot plate (see Figure 3.17). When this type of arrangement is used, the container being stirred should be placed directly in the center of the hot plate, so as to gain maximum efficiency of the magnetic flux. The spin rate of the bar or vane should be adjusted to obtain smooth mixing of the contents in the container. The insertion of a thermometer into the sand bath or aluminum block easily allows you to monitor the temperature of the container and its contents. The magnetic stir bar can be easily removed from the container using forceps or a magnetic wand.

Inexpensive micromagnetic stir bars may be constructed in the following manner. Take a Pasteur pipet (long tipped) and seal off the bottom of the tip. Drop a 0.5-cm long piece of paper clip down the open end of the pipet. Place the tip, just above the paper clip piece, in the flame of a microburner, and with tweezers, pull off the now sealed glass-enclosed magnetic stir bar. Several bars can be made from the same pipet/paper clip.

Stirring is a technique whose importance is often overlooked. When the mixed materials give off large quantities of heat (dilution of concentrated acids, for example), they should be efficiently stirred. Stirring also helps achieve rapid equilibrium, and is especially helpful when making pH measurements or taking kinetic data.

When the mixing of solutions or dissolution of a solid in a given solvent is desired and heating is not required, the use of a touch mixer (Figure 3.18) is most convenient. These mixers are commercially available, and are relatively inexpensive and durable. The mixer speed is adjustable, and the mixing operation is fast and efficient. Vortex mixing is recommended over the techniques of *swirling* or *shaking* by hand, since much greater control is maintained during the mixing process.

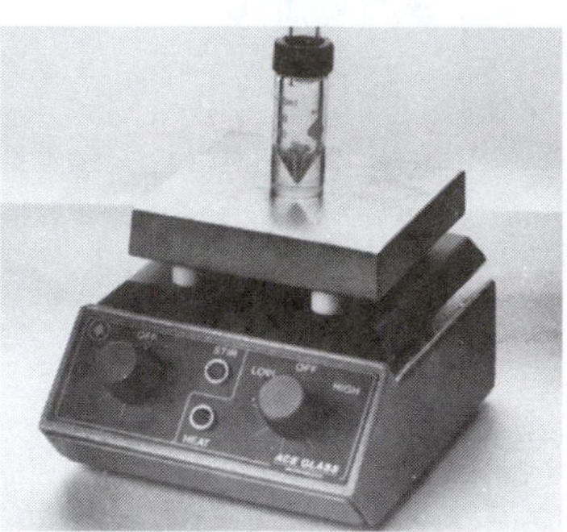

Figure 3.17 *Magnetic-stirring hot plate. (Courtesy of Ace Glass Inc., Vineland, NJ.)*

Figure 3.18 *Touch Mixer. (Courtesy of Fisher Scientific, 711 Forbes Ave., Pittsburgh, PA 15219.)*

3.7 FILTRATION

The process of collecting crystalline products at the microscale level usually involves the technique of filtration. The equipment used generally relates to the amount of material one is dealing with. The conventional procedures are discussed below.

Suction Filtration

When the amount of solid exceeds approximately 100 mg, a conventional porcelain Hirsch funnel with a filter paper disk, or a sintered glass or plastic Hirsch funnel fitted with a polyethylene frit, is used to collect the material. This operation can be done under vacuum, generally by use of a water aspirator, and is called "suction" or "vacuum" filtration. A typical arrangement is shown in Figure 3.19. When using this arrangement, **always securely clamp the filter flask to a support to prevent tipping.** Many times a valuable product has been lost due to not observing this simple rule.

A water trap must *always* be placed between the filter flask and the aspirator. This is to prevent water from backing up into the filter flask, and perhaps

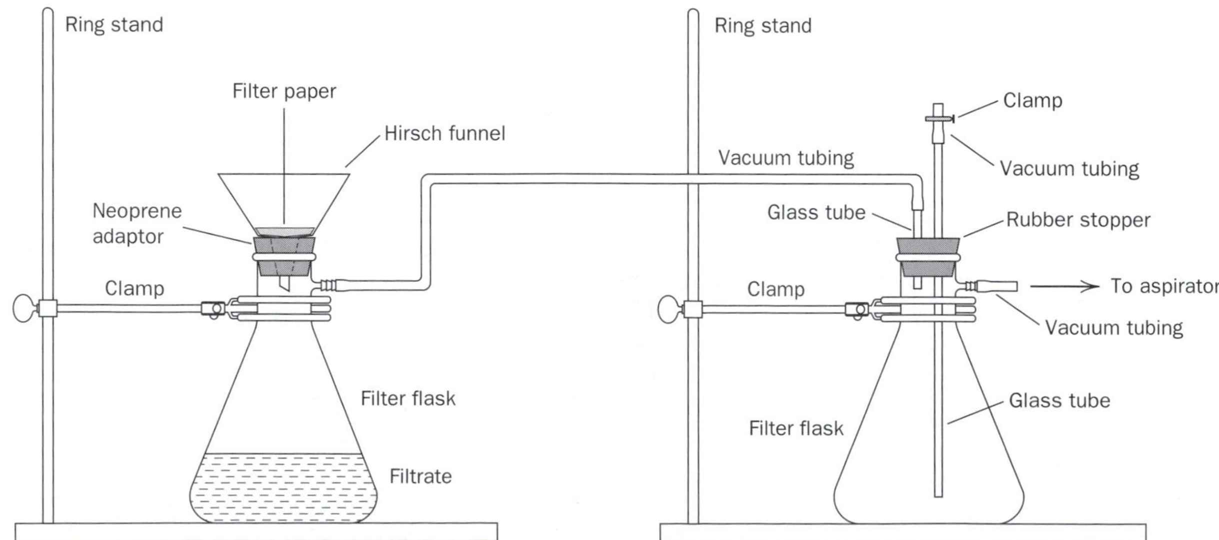

Figure 3.19 *Vacuum filtration apparatus with water trap.*

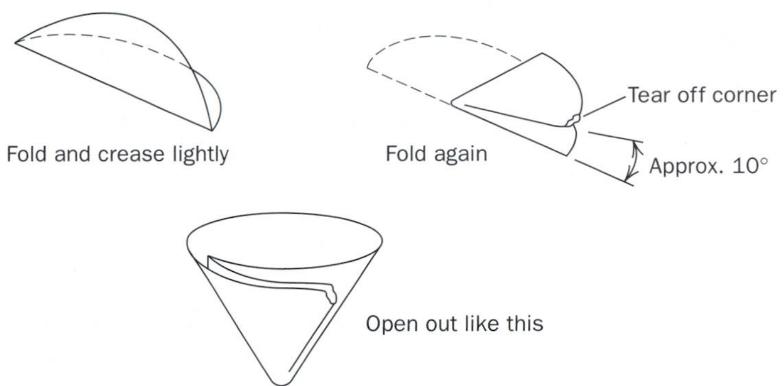

Figure 3.20 *Proper folding of filter paper.*

destroying the product. Be aware that when several persons are using the same water line, the water pressure can change at any time. It is recommended that the system be opened to the atmosphere by loosening the screw clamp on the trap or by removing the tubing from the filter flask before the water is turned off.

Gravity Filtration

Gravity filtration may be used with amounts greater than 100 mg. It requires a filter funnel (usually glass) with a stem to direct the filtrate (liquid) into the receiving container. A piece of filter paper is prepared (Figure 3.20) and inserted into the funnel. The filter paper should be wet with solvent before filtering, so that the paper adheres evenly to the inside surface of the funnel.

To prepare a piece of filter paper, fold it in half, and then in quarters with one slightly smaller than the other. Tear off a corner of the smaller quarter, and open the cone so that there are three layers on one side, and one on the other. The three layers should include the smaller quarter with the tear (see Figure 3.20). Folding the paper in this manner allows the filter paper to seal against the funnel, and allows for faster filtering. If desired, the filter paper can be further folded down to a very small size. This is known as fluted filter paper.

The funnel should be firmly supported using a clay triangle or cork ring. When decanting into the funnel, it is often helpful to use a glass rod (as described under "Transferring of Chemicals—Liquids," Sec. 3.2.). Do not fill the funnel more than two-thirds of the way to the top of the filter paper. A rubber policeman (glass stirring rod with a rubber spatula) can be used to scrape solids out of the decanting vessel onto the filter paper. Small loose particles can be swept into the filter paper by holding the decanting vessel in the pouring position over the funnel, while using a squeeze bottle of water (or other solvent) to wash the solids out.

3.8 SOLUTION PREPARATION

Work in the fields of synthesis, quantitative analysis, and qualitative analysis all depend on the ability to prepare solutions of known concentration. Virtually all aspects of life are affected, from food preparation to medical diagnostics, to manufacturing to the extraction of compounds from exotic plants.

Solutions are made up of two components. The **solvent** is the compound, usually present in large excess, in which the **solute** is dissolved. The solvent is usually a liquid at room temperature. Water is the most commonly used solvent. The solvent itself may be a solution, say 0.1 M hydrochloric acid (HCl). A solution such as 1.0 M barium chloride ($BaCl_2$) in 0.1 M HCl can also be thought of as a solvent (water) with two solutes (HCl and $BaCl_2$).

The units of concentration vary widely. The most common unit used by chemists is molarity (M, moles per liter of solution). Others include molality (m, moles of solute per kg of solvent), normality (N, equivalents per liter), and in the medical field, mg per deciliter (deciliter = 0.1 L). Less commonly used are percents—for solid solutes, weight:volume (w/v), and for liquids, volume:volume (v/v). For example, a 10% (w/v) sodium bicarbonate solution contains 10 g of sodium bicarbonate per 100 mL of solution, whereas a 5% (v/v) alcohol solution contains 5 mL of alcohol per 100 mL of solution.

Preparation of a Solution by Dissolving a Solid

Most of the solutions you will prepare will have the concentration expressed in units of molarity. Knowing the desired volume and concentration of a solution to be made up from a solid, the required mass of compound is calculated as follows:

$$\text{\# moles required} = \text{M} \times \text{V} = (\text{mol/L})(\text{L}) \tag{1}$$

$$\text{mass} = \text{moles} \times \text{MW} = (\text{mol})(\text{g/mol}) \tag{2}$$

Combining these two equations:

$$\text{mass needed} = \text{M} \times \text{V} \times \text{MW} = (\text{mol/L})(\text{L})(\text{g/mol}) \tag{3}$$

Example 1: *You are asked to make up 5.00 mL of 0.25 M NaOH.*

$$\text{M} = 0.25 \text{ mol/L}, \text{ V} = 0.00500 \text{ L}, \text{ MW of NaOH} = 40.00 \text{ g/mol}$$

$$\begin{aligned}\text{mass of NaOH needed} &= (0.25 \text{ mol/L})(0.00500 \text{ L})(40.00 \text{ g/mol}) \\ &= 0.0500 \text{ g NaOH}\end{aligned}$$

Thus, 50 mg of NaOH should be diluted to a volume of 5.00 mL to make up this solution.

Dilution

When a solution is to be made up by diluting another solution, the equation is slightly different. We must determine the number of moles of solute wanted in the dilute solution. We then calculate what volume of the more concentrated solution contains this number of moles of solute.

$$\text{M}_1\text{V}_1 = \text{M}_2\text{V}_2 \tag{4}$$

This equation states that the number of moles taken from more concentrated solution (M_1V_1) is the same as the number of moles in the dilute solution (M_2V_2). The molarity and volume have changed when the dilution is made, but the number of moles remains the same.

Example 2: *You are asked to prepare 5.0 mL of 0.10 M sulfuric acid (H_2SO_4) from 6.0 M H_2SO_4.*

The volume of 6.0 M acid required is calculated from Eq. 4:

$$(6.0 \text{ mol/L})(\text{V L}) = (0.10 \text{ mol/L})(0.0050 \text{ L})$$

$$\text{V} = \frac{(0.10 \text{ mol/L})(0.0050 \text{ L})}{6.0 \text{ mol/L}} = 8.3 \times 10^{-5} \text{ L or } 83\ \mu\text{L}$$

This volume may be measured using a microliter syringe or automatic delivery pipet. It is then diluted with deionized water to give a total of 5.0 mL of solution.

Serial Dilution

Serial dilution is used when a series of solutions spanning several orders of magnitude must be made, or when a small quantity of very dilute solution must be prepared where it would be difficult to accurately measure out the very small quantity of solute required. In the latter case, serial dilution often avoids waste of solute and solvent by starting with a small amount of more concentrated solution, diluting it in steps, until the concentration and volume desired is obtained.

Serial dilution starts with a concentrated solution (possibly made from a solid). A first dilution, typically 10:1 or 100:1, is made. This second solution is then used to make a third solution by diluting the second by 10:1, 100:1, or whatever is necessary and so on, until the desired volume and concentration are reached. How many steps are used in a serial dilution is up to the experimenter. Unless very large volumes are involved, dilutions of greater than 1:100 should be avoided, as the amount of concentrated solution needed to make a dilution becomes very small and thus difficult to measure accurately. (It is also true, however, that more steps introduce greater error.)

Example 3: *You wish to prepare 25.0 mL of 2.50×10^{-6} M potassium dichromate (MW = 294.18 g/mol).*

This requires

$$(0.0250 \text{ L})(2.50 \times 10^{-6} \text{ mol/L})(294.18 \text{ g/mol}) = 1.84 \times 10^{-5} \text{ g of potassium dichromate}$$

On a good analytical balance, the minimum amount that could accurately be weighed out would be about 25 mg, or 2.5×10^{-2} g, over 1000 times what you need. If you were to weigh out 250 mg and dilute it to the desired concentration directly, you would have 34 L of solution, when you only need 25 mL! Waste solutions containing chromium cannot be discarded down the drain, as chromium is poisonous and, in some forms, carcinogenic. It is unthinkable to generate 34 L of solution and then pay someone to haul away 33.975 L of it as waste.

Instead, you should weigh out 25 mg of potassium dichromate, and dissolve it in 100 mL of water. The resulting concentration is 8.50×10^{-4} M. Obtaining a solution which is 2.5×10^{-6} M requires a 340:1 dilution. This is conveniently done in two steps: a 100:1 dilution of the first solution (1.00 mL diluted to 100.0 mL, yielding a solution 8.50×10^{-6} M in potassium dichromate), followed by a 3.4:1 dilution as follows to get 25 mL of 2.5×10^{-6} M potassium dichromate. Using Eq. 4 again:

$$(8.50 \times 10^{-5} \text{ mol/L})(V \text{ L}) = (2.5 \times 10^{-6} \text{ mol/L})(0.025 \text{ L})$$

$$V = 7.4 \times 10^{-3} \text{ L or } 7.4 \text{ mL}$$

When 7.4 mL of the second solution is diluted to 25.0 mL, the concentration will be 2.5×10^{-6} M. There is some waste, 99 mL of the first solution and 92.6 mL of the second solution, but the amount is less than 0.2 L as opposed to 33.975 L. (The number of moles of chromium waste is the same either way, but disposal cost is based on the volume of waste.)

Multiple Solute Solutions

Solutions often contain more than one solute. In such cases, one calculates the amount of each solute needed, measures it in an appropriate container, mixes all the solutes together in the required volume of solvent, and stirs to dissolve and homogenize the solution.

Example 4: *Prepare 100 mL of solution that is 0.0100 M in $Cu(NO_3)_2 \cdot 3H_2O$, 0.0100 M in $Ni(NO_3)_2 \cdot 6H_2O$, with 1.00 M HNO_3 as the solvent. You have solid $Cu(NO_3)_2 \cdot 3H_2O$, solid $Ni(NO_3)_2 \cdot 6H_2O$, and 6.00 M HNO_3 available.*

Calculate the amount of each solute:

$$Cu(NO_3)_2 \cdot 3H_2O\text{: } (0.0100\ \text{mol/L})(0.100\ \text{L})(241.6\ \text{g/mol}) = 0.242\ \text{g}$$

$$Ni(NO_3)_2 \cdot 6H_2O\text{: } (0.0100\ \text{mol/L})(0.100\ \text{L})(290.8\ \text{g/mol}) = 0.291\ \text{g}$$

$$HNO_3\text{: } (1.00\ \text{mol/L})(0.100\ \text{L})/(6.00\ \text{mol/L}) = .0167\ \text{L or } 16.7\ \text{mL of } 6\ \text{M } HNO_3$$

Mixing these quantities in a 100 mL volumetric flask, and filling to the mark with deionized water yields the desired concentrations.

3.9 TITRATION

In a titration, a buret is filled with a solution of known concentration that will react with the analyte in question. The reaction is usually arranged to result in a color change once the reaction is complete, **the onset or disappearance of color indicating the end of the titration.** In most cases, a small quantity of an indicator is added, which will change color at the desired reaction end point, which is when the analyte has completely reacted.

If a strong acid is titrated with a strong base, for example, the pH near the equivalence point will change sharply from about 5.0 to about 9.0 upon the addition of just a few drops of base. The pH at the equivalence point is 7.0 for a strong acid–strong base titration. The standard indicator solution for this type of titration is **phenolphthalein** (pronounced *fee-nol-thay-leen*), which undergoes a color change from colorless in acid solution to pink in base solution at a pH of 8.3. Only a few drops of phenolphthalein are needed. Titrations of weak acids with strong bases have an endpoint pH greater than 7.0, while titrations of weak bases with strong acids have an endpoint pH less than 7.0. Indicators that change color in other pH ranges are used for these. Some common indicators are listed in Table 3.1. Various mixtures of indicators (called **universal indicators**) are also available.

Use of the Buret

A. Microburet

The microburet assembly is shown in Figure 3.13 (a), p. 39. The buret is mounted on a ring stand using a micro clamp. It is always rinsed with the solution to be used before a titration is carried out.

1. To rinse a microburet, dip the yellow tip in the titrant solution. Draw the syringe plunger up *gently*, to add 100–150 μL of the titrant to the buret.

Table 3.1 Common Indicators: Color and Range of pH Change

Indicator Name	Acid Color	Base Color	pH Range
Thymol blue	Red	Yellow	1.2–2.8
Methyl orange	Red	Orange	3.2–4.4
Methyl red	Red	Yellow	4.8–6.0
Litmus	Red	Blue	4.7–8.3
Phenolphthalein	Colorless	Pink	8.2–10.0
Alizarin yellow	Yellow	Red	10.1–12.0

Remove the buret from the titrant solution. Move the syringe plunger up and down *slowly* several times. This operation rinses the buret in one easy step. Dispense this wash into a waste container. Repeat this step two additional times.

2. Now mount the buret in the clamp and dip the end of the buret into the titrant solution. Draw the titrant solution into the buret by moving the syringe plunger up *slowly,* until the desired volume of tritant is collected.
3. Read the volume of solution in the buret and record this value on the report sheet. With aqueous solutions, the volume of should be read at the center of the liquid meniscus curve (see Figure 3.10, p. 37). In any microscale titration, the volume of solution taken should be above 1.000 mL. This ensures better precision (four significant figures) in your volume measurments.
4. To titrate, *gently* push down on the syringe plunger. Microdrops of titrant can be delivered, if this is done with care. This operation may take some practice at first, but once the technique is mastered, excellent titration results are obtained. The sample being titrated is generally held in a small Erlenmeyer flask and as the drops of titrant are added the flask is swirled to ensure mixing of the two solutions. Alternatively, a small magnetic stir bar may be placed in the flask and the stirring done by placing the flask on a magnetic stirrer.
5. It is often helpful to place a white sheet of paper under or behind the Erlenmeyer flask to ensure color contrast. The end of the titration is indicated when the color of the indicator changes, and the color lasts for at least 30 secs.
6. After the titration is complete, discard any remaining titrant to a waste container. *Never return excess solution to the original titrant container.*

B. Conventional Buret (50-mL)

A conventional buret is generally one of 50-mL capacity with a glass or teflon stopcock at the bottom end. The buret is clamped to a ring stand using a buret clamp.

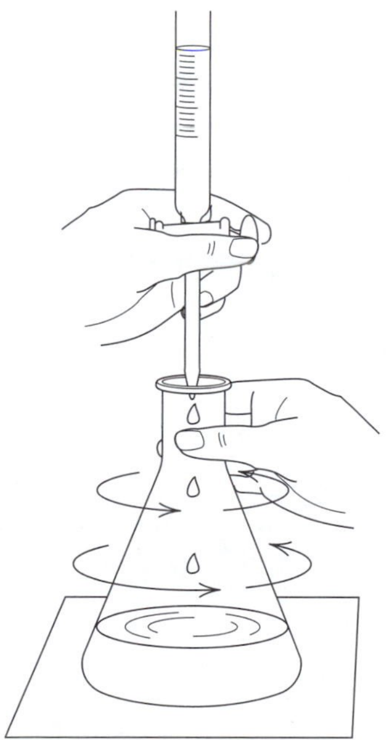

Figure 3.21 *Proper titration procedure.*

1. Prior to use, the buret should be rinsed a *minimum* of three times with a small amount (5–10 mL) of the solution to be used. **Make sure that the buret stopcock is in the closed position!** The rinse solution should be carefully added through a funnel. The buret is then gently rotated, making sure that the entire inner surface is moistened by the solution. The rinse solution is then discarded. The buret is then filled to the desired volumn.
2. To dispense the solution from the buret, slowly turn the stopcock towards the open position, until the liquid comes out slowly and dropwise. It is often convenient to titrate rapidly on a first titration, to get a "ballpark" value of solution needed to reach the endpoint. A second titration on a similar sample should then be performed, adding the solution rapidly until near the endpoint, and dropwise thereafter. The sample being titrated is usually held in an Erlenmeyer flask. The flask should be gently swirled as the solution is added, as shown in Figure 3.21. It is important to allow sufficient time for equilibrium to be reached, especially in the dropwise portion of the titration. It is often helpful to place a white sheet of paper under or behind the Erlenmeyer flask to provide color contrast. The titration endpoint is reached at the first permanent (30 sec. or longer) sign of color.

Chapter 4
Basic Concepts of Chemistry

Experiment 1 Basic Laboratory Measurements

OBJECTIVES

- To learn how to use balances, graduated cylinders, pipets, thermometers, microburners, and filtration apparatus.

PRIOR READING

Chapter 1: Safety in the Laboratory
Section 3.1: Glassware
Section 3.2: Transferring of Chemicals
Section 3.3: Weighing
Section 3.4: Liquid Volumes
Section 3.5: Heating Methods
Section 3.6: Stirring
Section 3.7: Filtration

INTRODUCTION

Chemistry is an experimental science. Any successful experiment depends upon careful measurements, good laboratory technique, an ability to observe phenomena and draw conclusions, and an eye toward safety. In this experiment, you will be introduced to some basic laboratory equipment and techniques. These methods will be used many times in your chemical career. Developing good laboratory technique is crucial.

Units of Measurement

Most scientific work is carried out using the **metric system.** Moves have been made to standardize basic units for various physical quantities. The most common set of units is the **SI** system (SI = *systeme internationale*, French for international system). The basic units are listed in Table E1.1, along with the more familiar English equivalents. Various prefixes indicating magnitude are also used together with the metric units, as listed in Table E1.2.

Measurement of Mass: Balances

The laboratory balance is used to obtain the mass of the object being investigated. The proper operation of a laboratory balance (Figures 3.6 and 3.7) is described in Section 3.3, p. 34–35. Most modern balances are capable of weighing accurately to at least the nearest milligram (mg). Never place a chemical

Table E1.1 Comparison of Metric, SI and English Unit Systems

Physical Quantity	Metric Unit	SI Unit	English Unit	Conversion Factors
Length	meter (m)	meter (m)	inch (in.), foot (ft)	1 in. = 2.54 cm 1 m = 100 cm 12 in. = 1 ft
Mass	gram (g)	kilogram (kg)	pound (lb)	1 lb = 0.454 kg 1000 g = 1 kg
Volume	liter (L)	cubic meter (m^3)	cubic foot (ft^3)	1 ft^3 = 0.0283 m^3 1 m^3 = 1000 L
Temperature	Celsius (°C)	Kelvin (K)	Fahrenheit (°F)	°C + 273.15 = K °C = (5/9) (°F − 32)
Energy	calorie (cal)	joule (J)	British thermal unit (BTU)	1 cal = 4.184 J 252 cal = 1 BTU

Table E1.2 Common Metric Prefixes

Prefix	Meaning	Abbreviation
pico-	$\times 10^{-12}$	p
nano-	$\times 10^{-9}$	n
micro-	$\times 10^{-6}$	μ
milli-	$\times 10^{-3}$	m
centi-	$\times 10^{-2}$	c
deci-	$\times 10^{-1}$	d
kilo-	$\times 10^{3}$	k
mega-	$\times 10^{6}$	M

directly on the balance pan—it can permanently damage the surface. Glazed weighing paper, small glassware, or a plastic weighing boat should be used.

Many balances are capable of taring out the weight of the container. The container should be placed on the balance pan. After a steady reading is obtained, the **Tare** button or lever should then be depressed. The material being weighed can now be added to the container, and its mass can be read directly.

Measurement of Liquid Volume: Graduated Cylinders and Pipets

Two types of glassware are used to measure liquid volumes. When only fair accuracy is needed, graduated cylinders (Figure 3.1) are used. For more exact work, pipets (Figure 3.8) are used. Glass pipets are commercially available to measure volumes from 1 mL to 100 mL. Their use is described in Chapter 3, Section 4. The liquid is drawn into the pipet using a bulb or a pipet pump (Figure 3.8). **Never pipet by mouth.** The smaller volumes that are necessary at the microscale level are measured using automatic delivery pipets (Figure 3.4). Your laboratory instructor will inform you as to the proper operation of these pipets.

Thermometers

Thermometers (Figure 3.2) are used to measure the temperature of a system. For exact work, a thermometer should be calibrated. The simplest way to do this is to determine the melting and/or boiling points of a series of pure materials. The measured temperature is then plotted against the literature values for the melting or boiling points. The resulting graph is known as a calibration curve.

Microburners

A microburner is a smaller version of the more familiar Bunsen burner, and is a convenient source of heat in the laboratory. Both are shown in Figure 3.14, and are operated in a similar fashion. A rubber hose is used to connect the microburner to a gas jet. Proper use and lighting of microburners is discussed in Chapter 3, Section 5. The gas flame can be adjusted by turning the air-flow adjustment (at the base of the microburner) to one side or the other, until a flame of the desired consistency is obtained. The best flames have a pale blue inner cone, with the hottest part of the flame being the region just above the cone.

Filtration Apparatus

Filtration is discussed in Section 3.7, and pictured in Figure 3.19. Many laboratory experiments have been ruined by accidentally knocking over of a filtration apparatus, scattering the product. This can be easily avoided by **clamping** the filtration apparatus to a ring stand.

A second common error when carrying out a suction filtration is to turn off the suction before the vacuum is released. Water is then sucked into the system, with potentially disastrous results to the filtrate. The filtration apparatus always should be connected to a water trap, which should in turn be connected to an aspirator. The connection is done at the side pipe of the aspirator, not at the water outlet. Turning on the water causes a partial vacuum to be formed in the system.

EXPERIMENTAL PROCEDURE

Part A: Determination of Volume Accuracy

Various pieces of apparatus are used to measure volume—some are more accurate than others. Graduated cylinders provide only fair accuracy of measurement, whereas calibrated pipets are far better for exacting work. Be sure to enter all measurements and calculations on the data sheets.

Use of a Graduated Cylinder

- Measure 2.0 mL of water in a 10 mL graduated cylinder. Be sure to read the liquid level from the bottom of the meniscus, as shown in Figure 3.10.
- Measure and record the water temperature.
- As completely as possible, transfer the water to a tared 10 mL Erlenmeyer flask.
- Determine the mass of the water. The densities of water at various temperatures are given in Table E1.3.
- Using the mass of the water and the densities (ρ) from the table, determine the volume of water actually transferred, using the density equation:

$$\rho = m/V$$

Use of a Pipet

- Now, measure 2.00 mL of water using a 2 mL glass pipet.

If using a pipet pump: Place a pipet pump (Figure 3.8) on top of the pipet, and carefully draw the water into the pipet, up to the mark.

If using a pipet bulb: Place the pipet bulb (Figure 3.9) on top of the pipet, and draw the water above the mark. Release the suction by removing the bulb and quickly place a finger on top of the pipet. Allow the water level to slowly drop to the mark.

- Transfer the water to a tared 10 mL Erlenmeyer flask, and determine the mass of the water and its temperature.
- Using the mass of the water and the densities (ρ) from Table E1.3, calculate the volume of water actually transferred.

Table E1.3 Density of Water at Various Temperatures

Temperature, °C	Density, g mL^{-1}	Temperature, °C	Density, g mL^{-1}
10	0.99973	20	0.99823
15	0.99913	25	0.99707
18	0.99862	30	0.99567

Use of an Automatic Delivery Pipet

- Measure 500 µL of water using an automatic delivery pipet (Figure 3.4). Your instructor will explain the operation of the particular brand of automatic delivery pipet used in your laboratory.
- Transfer the water to a tared 10 mL Erlenmeyer flask, and determine the mass of the water and its temperature.
- Using the mass of the water and the densities (ρ) from the table, determine the volume of water actually transferred.

Error Analysis

- Determine the % error (based on the determined mass of the water) for each type of apparatus that you used to measure volumes.

Part B: Calibration of a Thermometer

> ***A thermometer can be calibrated by making a plot of the observed temperature versus actual temperature. The actual temperatures can be obtained using known melting or boiling points of pure materials. Obviously, the more data points that are used, the more accurate the calibration curve will be. In this case, we will assume that the calibration is linear, and make do with only two points: the boiling point and freezing points of water. Additional materials can be used to obtain more accurate results. Be sure to enter all measurements and calculations on the data sheets.***

- Place a wire gauze on an iron ring attached to a ring stand. Place and clamp a 25 mL Erlenmeyer flask, containing 10 mL of water and a boiling stone on the wire gauze, as shown in Figure E1.1.

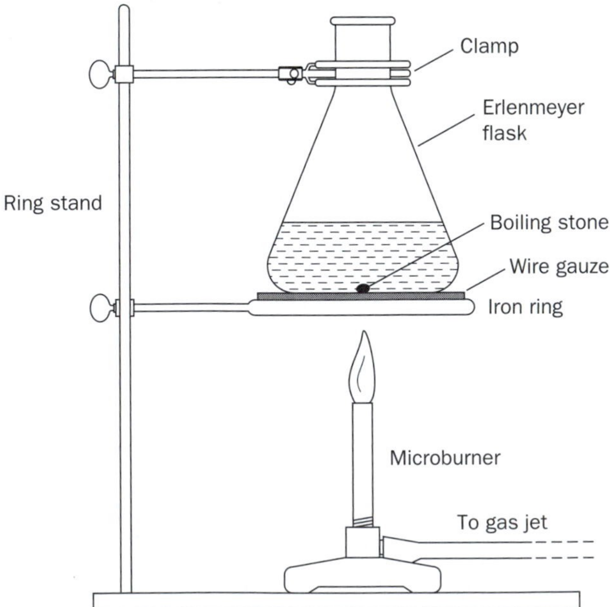

Figure E1.1 *Experimental apparatus for Experiment 1.*

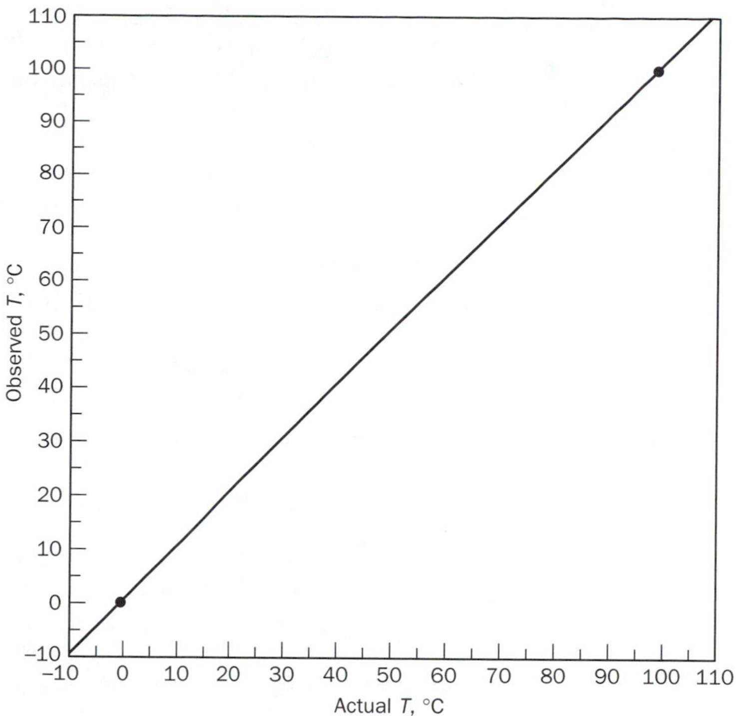

Figure E1.2 *Temperature calibration curve.*

- Light the microburner and heat the water to the boiling point. Record the observed temperature of the boiling water. Turn off the gas, and allow the water to cool back to room temperature.
- The actual boiling point temperature (T_{bp}) can be obtained from the equation

$$T_{bp} = 100.0°C - (760 \text{ torr} - \text{barometric pressure})(0.037°C/\text{torr})^*$$

  The barometric pressure will be given to you by your instructor.
- Using the same thermometer, measure the observed temperature of an ice–water mixture (The actual temperature is 0.0°C).
- Prepare a calibration curve graph of observed temperature (*y* axis) versus actual temperature (*x* axis) for your thermometer, as shown in Figure E1.2.

Part C: Solubility of KNO_3 at Ice Temperature

The object of this part of the experiment is to determine the solubility of KNO_3 in water at approximately 0°C (ice–water temperature). **Solubility** is defined as the mass of the material being dissolved (called the **solute,** KNO_3 in this case) that will dissolve in a given amount of **solvent** (water in this case). The normal units of solubility are g mL^{-1} or g L^{-1}. The solubility of a material depends on the temperature of the solvent it is being dissolved in. Most materials have higher solubilities at higher temperatures, although there are others for which the opposite is true. A good way of making a **saturated solution** (one containing the maximum amount of solute) is to add excess solid material to a hot solvent, and then allow the solution to cool. Any extra solute will then precipitate. A later experiment (Expt. 6) deals with this concept on a more quantitative basis. *Be sure to enter all measurements and calculations on the data sheets.*

*This equation is reasonably accurate down to about 650 torr. For a more exact equation, see Experiment 2.

- Obtain a weighing boat or piece of glazed weighing paper. Place it on the balance pan, and depress the "Tare" button or lever. If a taring balance is not available, the mass of the container must also be recorded on the data sheet.
- As shown in Figure 3.3, transfer and weigh 1.500 g (±10 mg) of potassium nitrate (KNO_3) to the nearest milligram. [For a non-taring balance, the mass shown on the balance should now be 1.500 g ***plus*** the mass of the container.]
- Carefully, transfer the KNO_3 into a 25 mL Erlenmeyer flask. Using a graduated cylinder, add 5.0 mL of water.
- Place a wire gauze on an iron ring attached to a ring stand. Place and clamp the Erlenmeyer flask on the wire gauze, as shown in Figure E1.1.
- Light the microburner and heat the solution with stirring until the KNO_3 has completely dissolved.
- Turn off the gas jet, and remove the microburner. Unscrew the clamp holding the Erlenmeyer flask from the ring stand **(NOT FROM THE FLASK!),** and using the clamp as a handle, set the flask in a cooling water bath. Allow the solution to cool to room temperature.
- Now place the flask in an ice-water bath, and allow the solution to sit for 10 minutes. Record the temperature of the cold solution using the calibrated thermometer. Some of the KNO_3 should precipitate from the cold solution.
- While waiting, assemble a suction filtration apparatus with water trap as shown in Figure 3.19. Be sure to clamp both filter flasks—it is very easy to tip the flasks over if this is not done. Place a piece of filter paper in the Hirsch funnel, and wet it with a few drops of water.
- Turn on the aspirator water, and with good agitation to loosen the precipitated KNO_3, rapidly transfer both the solid and liquid to the Hirsch funnel. Scrape any remaining solid into the Hirsch funnel using a rubber policeman. **Do not wash the filter cake.** Allow the filter cake to dry at the pump for 3–4 minutes.
- Transfer the solid KNO_3 to a tared container (a piece of weighing paper, a weighing boat, or a piece of glassware), and determine the amount of KNO_3 that precipitated. Since the original amount of KNO_3 added is known, the amount remaining in solution can be obtained by taking the difference.

$$KNO_3 \text{ (in solution)} = KNO_3 \text{ (original)} - KNO_3 \text{ (precipitated)}$$

Knowing the volume of water used, calculate the solubility of the KNO_3 in the ice-cold water.

$$\text{Solubility} = (\text{mass } KNO_3 \text{ in solution})/(\text{mL of solution})$$

Name: ____________________

Date: ________ Section: ____________________

Pre-Laboratory Questions: Experiment 1

1. Convert the following English measurements into their metric equivalents.

a. 13.15 in.

b. 2.35 ft^3

c. 0.655 lbs

2. Convert the following metric measurements into their SI equivalents.

a. 357 mm

b. 235 L

c. 1,355 g

3. What is the density of an object with a mass of 10.35 g and a volume of 2.95 mL? What volume of this substance would have a mass of 1.000 g?

4. What is the volume of a liquid that has a mass of 5.99 g and density of 0.755 g mL^{-1}?

Name: ______________________

Date: ________ Section: ______________________

Data Sheet: Experiment 1

Part A: Determination of Volume Accuracy

1. Graduated Cylinder

Volume of water (measured)	________	mL
Mass of water	________	g
Actual volume of water (calculated)	________	mL
Water temperature	________	°C

Calculate the % error.

$$\%\ \text{error} = \frac{(\text{measured volume}) - (\text{calculated volume})}{(\text{calculated volume})}(100)$$

________ %

2. Glass Pipet

Volume of water (measured)	________	mL
Mass of water	________	g
Actual volume of water (calculated)	________	mL
Calculate the % error.	________	%

3. Automatic Delivery Pipet

Volume of water	________	mL
Mass of water	________	g
Actual volume of water (calculated)	________	mL
Calculate the % error.	________	%

Part B: Calibration of a Thermometer

Barometric Pressure	________	torr
Observed Boiling Point of Water	________	°C
Actual Boiling Point of Water	________	°C
Observed Freezing Point of Water	________	°C
Actual Freezing Point of Water	0.0	°C

Prepare a calibration curve of observed temperature (*y* axis) versus actual temperature (*x* axis).

Part C: Solubility of KNO_3 at Ice Temperature

Original Mass of KNO_3	________	g
Volume of Water Added	________	mL
Temperature of Cold Solution	________	°C
Mass of KNO_3 Collected by Filtration	________	g
Mass of KNO_3 Remaining in Solution	________	g
Solubility of KNO_3 at Cold Temperature	________	g mL^{-1}

Name: ______________________________

Date: __________ Section: ______________________________

Questions: Experiment 1

1. Why is it necessary to calibrate a thermometer or to calibrate glassware?

2. Which of the three pieces of volume measurement apparatus was the most accurate? Explain.

3. An object is weighed on an analytical microbalance that has an accuracy of ± 0.1 mg. The mass of the object is precisely 4 g (infinite precision). To how many decimal places should the mass be written? Explain.

Experiment 2 Determination of the Physical Properties of a Chemical

OBJECTIVES

- To learn how to make basic physical measurements using laboratory glassware, electronic digital balances, syringes, and micropipets.
- To learn how to measure boiling points, melting points, solubilities, and densities as means of characterizing compounds.

PRIOR READING

Section 3.1: Glassware
Section 3.2: Transferring of Chemicals
Section 3.3: Weighing
Section 3.4: Liquid Volumes
Section 3.5: Heating Methods
Section 3.6: Stirring

INTRODUCTION

In many experiments, the goal is to prepare a new material. In order to be sure that the material has been prepared properly, its physical properties must be measured. Some physical properties that may be of interest are listed below.

Color

While many solids are white in color, and many liquids are colorless, others are distinctively colored. This is especially true of compounds containing the first row transition metals (elements 22–29). There are many gradations for a particular color as well. As will be seen in other experiments, color can be a very sensitive indicator of compound identity. Large crystals of a compound appear darker in color than the ground up compound. ***Colorless*** means having no color (like glass or water). The term ***clear*** has nothing to do with color. It tells whether or not you can see through something. Water is clear and colorless, while tea is clear and brown, and milk is cloudy and white (cloudy, meaning you cannot see through it).

Phase

Is the material a solid, liquid, or vapor?

Crystalline Form

Many materials form well-ordered crystals in the solid state. There are many different types of crystals: some are cubic, some look like needles, while others form "feathers." Any of these, when crushed, can be described as small crystals or, if finely ground, as powder. Other materials have no particular crystal structure, and are termed **amorphous.**

Melting Point

The melting point is the constant temperature at which a material changes from a solid to a liquid.

Boiling Point

The boiling point is that temperature at which the vapor pressure of a liquid is equal to that of the applied pressure. At this constant temperature, the liquid will change phase to form a vapor. When the applied pressure is 1 atmosphere (average sea level pressure, equal to 760 torr), the boiling point is termed the **normal boiling point.** At lower barometric pressure, the boiling point is lower than the normal boiling point. The relationship between the barometric pressure (in torr) and the boiling point of water is given by

$$T_{bp} = 4908/[19.786 - \ln(P_{barometric})] \tag{1}$$

For calculations involving the temperature of a gas or vapor, the absolute temperature (T, Kelvins) must be used (Note: no degree symbol is used with K).

$$T\ (\text{K}) = T\ (^\circ\text{C}) + 273.15^\circ$$

In Equation 1, T_{bp} is the boiling point in Kelvins—subtract 273.15 to get °C.

Example: What is the boiling point of water at a barometric pressure of 760.0 torr?

Answer:

$$T_{bp} = 4908/[19.786 - \ln(P_{barometric})]$$
$$T_{bp} = 4908/(19.786 - \ln(760))$$
$$T_{bp} = 373.15\ \text{K}$$
$$T_{bp} = 373.15\ \text{K} - 273.15 = 100.00^\circ\text{C}$$

Density

The density (ρ) of a material is defined as the mass per unit volume and has the units g mL^{-1}:

$$\rho = \frac{m}{V}$$

Example: A sample having a volume of 301 μL has a mass of 0.246 g. What is the density?

Answer: First, convert the volume to units of mL:

$$301\ \mu\text{L}\ (1\ \text{mL}/1000\ \mu\text{L}) = 0.301\ \text{mL}$$

Substituting into the density equation:

$$\rho = \frac{m}{V}$$

$$\rho = 0.246\ \text{g}/0.301\ \text{mL} = 0.817\ \text{g mL}^{-1}.$$

Solubility

Another convenient way of characterizing a material is by its solubility. The solubility is defined as the maximum mass of a given substance (the **solute**) that will dissolve in a given quantity of liquid (the **solvent**) at a particular temperature. The combination of solute and solvent is called a **solution.** While water is the most common of solvents, other materials are also extensively used, such as acid solutions (water with a dissolved acid), basic solutions (water with a dissolved base), ethanol (drinking alcohol), acetone, or hexane.

The solubility of a particular compound often varies greatly from one solvent to another. Materials that do not dissolve in a particular solvent are said to be **insoluble** in that solvent. Some materials will dissolve only in acidic solution, and some only in basic (alkaline) solution, some in both. Some compounds even react with the solvent (or the acid or base in the solvent). This would be indicated by obervations such as a color change, evolution of a gas, or formation of a new precipitate (solid).

There is no sharp dividing line for a material being soluble or insoluble—an arbitrary ratio of solute to solvent must be selected. A convenient ratio to use is that solutes with solubilities greater than 15 mg of solute per 0.5 mL of solvent are classified as soluble.

EXPERIMENTAL PROCEDURE

Record the color, phase, and crystalline form (as appropriate) for each material assigned by your laboratory instructor. Perform the appropriate tests from those listed below for each material. Fill in your observations and measurements on the laboratory data sheet.

Melting Point

With small quantities of chemicals, the melting point can be determined using a melting point capillary tube.

- Load the melting point capillary with a sample by placing a small quantity of the desired compound on a clay tile or filter paper. Gently, tap the tube into the material (open end down, see Figure E2.1), so that some material enters the tube. Invert the tube and tap it on the countertop to drive the

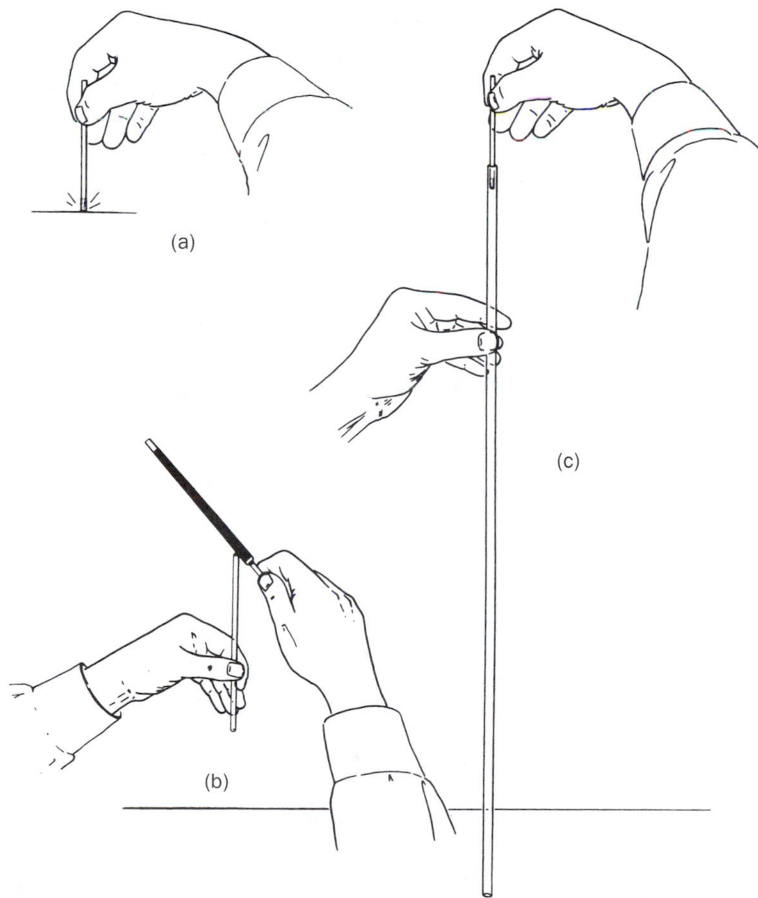

Figure E2.1 *Loading a melting-point capillary tube. (a) Tap the tube into the material. (b) File tube to pack material. (c) Drop tube down a hollow-bore glass tube.*

material to the bottom of the tube. An alternative tapping procedure is to drop the melting point tube down a long (2–3 ft), hollow bore glass tube, such that it bounces repeatedly off the countertop. Repeat this process until 3–4 mm in height of the material has been accumulated.

- Place the loaded melting point tube in a melting point apparatus. Follow the directions given by your laboratory instructor for the specific melting point apparatus that will be used. Note that most materials soften just before melting. Do not mistake this for the actual melting point, where the material liquefies. Note the temperature on your data sheet.

Boiling Point

- Using a pipet, place 0.5–1 mL of the desired liquid in a small (10 × 75 mm) test tube, along with a boiling stone.
- Clamp the test tube onto a ring stand, and insert a thermometer into the test tube, clamping it so that the bulb is just above (3–5 mm) the liquid surface (see Figure E2.2a).
- Heat the tube with a microburner, gently at first, then more vigorously. **Be careful not to melt the test tube.** Continue heating until liquid condenses on the thermometer bulb and drops back into the liquid. The temperature will reach a constant value at this point. With a little care and practice, boiling points accurate to within 1–3°C can be obtained in this way. Note the temperature on the data sheet.
- An alternative method uses a melting point capillary tube (see Figure E2.2b). Place the melting point capillary tube closed-end-up in a 10 × 75 mm test tube. Add about 0.5 mL of the experimental liquid to the test tube. Using an elastic band, attach the test tube to a thermometer.
- Insert the assemblage into a hot water or oil bath equipped with a magnetic stir bar. While stirring, increase the temperature of the bath until a rapid and continuous stream of bubbles comes out of the small capillary tube.
- Reduce the heat, and maintaining stirring, allow the bath to cool. The boiling point is the temperature at which bubbles cease to come out of the capillary, and liquid begins to enter it.

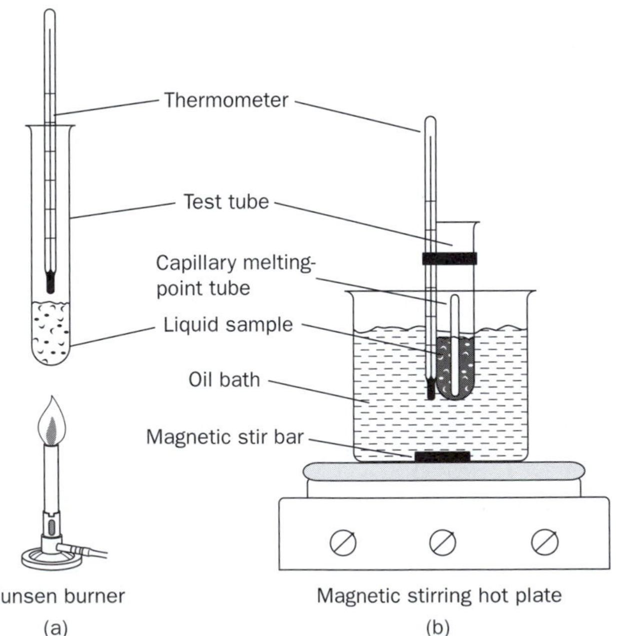

Figure E2.2 *Micro boiling point determinations. (a) Bunsen burner method. (b) Hot plate method.*

Density

In this experiment, the density of a solid will be determined by adding a sample to a weighed container that contains a known volume of water (the solid must not dissolve in water). The solid displaces a volume of water equal to the volume of the solid. The container is reweighed to obtain the mass of solid added. The mass divided by the change in volume gives the density. Liquid densities are determined by weighing a measured volume of the substance.

Density of a Solid

- Add about 5 mL of water to a 10 mL graduated cylinder (if the solid is soluble in water, some other solvent must be used that does not dissolve the solid). Read the volume of liquid in the cylinder to the nearest 0.1 mL, and weigh the cylinder and contents.
- Add the solid sample (an amount to displace 2–3 mL of water) to the graduated cylinder. If the solid to be added is heavy, tilt the graduated cylinder and let the solid slide into the liquid, so as not to break the cylinder. Make sure that all of the solid is below water, with none adhering to the sides. Also make sure that no air bubbles are present in the water.
- Weigh the charged graduated cylinder. Measure the new water level to the nearest 0.1 mL. Do as many trials as your instructor indicates.

Density of a Liquid

- Draw 200–400 μL of liquid into a micropipet, record the exact volume, and discharge it into a tared (or preweighed, record mass) 10 mL Erlenmeyer flask. Reweigh the flask and contents.
- Alternatively, draw 200–400 μL of the material into a previously weighed (record mass) 1 mL syringe. Record the exact volume and reweigh the charged syringe.

Solubility

In this experiment, the solubility of several compounds will be tested in deionized water, 3 M HCl (3 moles of hydrochloric acid in 1 L of solution) and 3 M NaOH (3 moles of sodium hydroxide in 1 L of solution). Solubility determinations can be carried out at room temperature in 10×75 mm test tubes.

- Place approximately 15 mg of the sample in the test tube, and add a total of about 0.5 mL of solvent using a pipet.
- Stir the mixture vigorously with a glass stirring rod for 1–2 minutes (or with a vortex mixer). If the amount of solid has visibly decreased, continue stirring until no further decrease is observed.

Name: ______________________________

Date: __________ Section: ______________________________

Pre-Laboratory Questions: Experiment 2

1. Using a micropipet, 275 μL of a particular liquid was added to a 10 mL Erlenmeyer flask. The weight of the empty flask was 8.471 g. After the liquid was added, the flask and contents weighed 8.785 g. What is the density (g mL^{-1}) of the liquid?

2. You should be familiar with the following common substances—describe their color, phase, and crystalline form, as appropriate. State whether the melting point is above or below room temperature.

Compound	*Color*	*Phase*	*Crystalline form*	*mp*
Water				
Quartz				
Oxygen				
Sugar				

3. Report the following information for the elements listed below. Obtain the data from your textbook or the *CRC Handbook of Chemistry and Physics*, Section B.

Compound	*Color*	*Phase*	*Crystalline form*	*mp*
Lead				
Chlorine				
Mercury				

Name: ____________________

Date: ________ Section: ____________________

Data Sheet: Experiment 2

Names and formulas of materials analyzed

A. ____________________

B. ____________________

C. ____________________

	Material A	*Material B*	*Material C*
Color	______	______	______
Phase	______	______	______
Crystalline form	______	______	______
mp range	______ °C	______ °C	______ °C
bp range	______ °C	______ °C	______ °C

Determination of Density, Solids

Compound ____________________

	Trial 1	*Trial 2*	*Trial 3*
Volume of water in grad. cylinder before adding solid	______ mL	______ mL	______ mL
Mass of grad. cylinder plus water	______ g	______ g	______ g
Volume of water in grad. cylinder after adding solid	______ mL	______ mL	______ mL
Mass of grad. cylinder, water and solid	______ g	______ g	______ g
Volume of solid	______ mL	______ mL	______ mL
Mass of solid	______ g	______ g	______ g
Density of the compound	______ g mL^{-1}	______ g mL^{-1}	______ g mL^{-1}
Average density		______ g mL^{-1}	

Name: ______________________________

Date: __________ Section: ______________________________

Data Sheet: Experiment 2, p. 2

Determination of Density, Solids (continued)

Compound ______________________________

	Trial 1	*Trial 2*	*Trial 3*
Volume of water in grad. cylinder before adding solid	______ mL	______ mL	______ mL
Mass of grad. cylinder plus water	______ g	______ g	______ g
Volume of water in grad. cylinder after adding solid	______ mL	______ mL	______ mL
Mass of grad. cylinder, water and solid	______ g	______ g	______ g
Volume of solid	______ mL	______ mL	______ mL
Mass of solid	______ g	______ g	______ g
Density of the compound	______ g mL^{-1}	______ g mL^{-1}	______ g mL^{-1}
Average density		______ g mL^{-1}	

Determination of Density, Liquids

Compound ______________________________

	Trial 1	*Trial 2*	*Trial 3*
Volume of liquid	______ mL	______ mL	______ mL
Mass of container	______ g	______ g	______ g
Mass of ctr. + liquid	______ g	______ g	______ g
Mass of liquid	______ g	______ g	______ g
Density of the compound	______ g mL^{-1}	______ g mL^{-1}	______ g mL^{-1}
Average density		______ g mL^{-1}	

Name: ______________________________

Date: __________ Section: ______________________________

Data Sheet: Experiment 2, p. 3

Determination of Density, Liquids (continued)

Compound ______________________________

	Trial 1	*Trial 2*	*Trial 3*
Volume of liquid	______ mL	______ mL	______ mL
Mass of container	______ g	______ g	______ g
Mass of container plus liquid	______ g	______ g	______ g
Mass of liquid	______ g	______ g	______ g
Density of the compound	______ g mL^{-1}	______ g mL^{-1}	______ g mL^{-1}
Average density		______ g mL^{-1}	

Solubility

If soluble, write **"S"**; if insoluble, write **"I"**; write **"R"** if it appears to react. Record what the mixture looks like. (Note, some samples may be soluble in more than one solvent.)

	Water	*3 M HCl*	*3 M NaOH*	*Observation*
A.	______	______	______	______________
B.	______	______	______	______________
C.	______	______	______	______________

Name: ______________________________

Date: __________ Section: ______________________

Questions: Experiment 2

1. Mountain climbers find that it takes longer to cook food in boiling water on top of a mountain than at its base. Explain why, in terms of the definition of boiling point.

2. In determining the density of a solid, a student does not notice some air bubbles trapped in with the solid. What measurement would be affected? Would the measurement be larger or smaller than it should be? Would the determined density be high or low? Explain.

3. The equation relating approximate barometric pressure (P, in torr) to altitude (h, in m) is (at room T):

$$\ln P = 6.63 - (h/8000)$$

Diethyl ether, $CH_3CH_2OCH_2CH_3$, a common organic solvent, has a normal boiling point of 34.6°C (and thus a vapor pressure of 760 torr at 34.6°C). At 20°C it has a vapor pressure of 440 torr. At what approximate altitude will diethyl ether boil at 20°C? Where on the earth would it be a problem to keep this solvent liquid?

Additional Literature Work

The solubility of most materials is quite different in one set of solvents compared to others. For example, common table salt is soluble in water, but insoluble in gasoline. Explain why the solubility depends on the natures of the solute and solvent.

Experiment 3 Generation and Identification of Gases

OBJECTIVES

- To prepare and identify the properties of several common gases.
- To identify an unknown gas from its properties.

PRIOR READING

Section 3.4: Liquid Volumes (Use of Pasteur Pipet)

Section 3.5: Heating Methods

INTRODUCTION

Gases are produced as products in many chemical reactions. Several common elements and a large number of compounds (both inorganic and organic) are found as gases at room temperature. To name a few:

Elements: Hydrogen, Helium, Nitrogen, Oxygen, Fluorine, Chlorine

Compounds: Ammonia, Carbon Dioxide, Methane, Ethylene

While all of the above materials are in the same phase at room temperature, they differ in their physical and chemical properties. These properties can be used to identify the particular gas produced in a given chemical reaction.

Hydrogen, H_2, is a diatomic, colorless, odorless gas, and is very flammable. It has a boiling point of −252.87°C. It forms explosive mixtures with oxygen. Hydrogen is usually produced due to the action of an acid on an active metal or by the electrolysis of water.

Helium, He, is a monotomic, colorless, odorless gas. It has a boiling point of −268.93°C. It undergoes no known reactions, and therefore is never formed in chemical reactions.

Nitrogen, N_2, is a diatomic, colorless, odorless gas. Its boiling point is −195.79°C. It is quite unreactive, but is commonly formed in chemical reactions involving nitrogen groups such as NO_3^- and NO_2^-. Nitrogen does not support combustion, i.e., few materials burn in an atmosphere of pure nitrogen.

Oxygen, O_2, is a diatomic, colorless, odorless gas. It supports combustion, i.e., many materials burn in oxygen. The common test for oxygen is to see if a candle or splint stays lit (and in fact, burns more brightly) in an atmosphere of oxygen. Oxygen has a boiling point of −182.95°C.

Fluorine, F_2, is a diatomic, pale yellow, pungent smelling gas boiling at −188.12°C. It is quite poisonous to breathe. It reacts with nearly anything, and is therefore quite dangerous. Many compounds will "burn" in fluorine, forming fluorides as products.

Chlorine, Cl_2, is a diatomic, yellow-green, pungent smelling gas boiling at −34.04°C. It is poisonous to breathe and very reactive, although not as reactive as fluorine. Many compounds will "burn" in chlorine, forming chlorides as products. Chlorine gas is used commercially as a bleaching agent. It was the first poisonous gas used during World War I.

Ammonia, NH_3, is a polyatomic, colorless, pungent smelling gas having a boiling point of −33.33°C. It has the distinctive smell of smelling salts, which are combinations of chemicals that form ammonia. It does not support combustion.

Carbon Dioxide, CO_2, is a triatomic, colorless, odorless gas boiling at −78.4°C. It does not support combustion, but in fact, is one of the end products of combustion. The common test for carbon dioxide is to bubble it through barium chloride ($BaCl_2$) solution, whereupon an insoluble white precipitate of barium carbonate ($BaCO_3$) will form.

In this experiment, we will prepare two known gaseous elements. An unknown gas will then be prepared and identified.

EXPERIMENTAL PROCEDURE

Preparation of a Gas Generator

- Obtain a 13 × 100 mm test tube, a #00 2-hole stopper, a short-stem Pasteur pipet and pipet bulb, a 1 ft length of 1/4 in. OD Tygon tubing, and a piece of glass tubing to fit the stopper and Tygon tubing.
- Insert the glass tubing into one of the holes in the stopper (see Figure E3.1), so that it extends just a little way below the stopper. *Be sure to wrap your hands in a towel during this operation.* Attach the Tygon tubing to the top of the glass tube. The Pasteur pipet and bulb will complete the assembly.

Part A: Preparation of Oxygen

$$\text{Reaction:} \quad 2\,H_2O_2 \xrightarrow{Fe^{3+}} 2\,H_2O + O_2(g)$$

- Place about 3 mL of 6% H_2O_2 (hydrogen peroxide) in the gas generator test tube and insert the stopper. Take up 0.5 mL of 0.5 M $FeCl_3$ [iron(III) chloride] in a clean Pasteur pipet (about 3/8 full) fitted with a pipet bulb. Holding the Pasteur pipet by the barrel, insert it firmly into the second hole of the stopper.
- Place a wire gauze pad on the iron ring of a ring stand. Add 150 mL of water to a 250 mL beaker, and place it on the pad. Fill a 13 × 100 mm test tube with water, block the opening with your thumb, invert the test tube, and place it in the beaker. Clamp the test tube, as shown in Figure E3.2. *Make sure that the water does not run out of the test tube.* Clamp a second test tube (inverted, filled with water identical to the first) in the beaker.
- Slowly, add the $FeCl_3$ solution a little at a time by squeezing the pipet bulb, allowing the gas being generated to flush the test tube and Tygon tubing. Now, run the open end of the Tygon tubing through the water in the 250 mL beaker, into the inverted test tube, as shown in Figure E3.2. and collect a test tube full of oxygen gas. When all of the water has been displaced from the first test tube, move the Tygon tube to the second test tube, and collect a second test tube full of oxygen gas.
- When full, allow the test tubes of oxygen to remain clamped upside down in the water container. Disassemble the gas generator and dispose of any leftover reaction solution into a Peroxide Waste Container in the **HOOD.**
- Using a microburner, ignite a 1 cm length of magnesium ribbon **(CAUTION: Do not look directly at the brilliant magnesium flame),** holding it with tongs. Using a test tube holder, remove one of the tubes of oxygen

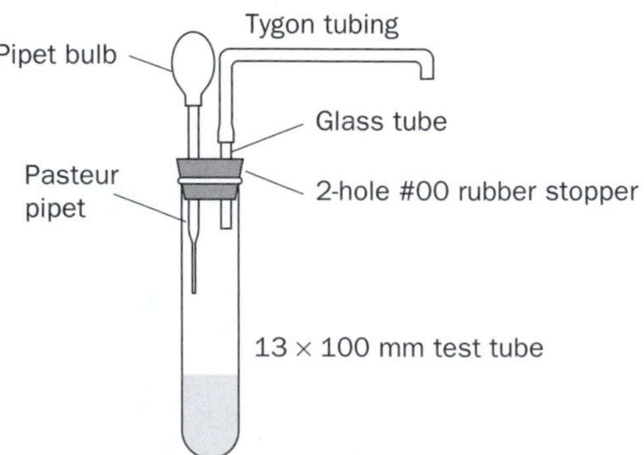

Figure E3.1 *Gas generator.*

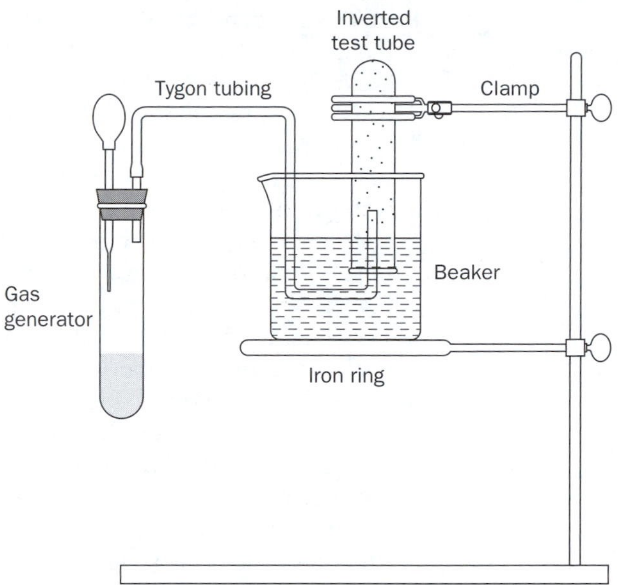

Figure E3.2 *Proper method of collecting gases.*

from the water container and immediately insert the burning magnesium. Record your observations on the data sheet.

- Take the second test tube of oxygen, still in the water container, to the **HOOD.** Place a small quantity (about 10 mg) of sulfur on the tip of a microspatula. Heat this over a microburner until the sulfur ignites.
- Using a test tube holder, remove the second test tube of oxygen from the water and immediately thrust the spatula containing the burning sulfur into the test tube. Record your observations on the data sheet.
- Wash and dry the gas generator thoroughly and remove the Pasteur pipet.

Part B: Preparation of Hydrogen

- Place about 150 mg of Zn powder in the gas generator test tube, and insert the stopper. Take up about 1 mL of 6 M HCl (hydrochloric acid) in a clean Pasteur pipet. **(CAUTION: CORROSIVE!)** Holding the Pasteur pipet by the barrel, insert it firmly into the second hole of the stopper.
- Place a wire gauze pad on the iron ring of a ring stand. Add 150 mL of water to a 250 mL beaker, and place it on the pad. Fill a 13 × 100 mm test tube with water, block the opening with your thumb, invert the test tube, and place it in the beaker. Clamp the test tube, as shown in Figure E3.2. *Make sure that the water does not run out of the test tube.*
- Release the acid a little at a time, allowing the gas being generated to flush the test tube and tubing. Collect a 13 × 100 mm test tube of hydrogen gas by displacement of water, as before.

$$\text{Reaction:} \quad 2\,H^{+}_{(aq)} + Zn^{0}_{(s)} \rightarrow H^{0}_{2(g)} + Zn^{2+}_{(aq)}$$

- When all of the water has been displaced from one of the test tubes, stopper it and remove it from the water. Clamp the test tube horizontally to a ring stand, **so that the mouth of the tube is not pointing at anyone (including yourself).**
- Strike a long kitchen match, remove the stopper from the hydrogen test tube, and place the burning tip cautiously alongside the mouth of the tube. Record your observations on the data sheet.
- Clean and dry the generator, and remove the Pasteur pipet.

Part C: Preparation of an Unknown Gas

- Place about 1 mL of saturated NH_4Cl (ammonium chloride) solution in the generator test tube. Insert the stopper. Take up about 1 mL of saturated $NaNO_2$ (sodium nitrite) solution in a clean Pasteur pipet. Holding the pipet by the barrel, firmly insert it into the second hole of the stopper.
- Place a wire gauze pad on the iron ring of a ring stand. Add 150 mL of water to a 250 mL beaker, and place it on the pad. Fill two 13 $\times$ 100 mm test tubes with water, block the opening with your thumb, invert the test tubes, and place them in the beaker. Clamp the test tubes, as shown in Figure E3.2. *Make sure that the water does not run out of the test tube.*
- Slowly add the $NaNO_2$ solution. Place the generator test tube in a 100 mL beaker containing about 65 mL of water. Heat the water to near boiling.
- Allow the gas produced to flush the test tube and Tygon tubing, and then collect two test tubes (13 $\times$ 100 mm) of the gas produced in the same manner as before. **NOTE:** The reaction may be stopped at any time by removing the generator test tube from the water bath.
- Test the gas you have collected and determine whether it is ammonia, nitrogen, oxygen, hydrogen or chlorine. Color, smell (See Safety Rules in the Laboratory, p. 4), whether combustion is supported, and any other tests from Parts A and B, should help in your identification of this unknown gas. Summarize your results on the data sheet.

Name: ______________________________

Date: __________ Section: ______________________

Pre-Laboratory Questions: Experiment 3

1. In the experimental procedure, what is the purpose of filling test tubes with water and inverting them in a water bath prior to collecting oxygen or hydrogen?

2. You will carry out a number of chemical reactions in this experiment. Several are given here. Balance each equation and name each compound and/or element.

a. $S + O_2 \rightarrow SO_2$

b. $Zn + HCl \rightarrow H_2 + ZnCl_2$

c. $NH_4NO_2 + heat \rightarrow H_2O + N_2$

3. Before collecting a gas sample by displacement of water, as in this experiment, it is prudent to allow gas to flush the gas generator and Tygon tubing for several seconds. Explain why.

4. Wet air contains monatomic, diatomic, and polyatomic gases. Give a formula and name for a gas that fits each of these types that is found in wet air.

Name: ______________________________

Date: __________ Section: ______________________________

Data Sheet: Experiment 3

1. Preparation of Oxygen: Write a balanced equation describing the reaction that takes place between magnesium metal and oxygen, assuming magnesium oxide, MgO, is formed.

Observation of combustion of sulfur:

2. Preparation of Hydrogen: Describe what occurs when the burning match is moved close to the hydrogen tube.

3. Preparation of Unknown Gas: Describe the procedure by which you identified the unknown gas, and eliminated other possibilities. What is the gas?

Name: ______________________________

Date: __________ Section: ______________________________

Questions: Experiment 3

1. Zeppelins were used as airships before their popularity was eclipsed by airplanes. Originally, zeppelins were filled with hydrogen gas because of its great lifting power. Later, hydrogen was replaced by helium. In terms of their chemical properties, explain why the change was made.

2. Most materials, when combusted in air, produce oxides rather than nitrides. Explain why, on the basis of the relative bond strengths of O_2 and N_2. Bond strength data may be found in any textbook on general chemistry.

3. When a diazonium salt is warmed to room temperature, it decomposes to form a gas. The gas is colorless and odorless, and does not support combustion. Which of the gases described in the introduction to this experiment might the decomposition product be? Explain your choice(s). The structure of diazonium salts can be found in any organic chemistry textbook.

4. Household bleach (NaOCl, sodium hypochlorite) should never be mixed with an acid solution. For example, the acid might come from household-based cleaning solutions that may contain $NaHSO_4$ (sodium bisulfate). The reaction between the two reagents generates a yellow-green, pungent, extremely toxic gas. What is the gas produced in this reaction?

Experiment 4 Periodic Trends via Internet Database (Dry Lab)

OBJECTIVE

- To recognize and plot periodic trends among the elements.

INTRODUCTION

The Periodic Table of the Elements arranges the elements in a pattern emphasizing the electronic similarities between related species. Since many physical and chemical properties arise from these electron configurations, the Periodic Table also emphasizes the elements' chemical and physical similarities. The Periodic Table is generally ascribed to Dmitri Mendeleev, a Russian chemist. While others had also noted that chemical properties repeated periodically among the elements (for example, lithium, sodium, and potassium are chemically similar), Mendeleev was the first to predict the existence and properties of hitherto unknown elements, by noting "gaps" in the periodic table due to undiscovered elements.

In the Periodic Table, the elements are arranged in vertical columns called **families** or **groups.** All members of a particular family (for example, the Alkali Metals, or Group 1 for Li, Na, K, Rb, Cs, Fr) have the same outer shell electron configuration, and hence similar properties. Horizontally, the elements were originally arranged by atomic mass, but are now arranged by atomic number. Another innovation of Mendeleev's was to deviate from strict atomic mass order whenever that was called for based on chemical behavior. At the time Mendeleev was developing his Periodic Table, the known element with the next highest atomic mass from zinc was arsenic. This would place arsenic under aluminum if strict mass order were followed. Arsenic has little chemical similarity to aluminum, but is quite chemically similar to phosphorus. Mendeleev therefore left two blank spaces for undiscovered elements under aluminum and silicon (which he called eka-aluminum and eka-silicon, later discovered and named gallium and germanium), and placed arsenic where it belonged chemically, under phosphorus. When atomic numbers and electron configurations were developed, a theoretical basis was established for Mendeleev's empirical correlations.

Various Internet websites (such as **webelements** at www.webelements.com) and computer database programs (such as KC? Discoverer) are available that allow one to list or plot various physical properties of the elements against the atomic number, or physical properties against each other. In this way, periodic trends can be observed and later understood. If Internet access or computer programs are not available, the data for the trends can be found in most general chemistry textbooks or in the *CRC Handbook of Chemistry and Physics*, and then plotted by hand or with the aid of a computer program.

EXPERIMENTAL PROCEDURE

Your laboratory instructor will show you how to access the particular website or to load the particular database program onto the computer. The discussion below is based on the website www.webelements.com, but other sites and programs are similar.

A. Atomic Weight versus Atomic Number

- Access the website as shown by your laboratory instructor. You will see a page containing a periodic table as its main feature. Select an element on the periodic table, and click on it.
- Under the element's name, you will see a section of the page labeled "The Essentials." Click on "atomic weight," and read its definition. Display a plot of **atomic weight** (y axis) versus **atomic number** (x axis) for all elements

by clicking on "bars" under the heading "Full Table Charts" on the left side of the page.

- What is the general trend that you observe? Are there any deviations from this trend among the naturally occurring elements? [This may be difficult to see—you may need to estimate where the deviations occur, and then check the elements individually to see if the deviation is there]. Under the heading "Groups Charts," select several groups and note any trends (and deviations) that you see. Do the same under "Periods Charts" for several periods.

B. Atomic Radius versus Atomic Number

- Click on the "back" button on the browser until you return to the element page you selected previously. Under the heading "Electronic Properties" on the left side of the page, click on "atom radii."
- Read the definitions of atomic radius. Plot **atomic radius** (y axis) versus **atomic number** (x axis) by selecting "bar" in the **select box** by "Atomic radius (empirical)," and then clicking "go." Select several periods, and find what is the general trend in atomic radius as you go from left to right. Explain the trend, on the basis of atomic structure. Now select several groups under the heading "Groups Charts," and determine if any trends are evident. You should look at families within the main group elements as well as within the transition metals.

C. Density versus Atomic Number

- Click on the "back" button on the browser until you return to the element page you selected previously. Under the heading "Physical Properties" on the left side of the page, click on "bulk properties."
- Plot **density** (y axis) versus **atomic number** (x axis) by selecting "bar" in the **select box** next to "density of the solid," and clicking "go." What is the general trend of density as the atomic number increases? What group of elements has the highest densities? Quite a few elements have very low densities—what physical property do they have in common? What is the trend for density across a period?

D. Melting Point versus Atomic Number

- Click on the "back" button on the browser until you return to the element page you selected previously. Under the heading "Physical Properties" on the left side of the page, click on "thermal properties."
- Plot **melting point** (y axis) versus **atomic number** (x axis) by selecting "bar" in the **select box** next to "melting point," and clicking "go." What is the general trend in melting points as you go down a particular group (not all groups show the same trend)? What group of elements has the highest melting points? Which one has the lowest melting points? You should be easily able to explain the lowest melting point family. Explain the melting point trend going down Groups 17 and 18.

E. Find Your Own Trends

- Investigate three more properties found on the website, and discuss whether a periodic trend is seen.

Name: ______________________________

Date: __________ Section: ______________________________

Pre-Laboratory Questions: Experiment 4

1. Define each of the following terms:

a. Atomic number

b. Atomic weight

c. Density

d. Isotope

e. Ionization energy

f. Electron affinity

g. Atomic volume

Name: ______________________________

Date: __________ Section: ______________________________

Data Sheet: Experiment 4

1. Write a laboratory report discussing any periodic trends you observe and answering each of the questions asked in the experimental procedure. Be sure to note and try to explain any trends across periods and down families.

2. Select three additional periodic properties of your own choice, and determine if a periodic trend is present. Discuss what the trend is, noting and trying to explain any trends across periods and down families.

Experiment 5 Investigation of Periodic Trends

OBJECTIVE

- To experimentally observe periodic trends.

PRIOR READING

Experiment 4: Introduction

INTRODUCTION

The arrangement of the various chemical elements in systematic manner, called the Periodic Table of the Elements, is generally ascribed to Mendeleev. Although the modern Periodic Table differs somewhat from his, the general principle of arranging the elements so that those with similar properties fall into the same column (family) remains the same.

The rows of the Periodic Table are called **periods,** and the columns are known as **families** or **groups.** Physical properties usually vary smoothly down a family. For example, we might wish to examine the boiling point of the **alkali metals,** Group 1 of the Periodic Table:

Li: 1317°C   Na: 892°C   K: 774°C   Rb: 701°C   Cs: 685°C

The trend is a decrease in the boiling point down the group. A major goal of chemists is to explain the periodic trends and to account for any deviations.

Other trends may be observed going across the periods. More anomalies are seen here, but once again, general patterns may be discerned. For example, if one looks at the electronegativities of the third period of elements:

Na: 1.0   Mg: 1.3   Al: 1.5   Si: 1.8   P: 2.1   S: 2.4   Cl: 2.9

The trend is toward an increase in electronegativity across the Periodic Table.

In this experiment, both a **family** and **period** trend will be investigated:

a. Oxides of the second and third period elements will be investigated to see if they form acidic or basic solutions with water. An aqueous acidic solution is one that contains more H_3O^+ than OH^-. When an oxide dissolves in water, it may react in such a way that it generates either H_3O^+ or OH^-, which can be determined by testing with litmus paper.

$$Al_2O_3 + 7\,H_2O \rightleftharpoons 2\,Al(OH)_4^- + 2\,H_3O^+$$

$$Tl_2O + H_2O \rightleftharpoons 2\,Tl^+ + 2\,OH^-$$

The results will be examined to see if there is a pattern across the periodic table.

b. The reactivities of the halogen family toward the halide ions will be investigated.

EXPERIMENTAL PROCEDURE

Record all observations on the Data Sheets. Include observations on production of heat.

Part A: Acid–Base Properties of the Second and Third Periods

Whether a solution is acidic or basic is determined by testing a drop of solution on litmus paper. Acidic solutions will turn blue litmus paper red, while basic solutions will turn red litmus paper blue. Alternatively, universal indicator paper may be used. After testing a drop of solution, compare the color of the paper with the color chart on the container to see whether the solution is acidic or basic.

Reaction 1: Sodium Oxide, Na_2O

Some instructors may elect to do this reaction as a demonstration. Sodium oxide reacts very violently with water. A related compound, sodium peroxide, Na_2O_2, is the major product of the reaction between sodium metal and oxygen. It undergoes almost the same reaction with water, but less violently.

- Place a small quantity (about the size of half of a pea) of Na_2O_2 in a 10×75 mm test tube. **Cautiously,** add one drop of water. When the reaction has subsided (give it at least 30 sec to occur), add another drop of water and so on, until any sign of reaction ceases. Test the resulting solution by dipping the tip of a glass rod into the solution, and then touching it to red and blue litmus paper, or universal indicator paper. Record your observations on the data sheet.

Reaction 2: Calcium Oxide, CaO

Calcium oxide will be used in this step instead of magnesium oxide, as MgO is insoluble in water and beryllium oxide (BeO) is quite toxic.

- Place approximately 10 mg of calcium oxide, CaO, in a 10×75 mm test tube. Dropwise, add 1 mL (about 20 drops) of water. Test the resulting solution with red and blue litmus paper, or universal indicator paper. Record your observations.

Reaction 3: Boron Oxide, B_2O_3

- Place about 10 mg of boron oxide (B_2O_3) in a 10×75 mm test tube. [If boron oxide is not available, boric acid, $B(OH)_3$, may be used instead.] Dropwise, add 1 mL of water. Test the resulting solution with red and blue litmus paper, or universal indicator paper. Record your observations.

Reaction 4: Carbon Dioxide, CO_2

> **CAUTION:** ***Dry ice is extremely cold. Do not allow it to contact your skin.***

- Using tweezers, place several small pieces of crushed dry ice (carbon dioxide, solid CO_2) in a 10×75 mm test tube. Quickly, add 1 mL of ice water to the tube, and stir rapidly using a spatula. Immediately, test the resulting solution with red and blue litmus paper, or universal indicator paper. Record your observations.

Reaction 5: Phosphorus Pentoxide, P_4O_{10}

> **NOTE:** ***Some instructors may elect to do this reaction as a demonstration.*** **CAUTION:** ***Phosphorus pentoxide is extremely corrosive. Do not allow it to contact your skin under any circumstances. Wear plastic gloves.***

- Place about 10 mg of phosphorus pentoxide, (actual formula is P_4O_{10}, tetraphosphorus decoxide) in a 10 × 75 mm test tube.
- Dropwise, add 1 mL of water to the test tube, with vigorous stirring. Be very careful not to splatter the solution that is formed. Test the resulting solution with red and blue litmus as before.

Reaction 6: Sulfur Dioxide, SO_2

CAUTION: *Do this reaction in the HOOD.*

- Place about 10 mg of sulfur in a 10 × 75 mm test tube, and clamp the test tube to a ring stand. Drape a moistened piece of blue litmus paper over the mouth of the test tube. Using a microburner, heat the sulfur until it first melts, and then begins to react with the oxygen within the test tube. **Be careful not to melt the test tube.** As the sulfur dioxide formed reaches the moistened litmus paper, it will dissolve in the water and the color will indicate the acid–base properties. Record your observations.

Reaction 7: Chlorine Oxides

Chlorine oxides are extremely dangerous, and cannot be safely handled in a general chemistry laboratory. The end result of the reaction of Cl_2O_7, dichlorine heptoxide and water is the formation of $HClO_4$.

- Obtain about 1 mL of very dilute (1 M) $HClO_4$. Test the solution with red and blue litmus paper. Record your observations.

Part B: Preparation and Reactivities of the Halogens

The reactivity of solutions of the elemental halogens (chlorine, Cl_2; bromine, Br_2; and iodine, I_2) will be tested against solutions of the halides (fluoride, F^-; chloride, Cl^-; bromide, Br^-; and iodide, I^-). The halogens, being neutral and nonpolar, dissolve better in nonpolar solvents such as toluene, whereas the ions dissolve better in water. The color of the toluene layer indicates what species are present. In toluene, chlorine is colorless, bromine is red-orange, and iodine is purple. The two solvents do not mix readily, so the mixture must be triturated (forced to mix by vigorous stirring), or by using an eyedropper or pipet to take up the lower layer (water) and drop it on top of the upper layer; as the drops fall through, the solutes can react. Elemental fluorine, F_2, is extremely dangerous and difficult to produce, so it will not be used. Fluoride, the halide form is much less dangerous.

Reaction 1: Preparation of Chlorine, Cl_2

- Place about 1 mL of chlorine laundry bleach in a 10 × 75 mm test tube. Chlorine bleach contains about 5% sodium hypochlorite, NaOCl, with the other 95% being made up of inert materials, mostly water. Add 750 μL (or 15 drops) of toluene. A two-layer system should form at this point. Note the color of the toluene (top) and water (bottom) layers.
- In the hood, acidify the system with 500 μL (or 10 drops) of 6 M HCl, and triturate with a spatula. What indication do you have that chlorine has formed, and dissolved in the toluene layer? Record your observations. Do not discard this test tube, it will be used later.

Reaction 2: Reactivity of Chlorine

Record your observations after each step.

- Place about 1 mL of 0.1 M potassium bromide (KBr) solution in a 10 × 75 mL test tube. Add one-third of the upper toluene layer (with chlorine in it) prepared in Reaction 1 above, using a Pasteur pipet. Agitate the test tube for a moment.
- Perform the same reaction again, but this time using about 1 mL of 0.1 M KI solution instead of KBr.
- Perform the same reaction again, but this time using about 1 mL of 0.1 M KF solution.

Reaction 3: Reactivity of Bromine

Record your observations after each step.

- Place about 1 mL of 0.1 M KCl solution, in a 10 × 75 mm test tube. Add about one mL of saturated bromine water **(HOOD),** using a Pasteur pipet. Add 0.5 mL of toluene, and agitate the test tube for a moment. Has a reaction taken place?
- Perform the same reaction again, but this time using about 1 mL of 0.1 M KI solution instead of the KCl.
- Perform the same reaction again, but this time using about 1 mL of 0.1 M KF solution.

Reaction 4: Reactivity of Iodine

Record your observations after each step.

- Place about 1 mL of 0.1 M KCl solution in a 10 × 75 mm test tube. Add 1 mL of 0.05 M iodine solution. Add 0.5 mL of toluene, and agitate the test tube for a moment. Has a reaction taken place?
- Perform the same reaction again, but this time using about 1 mL of 0.1 M KBr instead of the KCl.
- Perform the same reaction again, but this time using about 1 mL of 0.1 M KF solution.

Name: ____________________

Date: __________ Section: ____________________

Pre-Laboratory Questions: Experiment 5

1. Find the formula for the anhydride of each of the following acids. To obtain the formula of an anhydride, subtract H_2O's from the formula until all hydrogens have been subtracted. If the acid has an odd number of hydrogens, double its formula first.

a. H_2SO_4

b. NaOH

c. $HClO_4$

d. H_3PO_4

2. Determine the formula of the parent acid of each of the following anhydrides. To find the formula of the acid, add the indicated number of water molecules to the formula. Finally, divide through if the formula is divisible by some common factor.

	# of H_2O
a. SO_2	(1) __________
b. N_2O_5	(1) __________
c. CO_2	(1) __________
d. P_4O_{10}	(6) __________

3. Bond strengths can be correlated to such factors as bond length, bond order (single bonds, double bonds, triple bonds, etc.), electronegativity differences, and lone pair–lone pair electronic repulsions. Write Lewis dot structures for each of the halogen molecules, F_2, Cl_2, Br_2, and I_2. Which one would you expect to have the shortest bond length? Explain why fluorine is the most reactive of the halogen molecules.

Name: ____________________

Date: ________ Section: ____________________

Data Sheet: Experiment 5

Part A: Acid-Base Properties

	Acidic or Basic?	*Observations*
Sodium Peroxide Solution	______	____________
Calcium Oxide Solution	______	____________
Boron Oxide Solution	______	____________
Carbon Dioxide Solution	______	____________
Phosphorus Pentoxide Solution	______	____________
Sulfur Dioxide Solution	______	____________
Chlorine Oxide Solution	______	____________

Reactions

Write balanced equations for the reaction of each of the following oxides with water. One of the products should be H_3O^+ (for acids) or OH^- (for bases), in accordance with your results, above.

Sodium Peroxide

Calcium Oxide

Boron Oxide

Carbon Dioxide

Phosphorus Pentoxide

Sulfur Dioxide

Chlorine heptoxide

What is the common name of $HClO_4$? ____________

Name: ____________________

Date: __________ Section: ____________________

Data Sheet: Experiment 5, p. 2

Part B: Preparation and Reactivities of the Halogens

1. Observations for Preparation of Chlorine

2. Record your observations of the reactions of each halogen with each of the other halides. Include observations before and after mixing the layers (e.g., top layer was green before, colorless after mixing).

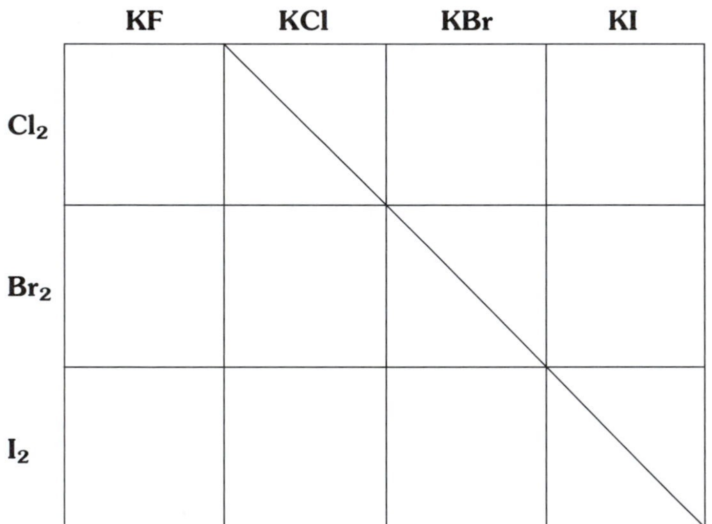

	KF	KCl	KBr	KI
Cl_2				
Br_2				
I_2				

Write balanced reactions for any of the above combinations that reacted. Use the color of the toluene layer to tell which elemental halogen is present (*NOTE*—colorless by itself does not mean that Cl_2 is present. But if it was purple, then turns colorless, that indicates a reaction occurred. Consider whether one of the products might be Cl_2.)

Name: ______________________________

Date: __________ Section: ______________________________

Questions: Experiment 5

1. Based on your experimental results, what general trend can be observed for the acidity or basicity of the oxides as one proceeds across a period on the Periodic Table?

2. Based on your experimental results, what conclusions can be drawn about the reactivities of the elemental halogens? Do you have conclusive evidence for fluorine? If not, can it be predicted by determining a pattern? Explain.

3. One of Mendeleev's great triumphs was predicting the atomic properties of the element gallium before it was discovered. On the basis of the physical properties of the then known aluminum and indium, explain Mendeleev's triumph. (Hint: how close are the properties of the middle element to the average of the properties of the elements above and below it in the same family?)

Extended Work

The structure of the atom (nucleus of neutrons and protons surrounded by an electron cloud) was unknown in Mendeleev's day. It was a stroke of genius that led him to arrange elements in groups according to behavior. Explain how the structure of atoms determines their behavior and properties and why those are periodic in nature.

Experiment 6 Determination of the Formula of a Salt by Job's Method

OBJECTIVES:

- To determine the empirical formula for a salt using gravimetric analysis.

PRIOR READING:

Section 3.3: Weighing
Section 3.6: Stirring
Section 3.7: Suction Filtration

INTRODUCTION

When a compound is synthesized, one of the first pieces of information needed is its molecular or empirical formula. No significant use can or should be made of a chemical in the laboratory until its identity is known. There are many methods of determining formulas, depending on the type of compound one has. These methods include gravimetric analysis (used in this experiment), titration (see Experiments 17–19), and spectroscopy (see Experiments 13–16).

In **gravimetric analysis**, a known reaction is carried out that produces a solid precipitate. For example, the amount of barium in a sample may be determined by dissolving the sample in water and precipitating the barium using sodium sulfate. The resulting solid, barium sulfate, $BaSO_4$, is dried and weighed. Since the formula is known, the amount of barium originally present may be calculated from the amount of $BaSO_4$ formed. Similar methods may be used to find the amounts of almost any material desired.

Other analytical techniques (such as atomic absorption spectroscopy, CHN analysis, mass spectral analysis, etc.) can also aid in determining elemental composition. These techniques are encountered in more advanced chemistry courses.

JOB'S METHOD FOR GRAVIMETRIC ANALYSIS

In 1928, a chemist named P. Job (rhymes with lobe) described a method for studying inorganic complexes in solution. This technique, also called the Method of Continuing Variation, has been used extensively as an analytical tool since that time.

This experiment introduces Job's Method. It provides a straightforward technique for determining the empirical formula of simple, relatively insoluble salts. The method involves mixing known amounts of the salt's **cation** (positive ion) and **anion** (negative ion) in varying ratios, and determining which ratio yields the greatest quantity of the salt, this ratio being the formula ratio.

As an example, suppose that we wish to determine the empirical formula of silver carbonate. Solutions of silver nitrate (providing a source of silver ions, Ag^+) and sodium carbonate (a source of carbonate ions, CO_3^{2-}) are prepared. Varying quantities of these two solutions are mixed, and the precipitate (silver carbonate) is collected, dried, and weighed after each trial. The mass of precipitate is plotted (y axis) versus the quantity of reagents used (x axis). The most precipitate is produced when the amounts of reagents mixed are in the same ratio as in the salt's empirical formula.

For this method to be effective, one must know the cation and anion present in the salt. For convenience, the concentrations of the initial cation and anion solutions should be the same. The total volume of the two solutions used must be the same for all trials (i.e., as more cation solution is used, less anion solution is used).

This process may be understood in terms of the silver carbonate example above. Suppose that the concentration of each of the two initial solutions is 0.100 M. In six different beakers, we place 1.0, 3.0, 5.0, 7.0, 9.0, and 11.0 mL of the $AgNO_3$ solution mixing with, respectively, 11.0, 9.0, 7.0, 5.0, 3.0, and 1.0 mL of the Na_2CO_3 solution. Note that the total volume is 12.0 mL in each case.

The number of moles of each reagent in a given beaker may be easily determined using the relationship

$$\text{moles} = (\text{volume, L})(\text{molarity, mol/L})$$

Thus, we can calculate that the total number of moles of silver and carbonate ions is constant in each mixture:

$$\text{moles} = (12\ \text{mL})(1\ \text{L}/1000\ \text{mL})(0.1\ \text{mol/L}) = 1.2 \times 10^{-3}\ \text{moles}$$

The amount of Ag^+ increases in a regular fashion, from 1.0×10^{-4} to 1.1×10^{-3} moles, and the amount of CO_3^{2-} decreases in the reverse order.

Job's method relies on the concept of the **limiting reagent,** i.e., the reagent that is present in insufficient amount for complete reaction of the other component present. In beaker number one, with 1.0×10^{-4} moles Ag^+ and 1.1×10^{-3} moles of CO_3^{2-} (a 1:11 ratio), it is likely that silver ion is the limiting reagent. We will run out of silver ion before all the carbonate has reacted, and therefore not produce the most possible product. In the second beaker, the ratio is 1:3 and silver ion is probably still limiting. In successive beakers the mole ratio of silver ion to carbonate ion is 5:7, 7:5, 3:1, and 11:1. With increasing amounts of silver ion from one mixture to the next, we would expect that more precipitate would be collected until the point is reached where silver ion is no longer limiting (remember that the amount of carbonate ion is decreasing). The amount of precipitate will reach a maximum at this point. From then on, carbonate ion becomes the limiting reagent. As the amount of carbonate ion continues to decrease, we would expect less and less precipitate to be collected in the remaining beakers. Specific instructions for this analysis using Job's method can be found in Pre-Laboratory Question 1.

A typical Job's plot is shown in Figure E6.1 for determining the formula of lead chloride. Note that the total number of moles of ions is constant at

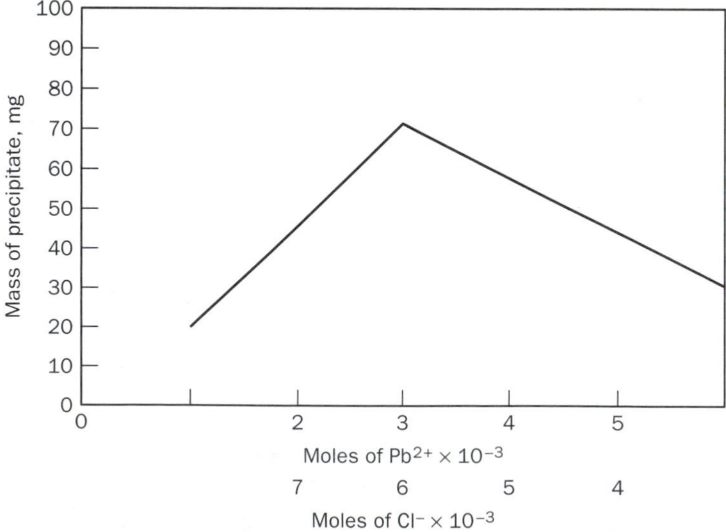

Figure E6.1 *Job's plot for lead chloride.*

9×10^{-3} moles, and that the maximum on the graph occurs at 3×10^{-3} moles of Pb^{2+} and 6×10^{-3} moles of Cl^-. Since there are two moles of Cl^- for every mole of Pb^{2+}, we can conclude that the empirical formula for lead chloride is $PbCl_2$.

In this experiment, the formula of the salt resulting from mixing ferrous ammonium sulfate hexahydrate [$Fe(NH_4)_2(SO_4)_2 \cdot 6H_2O$] and oxalic acid dihydrate ($H_2C_2O_4 \cdot 2H_2O$) will be determined. The ions involved in the reaction are Fe^{2+} and $C_2O_4^{2-}$ (oxalate). Note that the product is a dihydrate.

EXPERIMENTAL PROCEDURE

- Number six 25 mL beakers from 1 to 6 with a wax pencil. In a 50 mL beaker, place 30 mL of 0.1 M $Fe(NH_4)_2(SO_4)_2 \cdot 6H_2O$ solution. In a second 50-mL beaker, place 30 mL of 0.1 M $H_2C_2O_4 \cdot 2H_2O$ solution. Obtain two 10 mL graduated pipets (or short burets) to dispense each of the solutions above. **NOTE: Each solution should have its own pipet. The pipet should be pre-rinsed with a small amount of the solution.**
- In the six different 25 mL beakers, mix 1.00, 3.00, 5.00, 7.00, 9.00, and 11.00 mL (± 0.01 mL), of 0.100 M $Fe(NH_4)_2(SO_4)_2 \cdot 6H_2O$ solution with, respectively, 11.00, 9.00, 7.00, 5.00, 3.00, and 1.00 mL of 0.100 M $H_2C_2O_4 \cdot 2H_2O$ solution. Mix the solutions thoroughly by stirring with a clean glass rod (or a magnetic stir bar).
- Heat each of the beakers to approximately 70–80°C for 2 minutes on the hot plate. Leave the glass rod (or stir bar) in the solutions to prevent bumping. Allow the beakers to cool to room temperature.
- Assemble a suction filtration apparatus and water trap as shown in Figure 3.19, p. 42. Stir beaker #1 to suspend the precipitate and filter the contents using a Hirsch funnel. Use a rubber policeman and 2–3 mL of distilled water from a wash bottle as a rinse to ensure complete transfer. Wash the **filter cake** (the collected precipitate) with an additional 2–3 mL of distilled water. Discard the aqueous filtrate (liquid) and reassemble the filter. Wash the solid product twice with acetone (2–3 mL each). **NOTE: Acetone is extremely flammable. Keep it away from all flames and hot plates.**
- Draw air through the filter for 2 minutes to aid drying the filter cake. Discard the acetone washes in a designated waste container. Transfer the solid to a clay tile (or filter paper), and allow it to air dry.
- Repeat the filtering process for each of the five remaining mixtures. When each product is completely dry, transfer the solid to a tared piece of weighing paper, and weigh to the nearest mg. Dispose of the solid product in an appropriate waste container. Record your data on the Data Sheet.

The product of this reaction is a hydrate, which can be analyzed using the method shown in Experiment 7.

Name: ______________________________

Date: __________ Section: ______________________________

Pre-Laboratory Questions: Experiment 6

1. The following data were collected for the silver carbonate experiment described in the Introduction:

Beaker #	*1*	*2*	*3*	*4*	*5*	*6*
Moles Ag^+	1×10^{-4}	3×10^{-4}	5×10^{-4}	7×10^{-4}	9×10^{-4}	1.1×10^{-3}
Moles CO_3^{2-}	1.1×10^{-3}	9×10^{-4}	7×10^{-4}	5×10^{-4}	3×10^{-4}	1×10^{-4}
mg product	13.5	39.8	65.5	92.3	80.0	26.8

Construct a Job's plot using this data, and determine the formula of the product. The most accurate method of determining the correct ratio is to plot the moles of Ag^+ on the x axis, and mass of product on the y axis. To simplify determination of the ratio at any given point, it usually helps to have two x axis scales, moles of one ion increasing to the right, moles of the other ion increasing to the left. The data should generate two straight lines, one with a positive slope on the left and one with a negative slope on the right. The point where the two lines intersect is the point where the ratio of the two ions must be the stoichiometric ratio of the two components in the molecular formula. If both ions are represented on the x axis, this ratio may be read directly from the graph (see Figure E6.1).

2. The slope of either of the two straight lines on the graph you constructed for Question 1 may be used to calculate the empirical weight (EW). The units of the slope (on the left side, where Ag^+ is limiting) would be mg product/mol Ag^+. Thus, to convert to the units of EW (g/mol):

$$\text{EW} = \left[\frac{\text{(slope of line, mg product/mol silver)}}{\text{1000 mg/g}}\right]\left[\frac{\text{mol silver}}{\text{mol product}}\right]$$

Calculate the EW of the product.

Name: ______________________________

Date: __________ Section: ______________________________

Data Sheet: Experiment 6

Beaker #	*Moles Iron*	*Moles Oxalate*	*Mass of Product*
1.	________ mol	________ mol	________ g
2.	________ mol	________ mol	________ g
3.	________ mol	________ mol	________ g
4.	________ mol	________ mol	________ g
5.	________ mol	________ mol	________ g
6.	________ mol	________ mol	________ g

Construct a Job's plot (see Figure E6.1) on the supplied graph paper. Determine the formula of the product, as well as its empirical weight (see Pre-Laboratory Question 2).

Formula of Product ______________________________

Empirical Weight
(show calculation) ______________________________

Name: ______________________________

Date: __________ Section: ______________________________

Questions: Experiment 6

1. Explain why Job's method works in determining the formula of a salt.

2. What other types of experimental techniques (besides gravimetric) might Job's method be used for? (Hint: See other experiments in this book, and search the Internet under "Job's Method of Analysis").

3. An iron compound, when analyzed, gives the following elemental analysis:

Fe: 15.16%, K: 42.46%, N: 22.82% and C: 19.56%.

Determine the empirical formula of this compound.

4. Job's method can be used to analyze titrations of acid versus base. The Job's plot for the titration of citric acid with sodium hydroxide shows a maximum near the region of 0.75 mL of 0.100 M NaOH and 0.25 mL of 0.100 M citric acid. Predict the number of titratable acid protons per molecule of citric acid.

Experiment 7 Determination of the Empirical Formula of a Hydrate

OBJECTIVES

- To visually examine a group of hydrates and anhydrous salts, and observe their behavior.
- To establish the formula of a hydrate using gravimetric analysis.

PRIOR READING

Section 3.3: Weighing

Section 3.5: Heating Methods

INTRODUCTION

When aqueous solutions of inorganic salts are prepared, the ions that are formed are stabilized by a process called **hydration.** Each negative ion **(anion)** attracts the positive ends of many water molecules while each positive ion **(cation)** attracts the negative ends of many others. Positive ions, which are usually smaller than the negative ions, are usually strongly hydrated.

Divalent and trivalent transition metal ions are especially stable when hydrated. When aqueous solutions of these salts are concentrated, the forces between the water molecules and the hydrated cations often are of sufficient strength that the water molecules are retained in the crystal as the solid salt separates from solution. These hydrated salts are called **hydrates.** Hydrates can be isolated, and have definite compositions and structures. An example of a hydrated salt is $CoCl_2 \cdot 6H_2O$ [named cobalt(II) chloride hexahydrate], the cation of which is shown below.

OH_2
H_2O — Co^{2+} — OH_2
H_2O — OH_2
OH_2

In this salt there is one $[Co(H_2O)_6]^{2+}$ ion and two Cl^- ions present. The cation is bonded to six water molecules arranged in the shape of an octahedron. It should be emphasized that the lines in the figure therefore do not necessarily represent bonds—they merely indicate the shape of the ion. The waters associated with the salt are called **waters of hydration.**

The water bound in hydrates can often be removed at low temperatures (about 120°C in most cases) to produce water and the **anhydrous** (*an*—without, *hydro*—water) salt.

$$CoCl_2 \cdot 6H_2O \rightarrow CoCl_2 + 6\ H_2O$$

The mass percentage of water in a hydrate can usually be determined by heating a known amount of the hydrate, until dehydration occurs and the waters have been completely driven off. Since the waters of hydration are now no longer present, the remaining anhydrous salt has a lower mass than the original hydrate, or as an equation:

(mass of hydrate) = (mass of anhydrous salt) + (mass of waters of hydration)

The mass percentage of water in the hydrate can therefore be easily calculated:

$$\% \text{ water} = \frac{\text{mass of water lost}}{\text{mass of hydrate}} \times 100$$

Some hydrates lose their waters of hydration easily, by exposing them to the atmosphere. This loss of water is known as **efflorescence.** For example, when a sample of $MgSO_4 \cdot 7H_2O$ (known as Epsom salts) is placed on a watch glass on the laboratory bench top, the following process occurs:

$$MgSO_4 \cdot 7H_2O_{(s)} \rightarrow MgSO_{4(s)} + 7\ H_2O_{(g)}$$

The efflorescence reaction above produces seven molecules of water vapor per molecule of the hydrate. If the process is carried out in a closed container, the pressure of the water vapor (vapor pressure) can be measured. For this salt, the vapor pressure equals 11.5 torr at 25°C.

Some anhydrous salts such as $CaCl_2$ and $MgSO_4$ are **hygroscopic**—when exposed to the atmosphere, they absorb water to form hydrates. Such salts are used as drying agents or **desiccants.** Some compounds of this type actually absorb enough water to dissolve in, forming a solution. This process is termed **deliquescence.**

In this experiment, the mass percentage of water in a hydrate will be determined, using the relationships given above. The formula of the hydrate will then be determined by calculating the number of moles of anhydrous salt and the number of moles of water of hydration. Their molar ratio is then determined, giving the general formula of the hydrate. For example, suppose that the hydrate in question was sodium sulfate decahydrate, $Na_2SO_4 \cdot 10\ H_2O$. The original sample of the hydrate weighed 320 mg. After heating, the anhydrous salt weighed 141 mg. The mass of water lost would therefore be 320 − 141 = 179 mg. The number of moles of water lost would be

$$\text{moles water} = \frac{\text{mass of water lost}}{\text{MW of water}} = \frac{179 \text{ mg}}{18.0 \text{ g/mol}} \frac{1 \text{ g}}{1000 \text{ mg}} = 9.94 \times 10^{-3}$$

The number of moles of sodium sulfate would be

$$\text{moles sodium sulfate} = \frac{\text{mass of anhydrous salt}}{\text{MW of sodium sulfate}} = \frac{141 \text{ mg}}{142 \text{ g/mol}} \frac{1 \text{ g}}{1000 \text{ mg}} = 9.93 \times 10^{-4}$$

The ratio of the number of moles is

$$\frac{\text{moles water}}{\text{moles sodium sulfate}} = \frac{9.94 \times 10^{-3}}{9.93 \times 10^{-4}} = \frac{10}{1}$$

Thus, there are 10 moles of water associated with each mole of sodium sulfate. The mass percentage of water in the compound is:

$$\text{mass \% } H_2O = \frac{\text{mass of } H_2O \text{ lost}}{\text{mass of hydrate}} \times 100\% = \frac{179 \text{ mg}}{320 \text{ mg}} \times 100 = 55.9\%$$

EXPERIMENTAL PROCEDURE

Part A. Visual Inspection of a Series of Hydrates and Anhydrous Salts

- Inspect the group of hydrates and anhydrous salts provided by your instructor. A suggested list is given below. Comment, on the data sheet, on their physical appearance and color. Which show deliquescent properties? Which might find use as desiccants? Do any exhibit efflorescence?

Hydrates and Anhydrous Salts

Name	**Formula**
Copper sulfate, anhydrous	$CuSO_4$
Copper sulfate pentahydrate	$CuSO_4 \cdot 5\ H_2O$
Sodium sulfate, anhydrous	Na_2SO_4
Sodium sulfate decahydrate	$Na_2SO_4 \cdot 10\ H_2O$
Calcium chloride, anhydrous	$CaCl_2$
Sodium hydroxide	$NaOH$
Cobalt chloride, anhydrous	$CoCl_2$
Cobalt chloride hexahydrate	$CoCl_2 \cdot 6\ H_2O$

Part B. Determining the Formula of a Hydrate by Gravimetric Analysis

This analysis will be done in triplicate by using three micro-crucibles at once.

- Mark three micro-crucibles "A", "B," and "C" in pencil. While one crucible is cooling, the next can be heated.
- Attach an iron ring to a ring stand. Place a wire gauze square on the ring. (If clay triangles to fit the micro-crucibles are available, use them instead, as tipping is less of a problem.) Put the three marked micro-crucibles on the edge of the wire gauze (not on the matted section) and heat each to redness, with an intense flame, using a micro- or Bunsen burner for approximately 5 minutes. *Alternatively, each crucible may be held in the burner flame with crucible tongs. (If micro-crucibles are not available, use the technique shown in Figure E7.1.)* **NOTE: Use forceps or tongs to handle the crucibles, never your fingers.**
- Place each hot crucible in a 10 mL beaker, and then place all the beakers (with the crucibles in them) in a small desiccator. Allow them to cool to room temperature.
- Weigh each micro-crucible to the nearest mg. Repeat the heating, cooling, and weighing process until a constant mass (± 1 mg) for each of the crucibles is obtained. Note your results on the data sheets.
- Add a 100–150 mg sample of an unknown hydrate to each of the crucibles and weigh to the nearest mg.
- Place a small piece of aluminum foil (do not cover the opening completely) on each crucible to serve as a cover. Place one crucible (with cover) on the wire gauze. Heat the sample gently at first, and then increase the heat moderately. Do not permit the crucible to become red hot. Continue the heating for approximately 2–3 minutes.

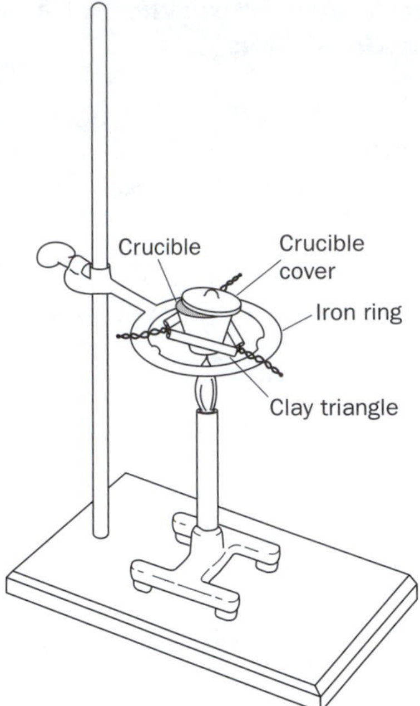

Figure E7.1 *Proper method of heating a crucible.*

- Remove the crucible (using forceps) and place it in a 10 mL beaker. Transfer the beaker and crucible to a desiccator to cool to room temperature. Weigh the crucible (without the cover) and record the results on the data sheet.
- Repeat this procedure for the second and third crucibles.
- To make sure that all the waters of hydration have been removed, reheat each sample, cool as before, and reweigh. Repeat this process until each trial achieves a constant mass (± 1 mg). Record your results as before.

Name: ______________________________

Date: __________ Section: ______________________

Pre-Laboratory Questions: Experiment 7

1. Consider the introduction to this experiment concerning the vapor pressure of the magnesium sulfate heptahydrate salt. Would you expect this hydrate to lose water when exposed to the atmosphere, if the partial pressure of water vapor in the air was greater than 11.5 torr? Explain.

2. Explain why water is a polar molecule, and how this allows it to bond to positively charged metal ions.

3. A 100 mg sample of the blue hydrate $CuSO_4 \cdot 5\ H_2O$ is heated in a crucible. The white, anhydrous salt ($CuSO_4$) is obtained, and weighs 64 mg after being cooled to room temperature. Calculate the % water of hydration for the hydrated salt. Confirm the empirical formula ratio of 1 $CuSO_4$:5 H_2O.

Name: ______________________________

Date: __________ Section: ______________________________

Data Sheet: Experiment 7

A. Visual Inspection of a Series of Hydrates and Anhydrous Salts
Write your observations for each of the following hydrates and anhydrous salts.

$CuSO_4$:

$CuSO_4 \cdot 5\ H_2O$:

Na_2SO_4:

$Na_2SO_4 \cdot 10\ H_2O$:

$CaCl_2$:

NaOH:

$CoCl_2$:

$CoCl_2 \cdot 6\ H_2O$:

Name: ______________________________

Date: __________ Section: ______________________________

Data Sheet, Experiment 7, p. 2

B. Gravimetric Analysis of an Unknown Hydrate

	Sample A	*Sample B*	*Sample C*	
1. Constant mass of crucible				
1st heating	_______	_______	_______	g
2nd heating	_______	_______	_______	g
3rd heating (if necessary)	_______	_______	_______	g
2. Mass of crucible and hydrated salt	_______	_______	_______	g
3. Constant mass of crucible and anhydrous salt				
1st heating	_______	_______	_______	g
2nd heating	_______	_______	_______	g
3rd heating (if necessary)	_______	_______	_______	g
4. Calculations				
Mass of the hydrated salt	_______	_______	_______	g
Mass of the anhydrous salt	_______	_______	_______	g
Mass of water lost	_______	_______	_______	g
Percentage of water in the hydrate	_______	_______	_______	%
Average of percent water in the unknown		_______ %		
Formula of the anhydrous salt (ask instructor)	_______________________			
Moles of water lost	_______	_______	_______	mol
Average moles of water lost		_______ mol		
Moles of anhydrous salt obtained	_______	_______	_______	mol
Average moles of anhydrous salt		_______ mol		
Mole ratio of anhydrous salt to water (based on average values)	_______________________			

Formula of the unknown hydrate _______________________

Name: ______________________________

Date: __________ Section: ______________________

Questions: Experiment 7

1. You are given a sample of the hydrate of $CaSO_4$. The average values of the data collected and recorded in your laboratory notebook were:

Mass of empty crucible	1.138 g
Mass of crucible and sample before heating	1.457 g
Mass of crucible and sample after heating	1.436 g

Calculate the formula of the $CaSO_4$ hydrate.

2. Sodium hydroxide pellets are very hygroscopic. Describe what happens chemically when a cover is left off a bottle of NaOH pellets. (Hint: carbon dioxide reacts with NaOH.)

3. A 120 mg sample of a lead acetate hydrate, $Pb(C_2H_3O_2)_2 \cdot xH_2O$ was heated to drive off the waters of hydration. The cooled residue weighed 103 mg. Calculate the value of x in the lead acetate formula.

Experiment 8 Ideal Gas Calculations (Dry Lab)

OBJECTIVES

- To learn how to use the Ideal Gas Laws.

PRIOR READING

Gas Laws chapter of your General Chemistry textbook
Introduction to Experiments 9–11

INTRODUCTION

Experiments 3, 9, 10, 11 and 12 deal with the properties of gases and the various gas laws that define their behavior. This dry lab is meant to give you practice in using the Ideal Gas Laws in various situations, so that you will become more familiar with the properties and applications of this important state of matter.

There are four variables that are used to define the state of a gas. These variables are temperature (T), pressure (P), volume (V), and the number of moles of gas (n). There are various simple relationships that relate these variables with reasonable accuracy. These relationships are collectively called the **Ideal Gas Laws:**

Boyle's Law:	$P_1V_1 = P_2V_2$	(T and n constant)
Charles' Law:	$V_1/T_1 = V_2/T_2$	(P and n constant)
Avogadro's Law:	$V_1/n_1 = V_2/n_2$	(P and T constant)
Combined Law:	$\frac{P_1V_1}{T_1} = \frac{P_2V_2}{T_2}$	(n constant)
Ideal Gas Law:	$PV = nRT$	
Dalton's Law:	$P_{\text{Total}} = P_1 + P_2 + P_3 + \cdots$	
Graham's Law:	$v_1/v_2 = \sqrt{(M_2/M_1)}$	

The quantities *pressure and volume* are related by **Boyle's Law;** *volume and temperature* by **Charles' Law;** *diffusion of gases* by **Graham's Law;** and *total pressure of gas components in a system* by **Dalton's Law.** These relationships are also expressed in the **Ideal Gas Equation,** $PV = nRT$. Temperatures must be in absolute units (K). If we can effectively use these relationships, we will develop an in-depth understanding of how gases behave under various conditions. It is important to study these relationships because we live in an atmosphere of gas, and many of the everyday processes of living, such as breathing, depend on how we interact or use a gas to our advantage.

For example, an individual at the dentist might be given nitrous oxide gas, (laughing gas, N_2O) as an anesthetic. It is pleasant, quick in its action, and without harmful after-effects. The inflation of automobile tires, cooking on a gas stove, and smelling of perfume are other examples of the gas laws in action.

In attacking a gas law problem it is critical to identify what conditions are operating on the gas. Once this is done, the law(s) involved can be applied and the problem set up and solved.

Example 1: A sample of gas occupies a volume of 500 mL at a barometric pressure of 755 torr. What volume would it occupy at a pressure of 760 torr?

Answer: The given information is in terms of pressure and volume, so we will use Boyle's Law.

$$P_1V_1 = P_2V_2$$

$$(755 \text{ torr})(500 \text{ mL}) = (760 \text{ torr})(V_2)$$

$$V_2 = 496.7 \text{ mL}$$

Example 2: A gas occupies a volume of 42.06 mL at 765.0 torr pressure and 20.0°C. What is the volume under standard conditions?

Answer: Recall that standard temperature and pressure (STP) refer to the conditions of 1 atm (760.0 torr) and 273.15 K. Since we are using pressure, volume, and temperature, the Combined Gas Law is called for. All temperatures must be in K, so we must add 273.15 to 20.0°C = 293.15 K.

$$P_1V_1/T_1 = P_2V_2/T_2$$

$$(765.0 \text{ mm})(42.06 \text{ mL})/(293.15 \text{ K}) = (760.0 \text{ mm})(V_2)/(273.15 \text{ K})$$

$$V_2 = 39.42 \text{ mL}$$

Example 3: A volume of 43.0 mL of a gas is collected at 25°C at a pressure of 763.0 torr by the displacement of water. How many moles of dry gas are present?

Answer: Since the gas is collected over water, it must contain water vapor. We must therefore look up the vapor pressure of water at 25°C in Table E8.1.

Averaging the values at 24 and 26°C, we obtain a value of 23.8 mm Hg. Substituting into Dalton's Law,

$$P_{\text{Total}} = P_{\text{dry gas}} + P_{\text{water}}$$

$$763.0 \text{ mm} = P_{\text{dry gas}} + 23.8 \text{ mm}$$

$$P_{\text{dry gas}} = 739.2 \text{ mm Hg}$$

$$P_{\text{dry gas}} = 739.2 \text{ mm Hg } (1 \text{ atm}/760 \text{ mm Hg}) = 0.973 \text{ atm}$$

Since we are solving for moles, we must use the Ideal Gas Law. Recall that to use the value of the R (the Universal Gas Constant) of 0.082 L atm/K mol, the volume must be in liters, the pressure in atmospheres, and the temperature in Kelvins.

Table E8.1 Vapor Pressure of Water

T, °C	*P* (mm Hg)	*T*, °C	*P* (mm Hg)	*T*, °C	*P* (mm Hg)
0	4.58	16	13.63	26	25.21
5	6.54	18	15.48	28	28.35
10	9.21	20	17.54	30	31.82
12	10.52	22	19.83	40	55.3
14	11.99	24	22.38	50	92.5

$$V = (43.0 \text{ mL})(1 \text{ L}/1000 \text{ mL}) = 0.0430 \text{ L}$$

$$T = 25°\text{C} + 273 = 298 \text{ K}$$

$$PV = nRT$$

$$(0.973 \text{ atm})(0.0430 \text{ L}) = n\ (0.0820 \text{ L atm/mol K})(298 \text{ K})$$

$$n = 1.71 \times 10^{-3} \text{ mol.}$$

Example 4: Calculate the ratio of the rate of diffusion of hydrogen to the rate of diffusion of oxygen.

Answer: This is a Graham's Law problem. The molecular weight of O_2 is 32 g mol^{-1} and that of H_2 is 2 g mol^{-1}. Substituting into Graham's Law:

$$v_1/v_2 = \sqrt{(M_2/M_1)}$$

$$v_1/v_2 = \sqrt{(32/2)} = \sqrt{16} = 4$$

Thus, hydrogen diffuses four times faster than oxygen.

The Question pages contain a number of additional problems for you to practice on. Since this a drill session, the answers are given so that you can follow your progress in mastering the art of solving gas law problems.

Name: ______________________________

Date: __________ Section: ______________________

Questions: Experiment 8

1. If 500 mL of hydrogen gas are cooled from 26 to $-10°C$ at a constant pressure, what is the volume at the lower temperature?

Ans. 440 mL

2. The pressure on 600 mL of a gas is 758 mm. The pressure is increased to a value of 774 mm. What is the new volume of the gas?

Ans. 588 mL

3. Wet hydrogen gas is confined over water under a pressure of 760 mm and a temperature of 26°C. What is the actual pressure of the dry hydrogen gas?

Ans. 735 mm

4. A sample of limestone is treated with acid and 98.7 mL of carbon dioxide is collected over water at 23°C and 761 mm pressure. What is the volume of dry gas at STP?

Ans. 88.6 mL

5. Here is a new twist. In the laboratory $KClO_3$ is decomposed by heating to give KCl and O_2 gas. If you heat 1.00 kg of pure $KClO_3$, how many liters of oxygen gas can be obtained if it is collected over water at 17°C and 777 mm pressure?

Ans. 290 L

6. If 130.5 mL of a gas is collected at 28.5°C under 380 mm pressure, what is the volume of gas at STP?

Ans. 0.905 L

7. At a temperature of 25°C and a pressure of 763.0 torr, a volume of 43.0 mL of a gas was collected by displacement of water. How many moles of dry gas are present?

Ans. 1.71×10^{-3} mol

Name: ______________________________

Date: __________ Section: ______________________________

Questions: Experiment 8, p. 2

8. A sample of butane gas occupies a volume of 405 mL at 298 K and 1.1 atm pressure. What is the volume of the gas at 467 K and 1.1 atm?

Ans. 635 mL

9. You have a 10.3 L tank of chlorine gas maintained at 21.2°C and a pressure of 633 mm. How many moles of chlorine gas are contained in the tank?

Ans. 0.355 mol

10. In a 220 mL gas cylinder you have 0.299 g of a gas at a pressure of 0.757 atm and 298 K. What is the molecular mass of the gas?

Ans. 43.9 g mol^{-1}

11. You have a balloon that contains the gases oxygen and helium. Which gas will effuse out of the balloon at a faster rate?

Ans. He is 2.83 times faster

12. In a 10.0 L glass container maintained at 35°C, what is the total pressure in atmospheres if the container has 3×10^{-4} moles of argon, 1.00×10^{-3} moles of oxygen and 2.5×10^{-3} moles of neon ?

Ans. 9.61×10^{-3} atm

13. You collected a gas at 760 mm and and 25°C. The measured density was found to be 5.37 g L^{-1}. What is the formula weight of the gas?

Ans. 131 g mol^{-1}

14. Here is the clincher! A certain hydrocarbon is found to have a vapor density of 2.550 g L^{-1} at 100°C and 760 mm pressure. Chemical analysis shows that the substance contains one atom of carbon to one atom of hydrogen. What is the molecular formula of the hydrocarbon?

Ans. C_6H_6

Experiment 9 The Gas Laws

OBJECTIVES

- To experimentally verify Boyle's, Charles' and Graham's Laws.
- To determine the value of absolute zero.

PRIOR READING

Section 2.6: Graphing
Experiment 8

INTRODUCTION

Experiments with a large number of gases have demonstrated that four variables can define the state of a gas. These variables are temperature (T), pressure (P), volume (V), and the number of moles of gas (n). There are various simple relationships that relate these variables with reasonable accuracy. These relationships are known as the **Ideal Gas Laws.**

Boyle's Law: Pressure and Volume

The first relationship found between these variables was established in 1662 by the Irish chemist, Robert Boyle. He found that the volume of a fixed quantity of gas, maintained at constant temperature, is inversely proportional to the gas pressure. That is, as the pressure increases, the volume decreases. This relationship is now known as **Boyle's Law.**

Boyle lived from 1627–1691. He was born in Ireland and at 11 toured Europe and eventually settled in England. He was founder of the Royal Society of London in 1660. He was known not only for his work on gases—he also introduced many chemical tests. Among these were the use of vegetable dyes as indicators, and flame tests to detect metals.

Boyle's experimental method with gases involved the trapping of air in a U-tube (called a **manometer**) as shown in Figure E9.1. Mercury was used as the manometer liquid. The pressure of the trapped gas can be increased by adding more mercury to the open arm of the manometer. The result is a decrease in the volume of the gas. If mercury is removed from the manometer, it follows that the gas pressure decreases and the gas volume increases.

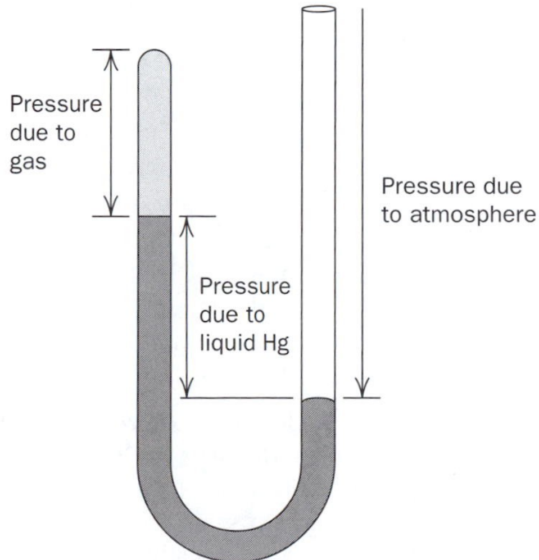

Figure E9.1 *Boyle's Law apparatus.*

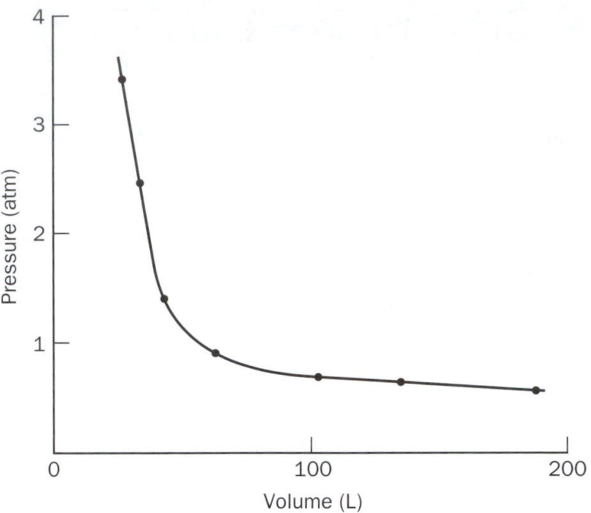

Figure E9.2 *Boyle's Law plot of P versus V at constant T.*

In this experiment, a sample of gas in a syringe (piston) will be subjected to different pressures by adding masses to the plunger and noting the change in volume. The Boyle's Law relationship can be expressed as (the symbol $\propto$ means "is proportional to")

$$P \propto 1/V$$

or as

$$P_1V_1 = P_2V_2$$

For a fixed amount of gas, the product of the pressure and the volume will be constant, if the temperature is not changed. A graphical interpretation of typical experimental results is shown in Figure E9.2. If Boyle's Law is obeyed, a hyperbola will connect the data points. The Boyle's Law relationship expresses the important fact that gases are compressible.

Charles' Law: Volume and Temperature

The relationship between the volume of a gas and the absolute temperature (defined below) at constant pressure is known as **Charles' Law.** The law states that the volume of a fixed quantity of gas at constant pressure increases in a linear fashion with the absolute temperature.

Jacques Alexander Cesar Charles, a French physicist, in about 1787 investigated the effect of changing the temperature on the volume of a confined gas at constant pressure. He found that equal amounts of oxygen, nitrogen, hydrogen, carbon dioxide, and air expand to the same extent over the same temperature interval. However, he did not publish his work. His study of this effect was brought about by his interest in hot-air ballooning. In those times, balloons were inflated with air heated by burning straw. Charles invented the hydrogen-filled balloon.

A fellow scientist, Joseph Louis Gay-Lussac, rediscovered the Law in 1802 and in his *published* paper made reference to Charles's earlier work. Thus, some people refer to this relationship as Gay-Lussac's Law. Controlled laboratory experiments led to results such as those plotted in Figure E9.3.

The plot illustrates the behavior of one mole of a gas at three different pressures. When the volume of the gas is plotted versus the temperature (°C),

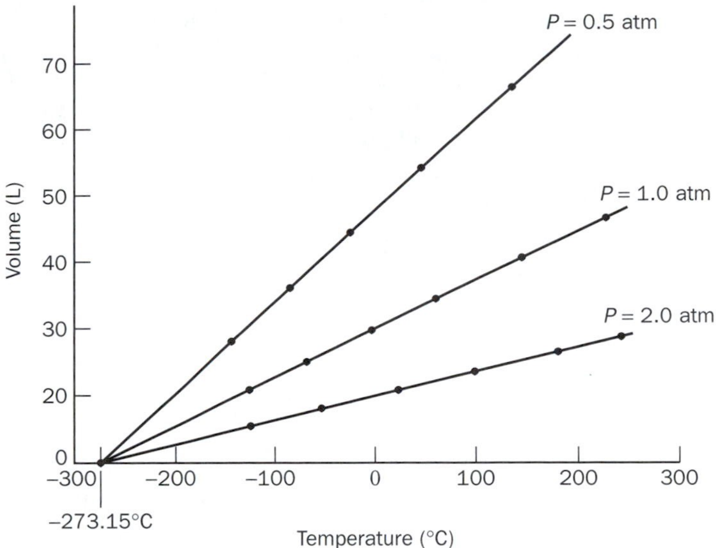

Figure E9.3 *Charles' Law experimental data.*

the data points fall on straight lines that intersect at −273.15°C. In actual practice, all gases condense at temperatures higher than −273.15°C, and the graph reflects the extrapolation of the data lines to a hypothetical volume of zero. This temperature is known as **absolute zero.**

A Scottish physicist, Sir William Thomson (whose title was Lord Kelvin), in 1848 proposed the idea of an **absolute temperature scale.** On the Kelvin scale −273.15°C = 0 K. Charles' Law can therefore be expressed as

$$V \propto T$$

or

$$\frac{V_1}{T_1} = \frac{V_2}{T_2}$$

where T is the absolute temperature in K, and V is the volume of the gas.

The relationship between the Kelvin scale and the Celsius scale is

$$K = °\text{C} + 273.15$$

In doing gas law problems, remember to convert all temperatures to the Kelvin scale. On the Kelvin scale, water freezes at 273.15 K and boils at 373.15 K. The only difference between the Kelvin scale and the Celsius scale is that zero is shifted—the size of the degree remains the same. One degree Celsius is equal in magnitude to one degree Kelvin. In this experiment, the volume of a sample of air will be determined at several different temperatures.

Graham's Law: Diffusion of Gases

The process by which a gas moves from a location of higher pressure to a location of lower pressure through a porous wall or through a small diameter tube is known as **diffusion.** If we are speaking about individual molecules, the process is known as **effusion.** The rate at which a given gas will diffuse is a characteristic property of the gas. This fact is extensively utilized in industry in contexts ranging from the production of foams to the separation of isotopes of uranium. The central premise of gas diffusion is a simple one: the relative rate of diffusion of a gas is proportional to the square root of its molecular weight. In simpler terms, the greater the molecular weight of the gas, the more slowly it moves. This is summarized by **Graham's Law of Diffusion** (also known as Graham's Law of Effusion), first developed by Thomas Graham in 1829:

$$\frac{v_1}{v_2} = \sqrt{\frac{M_2}{M_1}}$$

Here, v is the velocity or rate at which the gas is moving, and M is the molecular weight of the gas.

Graham (1805–1868) attended Glasgow University at the age of 14 and was fascinated by the study of chemistry. In 1830 he was elected chair of chemistry at Glasgow University. The year 1837 found him as professor at the University College, London, where he became a noted teacher. He was the first president of the Chemical Society of London.

In this experiment, the rates of diffusion of gaseous ammonia (NH_3) and chlorine (Cl_2) will be determined by measuring the time it takes the gas to react with an indicator that is a known distance away from the gas source.

EXPERIMENTAL PROCEDURE

Part A: Boyle's Law

The apparatus to be used is shown in Figure E9.4. Note: This procedure is adapted from the work of Prof. D. W. Brooks on microscale techniques.

- Obtain a U-100 Insulin plastic syringe (1 mL). Lubricate the plunger with a few drops of silicone oil, and insert the plunger into the body of the syringe. Set the plunger at the 90–100 unit mark. Remove the needle protector and insert the needle into a soft rubber stopper. (*As an alternative,*

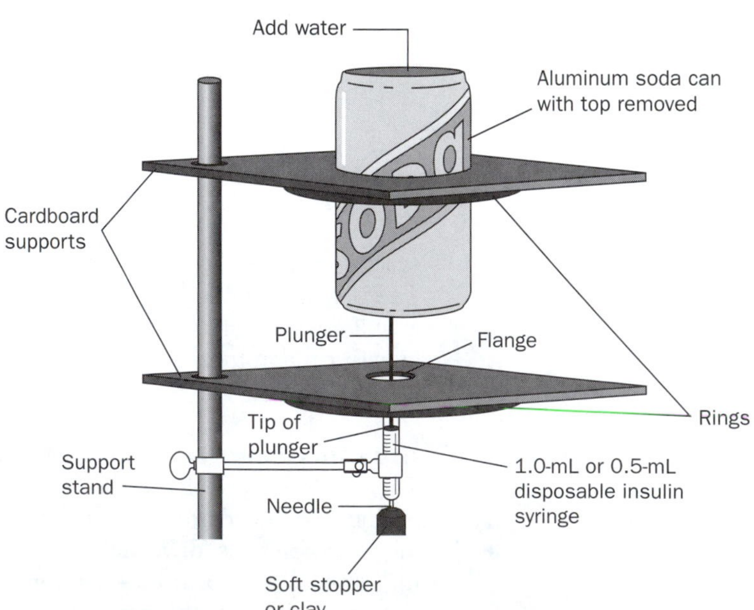

Figure E9.4 *Experimental apparatus.*

the needle can be cut and the small hole sealed with super-glue or modeling clay.) **NOTE: This must be a tight seal, since air leakage will lead to erroneous results.**

- Clamp the syringe to a ring stand as shown in Fig. E9.4. Record the volume of trapped air as that corresponding to room pressure. Record the room pressure.
- Determine the mass of the empty Al can. Add water to make the mass of the can plus water equal to 50 g. Now, place and secure the can so that it sits evenly on the plunger of the syringe. *Make sure the can is able to move easily in the clamp arrangement.*
- Press down on the can several times to force the plunger down and then let the system obtain equilibrium. Record the mass of the can and water and the volume reading on the syringe.
- Sequentially, add water to the can, adding 50 mL of water each time. Measure the water volume using a 50 mL pipet or a graduated cylinder. Press as before and after equilibrium is reached record the weight of water (volume) and volume of air reading on the syringe. Allow at least 1/2 minute for the system to stabilize after each addition. Record the volume of the air and the total mass of the added water in the data table. *Note: assume the density of water is one g/mL and thus the volume of water added in mL equals the mass of water added in grams.* Obtain information for at least 7 data points. The total mass will be 350 g at this point.
- Repeat the procedure adding water as before to determine the reproducibility of the experimental data.

Calculations

To obtain a true relationship related to Boyle's Law, the pressure (measured in grams, in this experiment) must be converted to absolute units. The calculation is illustrated for a selected data point from the following experimental data:

Mass (g)	*Volume (mL)*
150	37

- The area of the piston head is equal to πr^2. For the piston in Figure E9.4, $r = 0.9$ cm, therefore:

$$A = (3.14)(0.9\text{ cm})^2 = 2.54\text{ cm}^2$$

- The mass, 150 g, of the entry above, is divided by the area to give g/cm^2.

$$150\text{ g}/2.54\text{ cm}^2 = 59.06\text{ g cm}^{-2}$$

- A mass of 1 g on an area of 1 cm^2 exerts a pressure of 101.3 Pa (a Pascal is a metric unit of pressure). Thus,

$$P = (59.06\text{ g cm}^{-2})\,(101.3\text{ Pa cm}^2\text{ g}^{-1}) = 5982\text{ Pa} = 5.982\text{ kPa}$$

This is a gauge pressure, however. To obtain the total (absolute) pressure P_T, the atmospheric pressure (101.32 kPa) must be added:

$$P_T = P_{gauge} + P_{atm}$$

$$P_T = 5.982\text{ kPa} + 101.32\text{ kPa} = 107.3\text{ kPa}$$

Thus, the pressure (107.3 kPa) and volume (37 mL) would account for one data point on a Boyle's Law plot such as shown in Figure E9.2. The simplest test to confirm Boyle's Law behavior of a gas is to plot pressure (P) versus $1/V$. This results in a straight-line plot, if the gas obeys Boyle's Law.

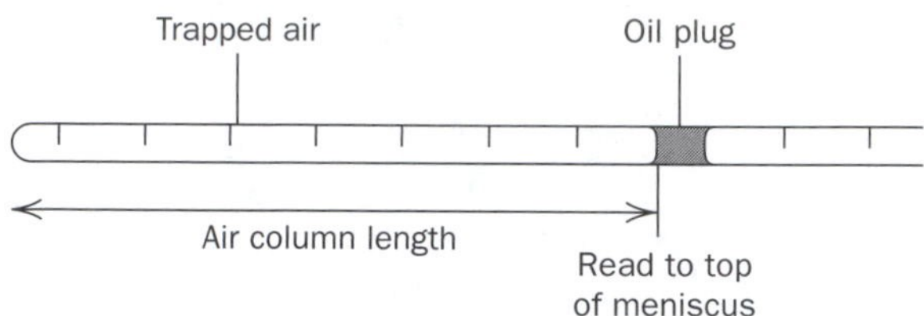

Figure E9.5 *Gas law tube in horizontal position.*

Part B: Charles' Law

> ***Your teacher will have prepared a series of baths set at various temperatures.***

- Obtain a blood serum sample tube (shown in Figure E9.5), which will serve as the gas law tube. **NOTE: The tube must be clean. No residual oil from a previous run should be present.**
- Using a Pasteur pipet, insert a plug of Nujol oil, 2–3 mm in length, approximately halfway down the tube. Record room temperature, and read the length of the air column (± 0.1 mm) entrapped in the tube, as shown in Figure E9.5.
- Place the tube vertically in the ice/water bath, making sure the tube is immersed from the bottom of the tube to the oil plug. Allow the tube to sit in the bath for 2–3 minutes, to establish thermal equilibrium. Record the temperature of the bath, as well as the length of the air column entrapped in the tube.
- Repeat the above procedure, but this time using an ice/salt bath.
- Holding the tube with a test tube clamp, carefully place the blood serum tube in the hot water bath(s). Make sure the tube is immersed from the bottom to the level of the oil plug. *Allow the tube to be immersed for only 2–3 sec* (**no longer,** or else the oil in the plug will bleed to the bottom of the tube), and then read the length of the entrapped air column. Note the temperature of the hot water bath.

If time allows, repeat these measurements.

Part C: Graham's Law

> ***Read the entire procedure before starting on this part of the experiment.***

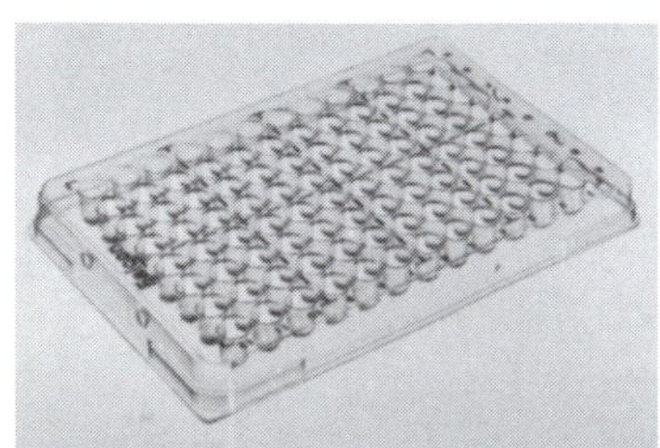

Figure E9.6 *96-Hole micro-well plate. (Courtesy of Fisher Scientific, 711 Frobes Ave., Pittsburgh, PA 15219.)*

- Obtain a 96-hole well plate with cover. Add 2 drops of phenolphthalein solution to every well in the plate *except one of the two corner wells* (where the cover fits). See Figure E9.6.
- Select three wells, and measure the distance from the empty well to them. As the ammonia diffuses to the wells, it will react with the phenolphthalein, which turns pink in the presence of the base.

- Place three drops of concentrated ammonia (NH_3 gas dissolved in water) in the empty well. Replace the cover on the well plate. Time how long it takes the ammonia to reach the three pre-selected wells. Record your data on the data sheet.
- Repeat the procedure, using laundry bleach instead of ammonia, and 0.1 M KI solution instead of phenolphthalein. Place the bleach in one well of a 96-hole well plate, and fill the other wells with the KI solution. Add two drops of 3 M HCl to the bleach and replace the cover immediately. The reaction taking place is

  $$NaOCl + HCl \rightarrow NaOH + Cl_2$$

  As the chlorine diffuses through the system, it will react with the KI, forming KCl and I_2:

  $$2KI_{(aq)} + Cl_2 \rightarrow 2KCl_{(aq)} + I_2$$

  The color change of the solution indicates reaction has taken place. Time how long it takes the chlorine to reach the pre-selected wells. Record your data on the data sheet.

Name: ____________________

Date: ________ Section: ____________________

Pre-Laboratory Questions: Experiment 9

Part A: Boyle's Law

1. Explain what it means to say, "the volume of a given amount of gas is inversely proportional to its pressure at constant temperature."

2. Explain why the compressibility of gases is so much greater than that of liquids.

3. By what factor must the pressure on a gas be decreased to triple the volume?

4. The raw data collected in this experiment consists of a series of masses added and of syringe volumes. What variables are actually plotted to obtain a straight line? How are these variables obtained from the raw data?

5. A balloon with a volume of 0.55 L is allowed to rise from sea level to a height of 4 miles, where the pressure is about 0.40 atm. Assuming the temperature remains constant, what is the final volume of the balloon?

Name: ______________________________

Date: __________ Section: ______________________________

Pre-Laboratory Questions: Experiment 9, p. 2

Part B: Charles' Law

1. In terms of Charles' Law, why is −273.15°C the lowest temperature possible?

2. In an experiment, the data collected consists of volumes and temperatures (°C). What must be plotted in order to obtain a straight-line graph going through the origin (0,0)?

3. At −196.0°C, the boiling point of liquid nitrogen, a sample of hydrogen gas has a volume of 2.25 L. The sample is allowed to warm to room temperature, measured as 27.0°C. What is the new volume of the gas, assuming no change in pressure?

Name: ______________________________

Date: __________ Section: ______________________________

Pre-Laboratory Questions: Experiment 9, p. 3

Part C: Graham's Law

1. A certain gas effuses at a rate 1/4 that of hydrogen. What is the molecular weight of that gas?

2. A normal balloon filled with helium will deflate much more quickly than one filled with air. In order to remain inflated with helium, Mylar balloons must be used. Explain this phenomenon.

3. Why must the 96-well plates be covered in order to obtain good results in this experiment?

Name: ______________________________

Date: __________ Section: ______________________

Data Sheet: Experiment 9

Part A: Boyle's Law

Trial #1

Mass, g	*V, mL*	*P, kPa*	*1/V, mL^{-1}*	*PV, kPa mL*

Trial #2

Mass, g	*V, mL*	*P, kPa*	*1/V, mL^{-1}*	*PV, kPa mL*

Name: ______________________________

Date: __________ Section: ______________________________

Data Sheet: Experiment 9, p. 2

Part A: Boyle's Law (Continued)

1. Write several concluding statements comparing your experimental data to that of Boyle's Law.

2. Construct a plot of P (y axis) versus $1/V$ (x axis). How linear is the plot? How does the linearity correlate with deviations from ideal behavior?

3. Comment on the consistency of the PV values.

Name: ____________________

Date: ________ Section: ____________________

Data Sheet: Experiment 9, p. 3

Part B: Charles' Law

1.

Condition	Temperature (°C)		Air-Column Length (mm)	
	Trial #1	*Trial #2*	*Trial #1*	*Trial #2*
Room Temperature	______	______	______	______
Ice-bath	______	______	______	______
Ice/salt bath	______	______	______	______
Hot water bath I	______	______	______	______
Hot water bath II	______	______	______	______

2. Prepare a graph, using the data tabulated above, to determine the value of absolute zero. Temperature should be plotted on the x axis, and air column length on the y axis (See Figure E9.3 as a guide). Attach the results to this data sheet. The data are easily plotted using Excel or a similar computer program.

3. Determined value of absolute zero: __________ °C __________ K

Name: ______________________________

Date: __________ Section: ______________________________

Data Sheet: Experiment 9, p. 4

Part C: Graham's Law

1. Ammonia Diffusion

Well #	*Time, sec*	*Distance, mm*	*Rate, mm s^{-1}*
______	______	______	______
______	______	______	______
______	______	______	______

2. Chlorine Diffusion

Well #	*Time, sec*	*Distance, mm*	*Rate, mm s^{-1}*
______	______	______	______
______	______	______	______
______	______	______	______

3. Determine if the relative rates for the two gases obey Graham's Law.

Name: ______________________________

Date: __________ Section: ______________________________

Questions: Experiment 9

Part A: Boyle's Law

1. There are deviations from Boyle's Law at low temperatures and high pressures. Why? (Hint: consult your textbook.)

2. A given mass of a gas has a volume of 150 mL under a pressure of 70 torr. What would be the volume of the gas if the pressure were increased to 150 torr at the same temperature?

Part B: Charles' Law

3. If it were desired to obtain data points at even lower temperatures, comment on the changes in the experimental procedure (apparatus and/or chemicals) that would be necessary to achieve this.

4. According to Charles' Law, how would the volume of a given quantity of gas, at constant pressure, vary with the temperature? Explain.

5. The blood sample tubes have a 3.0 mm inner diameter, and the height is graduated every 0.1 mm. The temperature is recorded to the nearest 0.1°C. Comment on the number of significant figures to which absolute zero (as −273.15°C) should be calculated.

Name: ______________________________

Date: __________ Section: ______________________________

Questions: Experiment 9, p. 2

Part C: Graham's Law

6. When ammonia and hydrogen chloride gas react, they form a solid ring of ammonium chloride. Ammonia and hydrogen chloride are introduced at opposite ends of a 1 m long tube. If the two gases obey Graham's Law, at what distance down the tube from the NH_3 end will the ring form?

7. Gas diffusion can be used to separate mixtures of gases. One of the more interesting uses is in separation of the isotopes of uranium, where $^{238}UF_6$ is separated from $^{235}UF_6$. What are the relative rates of diffusion for these two gases?

Experiment 10 Molar Volume of Hydrogen Gas

OBJECTIVES

- To determine the molar volume of a gas.
- To identify an unknown metal based on the volume of hydrogen gas that is generated (optional).

PRIOR READING **Introduction:** Experiment 9.

INTRODUCTION

Experiments conducted on gases have demonstrated that the variables of temperature (T), pressure (P), volume (V), and number of moles of gas (n) can be used to characterize this unique state of matter. Relationships have been established between the volume that a gas occupies and the three other variables. These relationships are collectively called the **Ideal Gas Laws:**

Boyle's Law:	$P_1V_1 = P_2V_2$	(T and n constant)
Charles' Law:	$V_1/T_1 = V_2/T_2$	(P and n constant)
Avogadro's Law:	$V_1/n_1 = V_2/n_2$	(P and T constant)
Combined Law:	$\frac{P_1V_1}{T_1} = \frac{P_2V_2}{T_2}$	(n constant)

These individual relationships have been shown to be special cases of a more general law, the **ideal gas equation.** If the proportionalities given above are combined into a single expression, we have

$$PV \propto nT$$

This proportionality expression may be written as an equality using the constant R, known as the **universal gas constant:**

$$PV = nRT$$

Any gas that behaves according to this relationship is defined as an ideal gas. The numerical value of R depends on the units chosen for n, V, P, and T. Table E10.1 lists several of these values.

The volume of 1.0 mole of a gas can be measured at standard temperature and pressure (abbreviated **STP:** 273 K and 1.0 atm). This volume is known as the **molar volume** (V/n). The average value of the molar volume of any gas at STP is 22.4 L/mol. The molar volumes of non-ideal **(real)** gases deviate somewhat from this value. Table E10.2 gives some experimentally determined values for a series of gases at STP.

Rearranging the ideal gas equation to solve for the molar volume gives

$$\frac{V}{n} = \frac{RT}{P}$$

Table E10.1 R in Various Units

Units	Numerical Value
Liter-atm/K-mol	0.08206
Calories/K-mol	1.987
Joules/K-mol	8.314
ft^3 psia/°R-lbmol	10.73

Table E10.2 Molar Volumes at STP

Gas	Molar Volume (L)
Oxygen	22.397
Hydrogen	22.433
Nitrogen	22.402
Helium	22.434
Carbon dioxide	22.260
Ammonia	22.079

Since R is a constant, we see that the molar volume of an ideal gas is dependent on its temperature and pressure.

The purpose of this experiment is to determine the molar volume of hydrogen gas, and to compare it to the value given in Table E10.2. Although the gas will be collected over water and at conditions other than STP, we can use the gas laws to calculate the desired molar volume.

The reaction of magnesium metal with hydrochloric acid will be used to generate the hydrogen gas:

$$Mg_{(s)} + 2\,HCl_{(aq)} \rightarrow Mg^{2+}_{(aq)} + 2Cl^{-}_{(aq)} + H_{2(g)}$$

The hydrogen gas will be collected by displacement of water from a eudiometer placed upside down in a beaker of water (see Figure E10.2). If the water level in the eudiometer is not the same as the water level in the beaker, the pressure of the wet vapor (water vapor + hydrogen gas) plus the pressure due to any liquid water in the eudiometer above the water level of the beaker is equal to the barometric pressure. Thus,

$$P_{\text{barometric}} = P_{\text{wet vapor}} + P\,(H_2O)$$

The difference in height (mm) of water levels is measured after the gas is collected and converted to torr (1 mm Hg = 13.6 mm water). This is subtracted from the barometric pressure to find the pressure of the wet gas (water vapor plus hydrogen gas).

Example: 38.5 mg of magnesium react with excess acid. 38.2 mL of gas is collected at $T = 19.5°C$, $P_{\text{barometric}}$ is 752.3 mm Hg, and the water is 30.0 mm higher in the eudiometer. Calculate the pressure of the dry hydrogen gas.

Answer: First, find the pressure of the wet vapor:

$$30.0 \text{ mm water } (1.00 \text{ mm Hg}/13.6 \text{ mm water}) = 2.21 \text{ mm Hg}$$

$$P_{\text{barometric}} = P_{\text{wet vapor}} + P\,(H_2O)$$

$$752.3 \text{ mm Hg} = P_{\text{wet vapor}} + 2.21 \text{ mm Hg}$$

$$P_{\text{wet vapor}} = 750.1 \text{ mm Hg}$$

We now solve for the pressure of the dry hydrogen using **Dalton's Law of Partial Pressures.**

$$P_{\text{Total}} = P_A + P_B + P_C + \cdots$$

$$P_{\text{wet vapor}} = P\,(H_2) + P\,(\text{water vapor})$$

Table E10.3 Vapor Pressure of Water

T, °C	*P* (mm Hg)	*T*, °C	*P* (mm Hg)	*T*, °C	*P* (mm Hg)
0	4.58	16	13.63	26	25.21
5	6.54	18	15.48	28	28.35
10	9.21	20	17.54	30	31.82
12	10.52	22	19.83	40	55.3
14	11.99	24	22.38	50	92.5

Vapor pressures of water at various temperatures are given in Table E10.3. If the measured temperature is not listed, interpolate to find the vapor pressure at the measured temperature. Since T = 19.5°C (3/4 of the way from 18 to 20°C), we take 3/4 of the difference in vapor pressures [3/4 (17.54 − 15.48) = 1.55] and add this to the vapor pressure at the lower T. Thus, the vapor pressure of water at 19.5°C = 15.48 + 1.55 = 17.03 torr.

$$P_{\text{wet vapor}} = P\,(\text{H}_2) + P\,(\text{water vapor})$$

$$750.1 \text{ mm Hg} = P\,(\text{H}_2) + 17.03 \text{ mm Hg}$$

$$P\,(\text{H}_2) = 733.1 \text{ mm Hg.}$$

The volume of dry hydrogen gas collected under the laboratory conditions is converted to STP conditions using the Combined Gas Law. This value, divided by the expected number of moles of hydrogen gas to be formed from the amount of magnesium metal used, gives the molar volume.

Example: For the example above, calculate the molar volume of hydrogen gas at STP.

Answer: Recall that STP is 1 atm (760 mm Hg) and 273.15 K. Substituting into the Combined Gas Law,

$$P_1V_1/T_1 = P_2V_2/T_2$$

$$(733.1 \text{ mm Hg})(38.2 \text{ mL})/(292.7 \text{ K}) = (760.0 \text{ mm Hg})(V_{\text{STP}})/(273.15 \text{ K})$$

$$V_{\text{STP}} = 34.4 \text{ mL}$$

This is the volume generated by 38.5 mg Mg (0.0385 g). From the reaction stoichiometry, we can see that one mole of H_2 was formed from the reaction of each mole of Mg. Thus, H_2 and Mg are in a 1:1 ratio.

$$V = 34.4 \text{ mL } (1 \text{ L}/1000 \text{ mL}) = 0.034 \text{ L}$$

$$n = 0.0385 \text{ g Mg } (1 \text{ mol}/24.3 \text{ g Mg}) = 1.584 \times 10^{-3} \text{ mol Mg} = 1.584 \times 10^{-3} \text{ mol H}_2.$$

$$V/n = 0.034 \text{ L}/1.584 \times 10^{-3} \text{ mol H}_2 = 21.7 \text{ L/mol H}_2$$

The value obtained is reasonably close to the theoretical value of 22.4 L/mol H_2 at STP.

Obtaining the Empirical Weight

In the above calculation, we used the known molecular weight of magnesium to find the molar volume of hydrogen. Another way to use the data is to use the ideal molar gas volume at STP (22,400 mL) to calculate the mass of the metal that is stoichiometrically equivalent to one mole of hydrogen. This quantity is equal to the equivalent weight of the metal.

Example: Solve for the equivalent weight of Mg using the data in the example above.

Answer: (34.4 mL H_2 at STP)(1 mol/22,400 mL at STP) = 1.54×10^{-3} mol H_2
Since H_2 and Mg are in a 1:1 ratio,

$$1.54 \times 10^{-3} \text{ mol Mg} = 1.54 \times 10^{-3} \text{ mol H}_2$$

$$\text{Equivalent weight} = (\text{g Mg})/(\text{mol Mg})$$

$$= (0.0385 \text{ g Mg})/(1.54 \times 10^{-3} \text{ mol Mg})$$

$$\text{Equivalent weight} = 25.1 \text{ g/mol}$$

The value obtained is close to the actual value (24.3 g/mol) for magnesium. Unknown metals can often be identified in this manner.

EXPERIMENTAL PROCEDURE

Record all measurements on the data sheets.

- Obtain a metal strip of magnesium (or unknown metal) approximately 4–5 cm in length, and scrape it with a spatula or rub it with fine emery cloth (or sandpaper) to remove the oxide/nitride film. Place the strip (with forceps) in a tared 10 mL beaker, and weigh it to the nearest mg.
- Obtain a 50 mL eudiometer. Pour 9–10 mL of dilute (6 M) hydrochloric acid into it, filling the rest of the eudiometer with water. Add the water slowly to avoid undue mixing of the acid.
- Insert the metal sample, held by a piece of thin copper wire, through a one-hole stopper, about 3–4 cm into the eudiometer. (**NOTE:** The hole in the stopper should not be sealed off.) Firmly press the stopper into the eudiometer; make certain that no air bubbles are trapped under the stopper (see Figure E10.1).

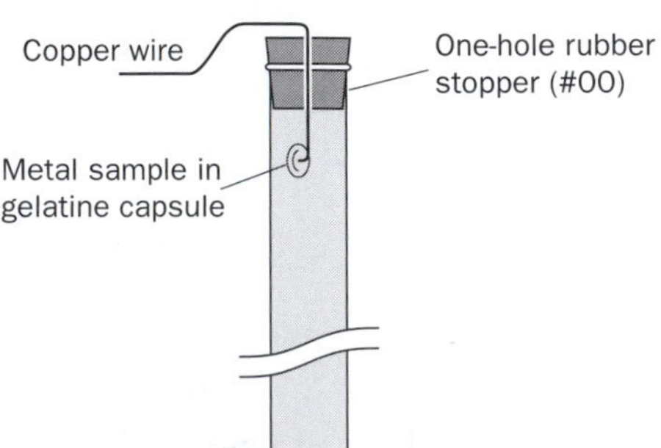

Figure E10.1 *Filled eudiometer with sample inserted.*

> **NOTE:** ***The metal sample may be held in the eudiometer in several ways:***
>
> ***a. If gelatin capsules are available, place the strip of metal into the capsule. Wrap the copper wire around the capsule several times, and insert the long wire end through the narrow end of the stopper. Insert this assembly into the eudiometer.***
>
> ***b. If gelatin capsules are not available, wrap the copper wire around the metal strip so as to entrap the sample in a cage-like assembly.***
>
> ***c. Bend the metal strip into a "V"-shaped wedge and then insert it into the end of the eudiometer, followed by the stopper.***

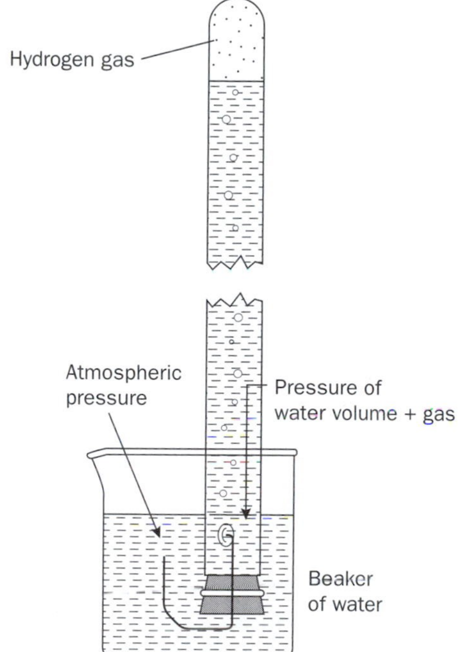

Figure E10.2 *Gas-collecting assembly (clamp and ring stand not shown).*

- Immediately place your finger over the hole in the stopper, invert the eudiometer, and immerse it in a 250 mL beaker of water (Figure E10.2).
- Remove your finger, and clamp the eudiometer so that the end is about 3–4 mm from the bottom of the beaker. If a gelatin capsule was used, the reaction will commence in 4–5 minutes as the capsule dissolves in the acid mixture; otherwise the reaction commences almost immediately as the acid settles to the bottom. Bubbles of hydrogen gas will be seen collecting in the eudiometer. **NOTE: If the metal is not inserted far enough into the eudiometer, small bubbles of gas may escape from the bottom of the eudiometer. If this occurs, the experiment must be repeated.**
- When all the metal has reacted, allow the equipment to cool to room temperature (heat is generated in the reaction). Tap the side of the eudiometer to free any adhering bubbles of gas. Raise or lower the eudiometer to make the water levels (inside the beaker and inside the eudiometer) equal. This equalizes the pressure of the collected hydrogen gas to the atmospheric pressure. If you cannot make the levels equal, measure and record the difference in the two levels in mm.
- Read and record the volume of gas in the eudiometer. Place a thermometer in the beaker of water, and record this value as the gas temperature. Record the barometric pressure.
- If time permits, repeat the experiment twice more with fresh samples of metal. If this is done, report the average of the results.

Name: ______________________________

Date: __________ Section: ______________________

Pre-Laboratory Questions: Experiment 10

1. A 45 mL sample of a dry gas is collected at 380 torr and 273°C. Calculate the volume of the gas sample at STP.

2. How is the calculated molar volume affected if the difference in water heights is not accounted for? Explain what piece of data is affected and how that affects the result.

3. How is the calculated molar volume affected if the partial pressure of water is not subtracted? Explain.

4. Use Table E10.3 to calculate the vapor pressure of water at 21.0°C.

Name: ____________________

Date: ________ Section: ____________________

Data Sheet: Experiment 10

Experimental Data

	Trial 1	Trial 2	Trial 3
1. Mass of container (0.0000 g, if tared)	_____ g	_____ g	_____ g
2. Mass of container + metal	_____ g	_____ g	_____ g
3. Difference in water levels (may be 0.0 mm)	_____ mm	_____ mm	_____ mm
4. Volume of gas in eudiometer	_____ mL	_____ mL	_____ mL
5. Temperature of the gas	_____ °C	_____ °C	_____ °C
6. Barometric pressure, mm Hg	_____ mm Hg	_____ mm Hg	_____ mm Hg

Calculations (show all work)

	Trial 1	Trial 2	Trial 3
1. Mass of metal, g	_____ g	_____ g	_____ g

If water levels are equal, omit step 2.

	Trial 1	Trial 2	Trial 3
2. Pressure of liquid water in eudiometer	_____ mm Hg	_____ mm Hg	_____ mm Hg
3. Total pressure of wet vapor	_____ mm Hg	_____ mm Hg	_____ mm Hg

($P_{barometric}$—result of step 2)

	Trial 1	Trial 2	Trial 3
4. Vapor pressure of water at experimental *T*	_____ mm Hg	_____ mm Hg	_____ mm Hg

Name: ____________________

Date: __________ Section: ____________________

Data Sheet: Experiment 10, p. 2

Calculations (continued)

Do 8–10 if finding molar volume **or** 11–12 if finding equivalent weight.

	Trial 1	*Trial 2*	*Trial 3*
5. Pressure of dry hydrogen gas	_____ mm Hg	_____ mm Hg	_____ mm Hg
6. Temperature, K	_____ K	_____ K	_____ K
7. Volume of dry gas at STP	_____ mL	_____ mL	_____ mL
8. Molar volume of hydrogen gas	_____ L mol^{-1}	_____ L mol^{-1}	_____ L mol^{-1}
9. % difference from literature value	_____ %	_____ %	_____ %
10. Average molar volume		_____ L mol^{-1}	
11. Equivalent weight of metal	_____ g mol^{-1}	_____ g mol^{-1}	_____ g mol^{-1}
12. Average equivalent weight		_____ g mol^{-1}	

Name: ____________________

Date: ________ Section: ____________________

Questions: Experiment 10

1. The magnesium strip was cleaned prior to weighing and reacting. If this had not been done and the MgO film was left on the ribbon, would the calculated molar volume of hydrogen gas collected be higher or lower than if the weighed mass were all magnesium metal? Explain which experimental data would be incorrect for the measured mass.

2. Give an explanation of why gases are not ideal at low temperature or high pressure. (Hint: see your textbook.)

3. In a fire emergency, a pumper truck can refill by sucking water up from a stream. The pump pulls a vacuum (~0.0 mm Hg) at the top of the hose, but it is atmospheric pressure pushing down on the stream that pushes water up the hose. What height of water (in meters) is equivalent to one atmosphere (760 mm Hg)? If the stream is 28 feet below the truck's pump, will the pump be able to refill the truck? What if the truck is in the mountains at 3000 m? The equation relating approximate barometric pressure (P, in torr) to altitude (h, in m) is

$$\ln P = 6.63 - (h/8000)$$

Find the barometric pressure at 3000 m, then the equivalent height of water.

Experiment 11 Formula Weight of a Volatile Liquid

OBJECTIVES

- To determine the formula weight, density, and molar volume of a volatile liquid.

PRIOR READINGS

Experiment 8

Introduction: Experiments 9–10

INTRODUCTION

In Experiment 10, the molar volume of any ideal gas at STP was stated to be 22.4 L. Therefore, one can experimentally determine the formula weight of a gas by weighing 22.4 L of the substance at STP. Fortunately, we can avoid this cumbersome procedure by invoking the Ideal Gas Law relationship:

$$PV = nRT$$

where P is the pressure (atm), V is the volume (liters), n is the number of moles (mol), R is the universal gas constant (0.082 L atm K^{-1} mol^{-1}), and T is the temperature (K).

A mole (n) may be defined as mass (m) per formula weight (M). Thus, $n = m/M$. If $\boldsymbol{m}$ equals the mass of the gas in grams, and $\boldsymbol{M}$ represents the formula weight of the gas, the Ideal Gas Law may be rewritten as

$$PV = \frac{m}{M}RT$$

Rearrangement of this equation shows that the density of a gas (the mass in grams per unit volume, $\rho = m/V$, can be determined using the relationship given below:

$$\rho = \frac{m}{V} = \frac{MP}{RT}$$

Further rearrangement leads to the relationship for the calculation of the formula weight of a gas:

$$M = \frac{mRT}{PV} = \rho\frac{RT}{P}$$

Thus, the formula weight of a gas can be determined from the measured density at any given temperature and pressure.

In this experiment, the Dumas method will be used to obtain the experimental data. This is a classical method, having been developed by Dumas in 1826. It provides a fairly accurate measurement of the formula weight of a gas.

It will be assumed that the vapor obeys the ideal gas law. The unknown liquid is vaporized at a known temperature **($\boldsymbol{T}$)** on a hot water bath. The volume **($\boldsymbol{V}$)** of the vapor corresponds to the volume of the round-bottom flask used in the experiment. The pressure **($\boldsymbol{P}$)** is measured using a barometer. The weight of vapor **($\boldsymbol{m}$)** is equal to the weight of condensed vapor after the flask has been removed from the hot water bath. The density and formula weight are then calculated using the equations outlined above. From the density and formula weight, the molar volume can be determined.

The pressure is measured by reading a barometer. The values are in mm Hg, which should be converted to atmospheres when using the Ideal Gas Law (760 mm Hg = 1 atm). Weather reports often give the pressure in inches of mercury (29.92 inches of Hg = 760 mm Hg).

EXPERIMENTAL PROCEDURE

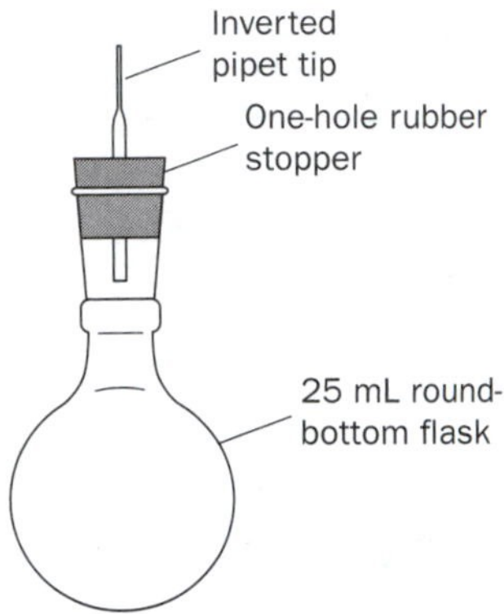

Figure E11.1 *Flask for formula weight determination.*

- Obtain a clean, dry 25-mL round-bottom flask and fit it with a one-hole rubber stopper in which a glass or plastic pipet tip has been inserted (see Figure E11.1). Note: *Be careful when inserting the pipet tip. Be sure to wrap your hands in a towel. A few drops of lubricant such as glycerin will assist this process.* Add a small boiling stone, and weigh the assembly to the nearest mg.
- Add approximately 1.5 mL of the unknown liquid to the flask and replace the stopper assembly. *Your unknown is a volatile organic liquid. It is flammable and could be an irritant. Do not inhale or get the liquid on your skin.*
- Attach an iron ring to a ring stand, and place a wire gauze pad on the ring. Prepare a hot water bath using a 250 mL beaker supported on the pad. Add a boiling stone to the beaker to prevent bumping and heat the water to nearly boiling.
- Remove the heat source, and clamp the flask in place as shown in Figure E11.2. *Be careful, the water bath is hot! Have the flask already clamped, and lower the clamp and flask into the water in one operation. Adjust the clamp as necessary.* Adjust the water level in the beaker by adding or removing water as required, so that the level is about 2–3 cm below the top of the flask.
- Heat the water in the beaker to a gentle boil, and record the temperature of the water bath. Continue to heat the flask until vapor can no longer be seen streaming from the capillary or until liquid can no longer be seen boiling in the flask. The boiling stone aids in seeing the removal of liquid from the flask. Heat the flask for an additional 3–5 minutes.

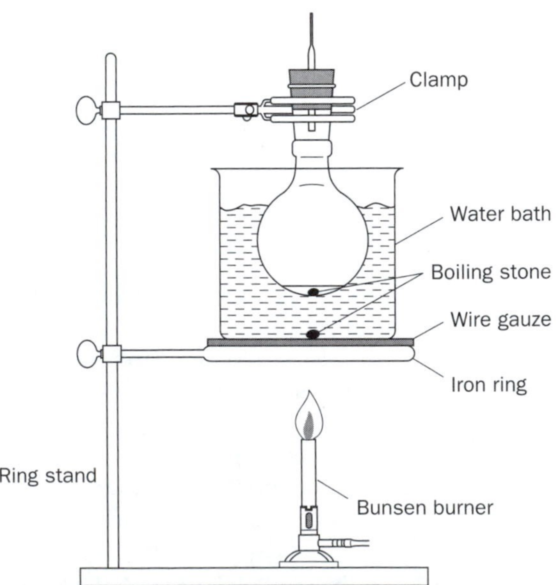

Figure E11.2 *Assembly for determining formula weight of a volatile liquid.*

- Remove the flask from the heat source, and allow it to cool to room temperature. **Do not remove the stopper and capillary.** Carefully dry the outside of the flask. Weigh the assembly containing the condensed vapor to the nearest mg.
- **Discard any condensed vapor as directed by your instructor, and the boiling stone.** If time permits, repeat the procedure using a fresh sample of the unknown.
- Fill the empty flask to the brim with water and insert the stopper containing the pipet tip. Make sure the water fills the tip. Remove the stopper and capillary. Measure the volume of water in the flask, using a 50 mL graduated cylinder. This corresponds to the volume of the flask, and hence the volume of the vapor.
- Record the barometric pressure in mm Hg or in inches of Hg. Calculate the formula weight of the volatile unknown liquid. If possible, obtain a copy of the MSD sheet for your liquid and attach it to your report.

Name: ______________________________

Date: __________ Section: ______________________________

Pre-Laboratory Questions: Experiment 11

1. Calculate the volume (in liters) of 0.15 moles of nitrogen gas at STP.

2. Calculate the number of moles of carbon dioxide when a lump of dry ice completely sublimes in an evacuated 5.0 L flask. The temperature of the gas is 25°C and the pressure 675 mm Hg.

3. A 20 mL vessel contains 70.7 mg of organic vapor at 1.5 atm and 27°C. What is the formula weight of the organic substance?

4. Calculate the density of ammonia gas in g L^{-1} at 100.0°C and 2.00 atm.

5. A gas was observed to have a density of 5.37 g L^{-1} at 1.00 atm and 25.0°C. What is the formula weight of the gas?

Name: ______________________________

Date: __________ Section: ______________________________

Data Sheet: Experiment 11

Unknown Number ______________

	Trial #1	*Trial # 2*
Mass of empty flask, stopper, capillary tube, and boiling stone	_______ g	_______ g
Mass of flask, stopper, capillary tube, boiling stone and condensed vapor	_______ g	_______ g
Temperature of hot water bath,°C	_______ °C	_______ °C
Volume of the round-bottom flask and tip (mL)	_______ mL	_______ mL
Barometric pressure (atm)	_______ atm	_______ atm

Calculations

Mass of unknown vapor	_______ g	_______ g
Moles of unknown vapor	_______ mol	_______ mol
Formula weight of unknown compound	_______ g mol^{-1}	_______ g mol^{-1}
Density of the gas at STP	_______ g L^{-1}	_______ g L^{-1}
Molar volume of the gas at STP	_______ L mol^{-1}	_______ L mol^{-1}
Deviation from ideal Molar volume	_______ %	_______ %

Show your calculations for the determination of the formula weight and the molar volume of your unknown. Hint: be careful to use the correct units.

Name: ______________________________

Date: __________ Section: ______________________________

Questions: Experiment 11

1. A chemist does not have access to a barometer, and therefore uses the value of 760 mm for the atmospheric pressure in a formula weight determination. Would the value of the formula weight be high, low, or the same as compared to your reported value, calculated with the actual barometric pressure? Explain.

2. If the unknown liquid does not completely vaporize during the heating process, will the formula weight be too high, low, or the same? Explain.

3. A chemist is given an unknown liquid having a boiling point of 155°C. What changes would have to be made in the experimental procedure to determine the formula weight of this compound? (Dumas determined the vapor density of numerous high boiling species and even some solids!)

4. You call the weather station and they tell you the barometric pressure for the day is 28.5 inches of Hg. Convert this value to atmospheres.

Experiment 12 Gas Phase Reactions

OBJECTIVES

- To construct a microscale gas generator
- To produce hydrogen and chlorine gas
- To photolytically prepare hydrogen chloride

PRIOR READINGS Experiment 8

Introduction: Experiments 9–10

INTRODUCTION

Note: This experiment is based on the work of Viktor Obendrauf (University of Graz, Austria, who can be contacted at v_obendrauf@styria.com). The syringe needles used are 1.2/40 mm, with the tips cut off in order to blunt the needle, converting them into steel tubes. Thus, they are useless as drug delivery devices.

Performing gas-phase reactions in the laboratory is normally difficult. Traditionally, gases were prepared in large-scale traditional glassware. The best known of such apparatus was the Kipp Generator, invented by Petrus Jacobus Kipp in 1844. Kipp generators were once quite common in qualitative analysis laboratories (see Experiments 30–33), where they were used to prepare hydrogen sulfide that was used to precipitate various metals. Other gases could also be prepared. The generator consisted of four parts: a base, an inner separator, a gas reservoir, and an acid reservoir, as shown in Figure E12.1.

To prepare hydrogen, for example, zinc chunks are packed around the inner separator in the gas reservoir. Acid is then added to the acid reservoir, where

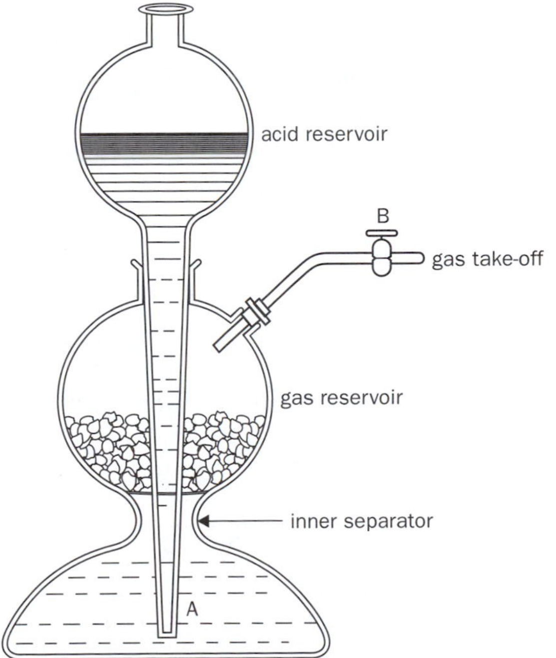

Figure E12.1 *Kipp gas generator.*

it flows through the long tube into the base. As the acid fills, and then overflows the base and inner separator, it reacts with the zinc to produce hydrogen:

$$Zn_{(s)} + 2\ HCl \rightarrow Zn^{2+}_{(aq)} + H_{2(g)} + 2\ Cl^{-1}_{(aq)}$$

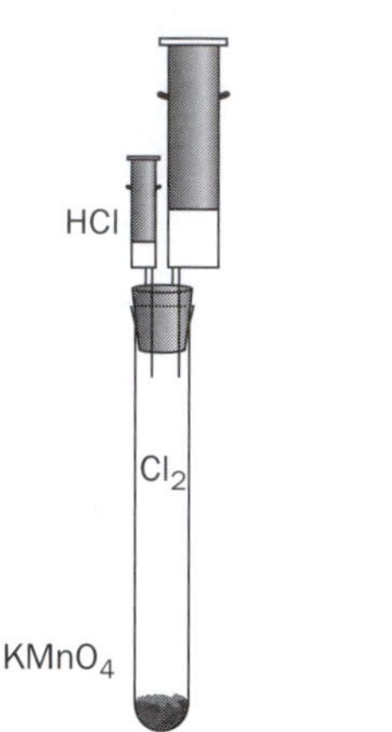

Figure E12.2 *Experimental apparatus for producing Cl_2.*

As the hydrogen accumulates, its pressure forces the acid out of the gas reservoir, back into the base and up into the acid reservoir. The system is self-regulating, since it is only when hydrogen is removed through the gas take-off that the pressure drops, more acid flows to react with more zinc, and produces more hydrogen.

In this experiment, we will prepare a micro-scale gas generator that has many similarities to the Kipp generator. A 2 mL disposable syringe (without inner washer) holds the acid solution, and its steel tube is inserted into a soft rubber stopper. A second, large syringe serves as the gas collection device, and is inserted through its steel tube into the rubber stopper as well. The stopper is inserted into a 16 × 160 mm wide-mouth test tube completing the gas generator (Figure E12.2).

The gas generator will be used to prepare hydrogen and chlorine gas, which will react to form hydrogen chloride through the photolytic reaction.

$$H_{2\,(g)} + Cl_{2(g)} \rightarrow 2\ HCl_{(g)}$$

The reaction is activated using a camera flash.

EXPERIMENTAL PROCEDURE

Preparation of Chlorine Gas Generator

- Obtain a 2 mL syringe, 20 mL syringe, 16 × 160 mm wide-mouth test tube, rubber stopper, and two 1.2/40 mm steel tubes. Attach one end of the steel tubes to the luer opening on the syringes. Label the 20 mL syringe "Cl_2."
- Draw up 2 mL of concentrated HCl **(Caution: corrosive)** into the 2 mL syringe, and insert its steel tube firmly through the rubber stopper so that a small amount protrudes beneath it. Attach the 20 mL syringe to the stopper as well, as shown in Figure E12.2.
- Place about 500 mg of solid $KMnO_4$ **(Caution: strong oxidizer,** wear gloves) in the test tube, tapping the tube to ensure that the solid falls to the bottom. Insert the stopper into the test tube, and set the complete gas generation apparatus aside in a test tube rack.

Preparation of Charcoal Filter

- Obtain a 10 mL syringe (without a plunger), and fill it with granular activated charcoal. Insert a rubber stopper in the top to seal it. Insert a single steel tube through the stopper, so that the tip protrudes a bit below the stopper. Make a second filter as well.

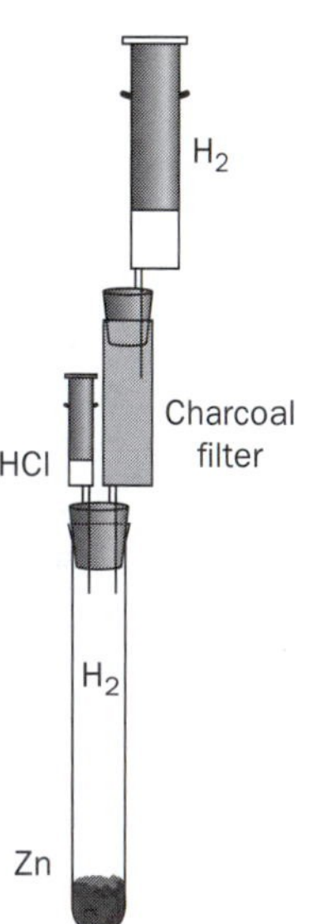

Figure E12.3 *Experimental apparatus for producing H_2.*

Preparation of Hydrogen Gas Generator

- Obtain a 2 mL syringe, 16 × 160 mm wide mouth test tube, 20 mL syringe, rubber stopper, and two 1.2/40 mm steel tubes. Attach one end of one steel tube to the luer opening of the 2 mL syringe. Attach one end of the other steel tube to the luer opening of the charcoal filter.
- Draw up 2 mL of concentrated HCl **(Caution: corrosive)** into the 2 mL syringe, and insert its steel tube firmly through the rubber stopper so that a small amount protrudes beneath it. Attach the charcoal filter in the same way, as shown in Figure E12.3. Attach the 20 mL syringe (label it "H_2") to the steel tube at the top of the charcoal filter.

- Place about 500 mg of solid Zn in the test tube, tapping the tube to ensure that the solid falls to the bottom. Insert the stopper into the test tube, and set the complete gas generation apparatus aside in a test tube rack.

Preparation of Chlorine

Do this reaction in the HOOD.

- Take the chlorine gas generator, and add a few drops of HCl from the 2 mL syringe by pushing down *gently* on the plunger. Chlorine gas should form immediately. As it does, draw up *gently* on the plunger of the 20 mL "Cl_2" syringe, filling it with the gas as it forms. Remove the "Cl_2" syringe (without its steel tube) from the stopper, and replace it with one of the charcoal filters. Expel the gas so that it bubbles through a beaker of water. The gas expelled will be a mixture of chlorine and air. Dry the luer on the "Cl_2" syringe, remove the charcoal filter and reattach the 20 mL syringe to the steel tube. Collect 10 mL of chlorine in the "Cl_2" syringe. It may be necessary to add another drop or two of the acid. Set the gas generator aside in a test tube rack, with the "Cl_2" syringe still attached.

Preparation of Hydrogen

- Take the hydrogen gas generator, and add a few drops of HCl from the 2 mL syringe by pushing down *gently* on the plunger. Hydrogen gas should form immediately as the acid reacts with the zinc. As it does, draw up *gently* on the plunger of the 20 mL "H_2" syringe, filling it with the gas as it forms. Remove the "H_2" syringe (without its steel tube), and expel the gas (a mixture of hydrogen and air) into the air. Reattach the "H_2" syringe to the steel tube at the top of the charcoal filter. Draw 10 mL of hydrogen gas in the "H_2" syringe, thereby flushing the charcoal filter with hydrogen. It may be necessary to add another drop or two of the acid.

Reaction of Hydrogen and Chlorine

- Remove the "H_2" syringe from the charcoal filter and set it aside. Now, remove the "Cl_2" syringe from the chlorine generator, and attach it to the steel tube on top of the charcoal filter on the hydrogen generator. Use the other charcoal filter to replace the syringe on the chlorine generator.
- Draw up 10 mL of hydrogen into the "Cl_2" syringe, which will now contain 10 mL of chlorine (from before) and 10 mL of hydrogen. Remove the "Cl_2" syringe, with its metal tube, from the charcoal filter, and place the steel tube vertically in a Styrofoam block. Place a camera flash alongside the syringe body, and fire the flash. The reaction forming HCl will start as the flash is fired.

Proving that HCl has Formed

- After the reaction has taken place, remove the syringe and steel tube from the Styrofoam block, and *slowly*, bubble the gas that has formed through 10 mL of water in a 10 mL beaker. Test the solution that is formed to confirm its acidity.

Name: ____

Date: __________ Section: ________________________________

Pre-Laboratory Questions: Experiment 12

1. In this experiment, chlorine gas is produced through the reaction of HCl with $KMnO_4$. Describe a two ways that we can confirm that the gas formed is chlorine. (Hint: See Experiment 3.)

2. In this experiment, hydrogen gas is produced through the reaction of HCl with Zn. Describe two ways that we can confirm that the gas formed is hydrogen.

3. When the hydrogen and chlorine gases react to form HCl, would you expect to see a volume change for the gas within the syringe? Explain why or why not.

Name: ______________________________

Date: __________ Section: ______________________

Data Sheet: Experiment 12

Sketch the apparatus that you prepared, and write your observations of the gas generation reactions and the reaction to form HCl.

Write a report detailing your experimental results for this experiment. Be sure to include how you prepared the experimental apparatus, the reaction procedure, and your observations at every step.

Name: ____________________

Date: ________ Section: ____________________

Questions: Experiment 12

1. The gas generator used in this experiment could be used to prepare gases other than hydrogen and chlorine. Choose any two of the following gases, consult your general chemistry book (or any other chemistry source, including the internet) and describe how this might be done.

a. Acetylene (C_2H_2)

b. Carbon dioxide

c. Oxygen

d. Nitrogen

Chapter 5
Spectroscopy

Introduction to Visible Spectroscopy

INTRODUCTION

Many familiar compounds in the chemistry laboratory form colored solutions. Potassium permanganate solutions are purple, many copper(II) solutions are blue, and most nickel(II) solutions are green. If a solution is colored, a species within it is absorbing light in the visible part of the spectrum. A graph of the amount of light absorbed by the species at different wavelengths is called an **absorption spectrum,** and can be used to identify the species or determine its concentration. A similar graph may be obtained showing the amount of light that is transmitted at various wavelengths, called a **transmittance spectrum.**

An example of an absorption spectrum is shown in Figure 5.1. The wavelength where the most light is absorbed is referred to as λ_{max} (read as lambda max), in this case occurring at 600 nm.

Spectroscopy is an important tool in analytical chemistry, where chemists seek to determine whether a particular species is present at all **(qualitative analysis),** the identity of the species, and/or how much of the material is present **(quantitative analysis).** Instruments used to measure either the transmittance or absorbance spectra of compounds are called **spectrophotometers.** In these instruments, a beam of energy (light) passes through the sample and is picked up by a detector. The percent transmittance ($\%T$) is defined as 100 times the ratio of the energy (I) reaching the detector relative to the original amount of energy (I_o):

$$\%T = 100\ (I/I_o)$$

The absorbance (A) is the logarithm of the inverse of the transmittance:

$$A = \log\ (I_o/I) = \log\ (100/\%T)$$

The visible region is only a small part of the overall electromagnetic spectrum (see Table 5.1). Spectrophotometers are available that can be used to investigate any portion of the overall electromagnetic spectrum, including the visible (see Experiments 13 and 14), the infrared (see Experiment 16), and the ultraviolet regions. When a substance absorbs in any region of the spectrum, the species undergoes a change due to that absorbance of energy. This change can be a bond vibration, the rotation of the molecule, an electron being promoted, etc. The amount of energy necessary for the change determines the region in which that change can be observed.

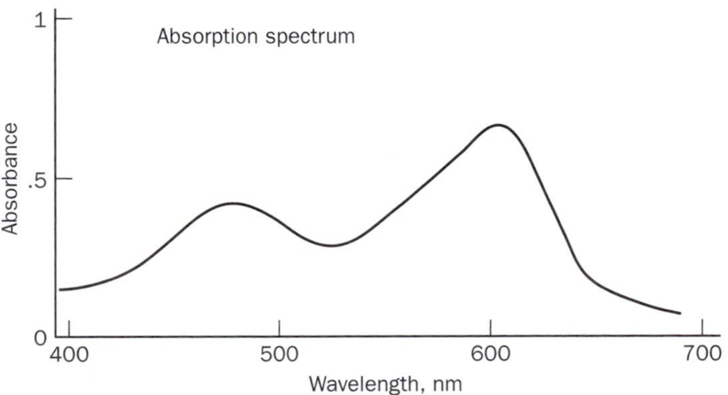

Figure 5.1 *Visible absorption spectrum.*

Table 5.1 Regions of the Electromagnetic Spectrum

Region	Wavelength (m)	Energy ($kJ\ mol^{-1}$)	Change Excited
Gamma Ray	$<10^{-10}$	$>10^{6}$	Nuclear Transformation
X-Ray	10^{-8}–10^{-10}	10^{4}–10^{6}	Inner-Shell Electrons
Ultraviolet (UV)	4×10^{-7}–10^{-8}	10^{3}–10^{4}	Outer-Shell Electrons
Visible (VIS)	8×10^{-7}–4×10^{-7}	10^{2}–10^{3}	Electronic Transitions
Infrared (IR)	10^{-4}–2.5×10^{-6}	1–50	Bond Vibrations
Microwave	10^{-2}–10^{-4}	0.01–1	Molecular Rotations
Electron Spin Resonance (ESR)	10^{-2}	0.01	Electron Spin Reversals
Nuclear Magnetic Resonance (NMR)	10	10^{-5}	Nuclear Spin Reversals

The Spectronic 20, shown in Figure 5.2, is a spectrophotometer that works largely in the visible portion of the spectrum (400–700 nm, the wavelengths of light that humans can see). It can be used to analyze colored solutions that are clear (not cloudy). In order to be detected by a visible spectrophotometer, the material being analyzed must be colored. This does not mean that colorless or faintly colored species cannot be analyzed in this manner. In many cases, a reaction can be carried out between a colorless analyte and some other material, producing a colored product, which can then be analyzed. As an example, aqueous solutions of iron(II) generally contain the $[Fe(H_2O)_6]^{2+}$ ion, which is pale blue-green in color and difficult to analyze. To improve the analysis, the iron(II) is often treated with 1,10-phenanthroline (phen), with which it forms a highly colored complex, $[Fe(phen)_3]^{2+}$. It is the iron–phenanthroline complex that is then detected and analyzed. This method is used in Experiment 14.

For quantitative analysis, the Spectronic 20 is set at a fixed wavelength, usually λ_{max} (unless an interfering species present in solution also absorbs at that wavelength). This allows for maximum light absorption. It is especially useful if the absorption spectrum is reasonably flat at λ_{max}. In this case, if the wavelength drifts slightly during an analysis, the readings will be minimally affected.

At a fixed wavelength, the amount of light absorbed is proportional to two parameters:

1. the concentration of the absorbing species, and
2. the pathlength of light through the solution.

The higher the concentration, the more compound is present to absorb light, and the more light will be absorbed. If the reading is made in cells of different

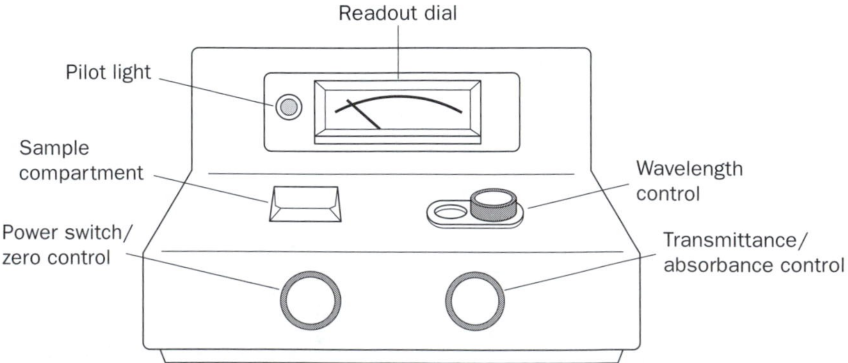

Figure 5.2 *Milton Roy Spectronic 20 spectrophotometer.*

pathlengths, different readings will be obtained. As the pathlength increases, the light has to pass through more solution and more compound, hence more light will be absorbed. These two factors are summarized in **Beer's Law:**

$$A = \varepsilon\, b\, c$$

where A is the amount of light absorbed **(absorbance),** ε is the molar extinction coefficient (a proportionality constant), b is the pathlength of light through the solution, and c is the concentration of the solution. In order to use Beer's Law to determine an unknown concentration, it is simplest to hold ε and b constant. By working at a fixed wavelength, ε is constant, and if all measurements are recorded in the same cell, then b is constant. Using these conditions, the amount of light absorbed by the solution is directly proportional to the concentration of compound in the solution.

The first part of a quantitative analysis experiment involves establishing the relationship between the amount of light absorbed and the concentration of the compound of interest in solution. This is accomplished by preparing a series of solutions of known concentration, called **standards.** One of these is a **blank** solution. It should contain all the components of the standards except for the compound of interest.

The amount of light absorbed by each of the standard and blank solutions is read on the spectrophotometer. A Beer's Law plot is then constructed, being a graph of absorbance (y axis) versus concentration (x axis). This plot is then used to determine the concentration of an unknown solution from the amount of light absorbed by that solution. For dilute solutions, such a plot is usually linear. Computer graphing programs are useful for this purpose, as they not only plot the graph, but also determine the slope of the Beer's Law data. A typical computer generated Beer's Law plot is shown in Figure 5.3.

It is readily seen that the slope of the line is equal to 1.12 L mol^{-1} in this example. If the absorbance of a solution of unknown concentration is obtained, the concentration of that solution can then be read from the Beer's Law plot. In the example in Figure 5.3, if the absorbance of the unknown solution was 0.080, the concentration could be read from the plot by moving horizontally from the 0.080 absorbance until the slope line is reached, and then moving vertically to the x axis, obtaining a concentration value of

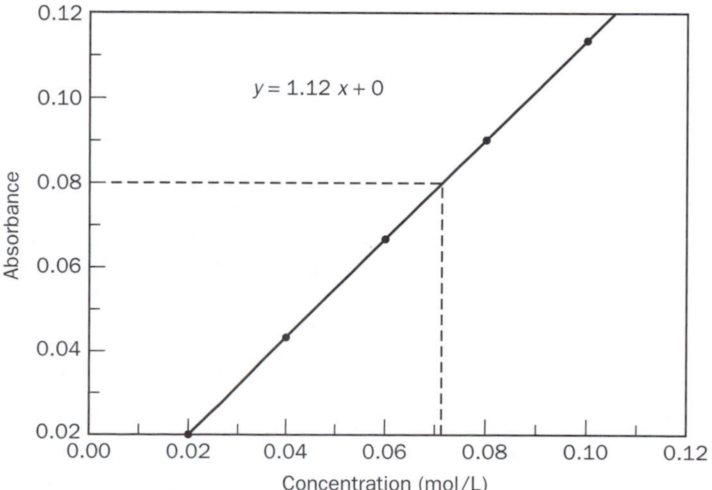

Figure 5.3 *Beer's Law plot.*

0.072 mol L^{-1}. Alternatively, the absorbance value of 0.080 could be substituted for y in the slope equation

$$y = 1.12x + 0$$

Solving for x gives the concentration as 0.071 mol L^{-1}. A third way of finding m and b is to perform a least squares calculation either by hand, using a spreadsheet or with a statistics program.

Operating instructions for the Milton Roy Spectronic 20 and 21 spectrophotometers are outlined below. If a different piece of equipment is used in your laboratory, the laboratory instructor will advise you as to its operation.

Operating Instructions for the Spectronic 20 Spectrophotometer

The Spectronic 20 is extremely easy to use, and gives accurate results. Refer to Figure 5.2 for the location of the various controls described below.

1. Turn on the Spectronic 20 by turning the **power switch/zero control** knob clockwise until you feel a "click." The **pilot light** will glow red at this point. Allow the unit to warm up for at least 15 minutes prior to making any measurements.
2. Set the desired wavelength (usually λ_{max}) using the **wavelength control** knob.
3. With the sample compartment empty and the lid closed, adjust the **readout dial** to 0% T by turning the **power switch/zero control** knob. Be careful not to inadvertently turn off the instrument.
4. Rinse and fill a matched pair of clean cells with the blank solution. Always rinse three times (about 1 mL per rinse) before filling. A half-full cell is usually sufficient for a reading. Then, wipe and blot the cells clean with a nonfibrous towel. Make sure that no fingerprints are on the cell. Place one of them in the **sample compartment,** aligning the guide mark on the cell with the guide mark on the sample compartment. If microcuvettes are used, an insert must be placed in the sample compartment first, to hold the microcells.
5. Adjust the **readout dial** to 100% T (0.00 Abs) with the **transmittance/absorbance control** knob.
6. Insert the second cell of blank solution, aligning the guide marks, and take a reading. If you get something close to 100% T (zero or a small positive reading, say 0.012 on the absorbance scale), record this value as the reading for your blank solution. Record both the % T and absorbance readings. If you get a value off scale to the right (>100% T, <0 absorbance), then you need to swap cells. Set the 100% T with this second cell. Then reinsert the first cell and you should get an on-scale reading. Record both the % T and absorbance.
7. Retain the cell you used to set 100% T. Label it **zero cell,** so you won't discard it by mistake. The <u>other</u> cell will be referred to as the **sample cell.** The zero cell will always be full of blank solution and is used only to set 100% T. The sample cell is the one in which you record the readings for all your solutions, including the blank solution. Discard the solution in the sample cell.
8. Rinse the sample cell three times with the least concentrated standard solution. Fill the cell about half full, wipe dry, and insert it into the **sample holder,** once again aligning the guide marks. Record the % T and absorbance readings and the concentration of the standard.

9. Recheck the 100% T reading using the zero cell, and readjust if necessary. Repeat step 8 for the other standards, in order of increasing concentration, and for your unknown solution.

Operating Instructions for the Spectronic 21 Spectrophotometer

The Spectronic 21 controls differ from those on the Spectronic 20, mainly due to the absence of a zero control knob.

1. Turn on the Spectronic 21 by turning the **power switch**. The **pilot light** will glow red at this point. Allow the unit to warm up for at least 15 minutes prior to making any measurements.
2. Set the desired wavelength (usually λ_{max}) using the **wavelength selector** knob.
3. Rinse and fill a matched pair of clean cells with the blank solution. Always rinse three times (about 1 mL per rinse) before filling. A half-full cell is usually sufficient for a reading. Then, wipe and blot the cells clean with a nonfibrous towel. Make sure that no fingerprints are on the cell. Place one of them in the **sample compartment,** aligning the guide mark on the cell with the guide mark on the sample compartment. If microcuvettes are used, an insert must be placed in the sample compartment first, to hold the microcells.
4. Adjust the **meter readout** to 100% T (0.00 Abs) by adjusting the **100% T control.**
5. Insert the second cell of blank solution, aligning the guide marks, and take a reading. If you get something close to 100% T (zero or a small positive reading, say 0.012 on the absorbance scale), record this value as the reading for your blank solution. Record both the % T and absorbance readings. If you get a value off scale to the right (>100 % T, <0 absorbance) then you need to swap cells. Set the 100% T with this second cell. Then reinsert the first cell and you should get an on-scale reading. Record both the % T and absorbance.
6. Retain the cell you used to set 100% T. Label it **zero cell,** so you won't discard it by mistake. The other cell will be referred to as the **sample cell.** The zero cell will always be full of blank solution and is used only to set 100% T. The sample cell is the one in which you record the readings for all your solutions, including the blank solution. Discard the solution in the sample cell.
7. Rinse the sample cell three times with the least concentrated standard solution. Fill the cell about half full, wipe dry, and insert it into the **sample holder,** once again aligning the guide marks. Record the % T and absorbance readings and the concentration of the standard.
8. Recheck the 100% T reading using the zero cell, and readjust if necessary. Repeat step 7 for the other standards, in order of increasing concentration, and for your unknown solution.

Experiment 13 Spectrophotometric Determination of Copper(II)

OBJECTIVES

- To determine the concentration of an unknown copper(II) solution using a Beer's Law plot.

PRIOR READING

Chapter Five: Introduction

Section 3.4: Liquid Volumes

INTRODUCTION

Visible spectroscopy most often is used to make quantitative measurements, for example, to determine the amount of a particular species present in solution. This is generally done by obtaining a **Beer's Law** plot for a series of standards of the species of interest. The unknown concentration is then read directly from the Beer's Law plot.

Suppose you are interested in obtaining the concentration of copper(II) in a particular solution. The Beer's Law method would consist of preparing a series of standards from a stock solution of copper(II). One might take a 0.100 M solution of $CuSO_4{\cdot}5H_2O$, and by dilution in volumetric flasks, obtain a series of solutions that are 0.0100, 0.0200, 0.0500, and 0.0800 M.

The dilution equation is

$$M_1V_1 = M_2V_2$$

where M is the molarity of the solution, and V is its volume.

Example: We wish to prepare 10.0 mL of 0.0800 M solution from a 0.1 M stock solution of $CuSO_4{\cdot}5H_2O$. How do we do it?

Answer: Substituting into the dilution equation,

$$M_1V_1 = M_2V_2$$

$$(0.100\text{M})(V_1) = (0.0800\text{M})(10.0 \text{ mL})$$

$$V_1 = 8.00 \text{ mL of the stock solution}$$

We would dilute 8.00 mL of the stock solution to 10.0 mL with solvent in a 10 mL volumetric flask.

The absorbance of the standards would then be obtained at λ_{max}, the wavelength of maximum absorption, and plotted (y axis) as a function of concentration (x axis). A straight line Beer's Law plot will be obtained if the concentrations are reasonably low. The absorbance of the unknown would then be obtained, and the concentration determined directly from the Beer's Law plot. The introduction to Chapter 5 details the use of the Spectronic 20 or 21 visible spectrophotometer, and the graphing and use of Beer's Law plots. In this experiment, the concentration of an unknown Cu(II) solution will be determined.

EXPERIMENTAL PROCEDURE

Preparation of Standards

The ion of interest in the solutions is $Cu(NH_3)_4^{2+}$, tetramminecopper(II), which has a deep royal blue color in aqueous solution. All solutions should be prepared from a 0.0200 M stock solution of copper sulfate pentahydrate, $CuSO_4{\cdot}5H_2O$, and a 2.0 M stock solution of aqueous ammonia.

- Rinse a clean 10 mL graduated pipet with a small amount of the copper sulfate stock solution. Rinse five 10 mL volumetric flasks with distilled water.
- Pipet exactly 5.0 mL of the copper sulfate solution into the first volumetric flask, 4.0 mL into the second flask, 3.0 mL into the third, 2.0 mL into the fourth, and 1.0 mL into the fifth. Label each flask with the amount of copper sulfate stock solution that it contains.
- Dilute each of the five flasks to the mark with the 2.0 M solution of aqueous ammonia. Stopper the flasks, and mix thoroughly. Each volumetric flask contains enough solution for two students.
- Use the 2.0 M aqueous ammonia as the blank solution. Obtain a sample of unknown concentration (already mixed with ammonia) from your instructor. ***NOTE: These solutions all contain aqueous ammonia, which is an irritant. Avoid inhalation. Keep discarded solutions in a stoppered flask.*** Following the procedure in the introduction to Chapter 5 (Introduction to Visible Spectroscopy), obtain absorbance readings for the blank, standards, and unknown at 615 nm using a Spectronic 20 or 21 spectrophotometer. If a scanning visible spectrophotometer is available, verify that λ_{max} occurs at this wavelength.

Graphing the Beer's Law Plot

A final correction needs to be made before the linear relationship can be used to calculate the concentration of the unknown solution. Ideally, the reading for the sample cell should be 0.00 absorbance, 100.0% T (the same as for the zero cell), when filled with the blank solution. In most cases it is not, indicating a slight difference in the two cells. As an example, suppose that the sample cell absorbance is 0.012. This means that all the readings taken in the sample cell (the blank, the five standards and the unknown) are slightly incorrect. They all differ by the same amount, the difference between the zero and sample cells. This difference is called the **cell-to-cell correction factor.** All readings must be corrected before calculations can be done.

- Take the difference between the zero cell (0.00 absorbance) and the sample (0.012 when filled with the blank solution in our example) and subtract it from all the absorbance readings (including the blank). In this example you would subtract 0.012 from all the absorbance readings. This will give the blank a reading of 0.00, and the others will be similarly corrected to what those readings would have been had they all been read in the zero cell.
- Construct a graph of concentration of standards (including the blank) on the x axis versus absorbance on the y axis. Draw a straight line through the data. Show the calculation of the slope and y intercept for your corrected data. Write the equation of the line ($y = mx + b$) using the slope and the y intercept calculated.
- Use this equation and the absorbance recorded for the unknown to calculate the concentration of the unknown solution. Alternatively, use a graphing program to construct the Beer's Law plots and obtain the line equation and unknown concentration.

Name: ______________________________

Date: __________ Section: ______________________________

Pre-Laboratory Questions: Experiment 13

1. What is the relationship between absorbance and transmittance?

2. Normally, calibration curves are plots of absorbance (y axis) versus concentration (x axis). Why couldn't a plot of transmittance versus concentration be used as easily?

3. Draw a sketch of the optical pathway in a visible spectrophotometer. Be sure to include the source, sample, and detector.

Name: ____________________

Date: ________ Section: ____________________

Data Sheet: Experiment 13

1. Zero Cell Reading ____________

2. Sample Cell Reading with Blank Solution ____________

3. Cell-to-Cell Correction Factor ____________

4. Standard Solution Readings

Solution #	*Concentration*	*%T*	*Absorbance*	*Corrected Abs.*
1	________ mol L^{-1}	____	________	________
2	________ mol L^{-1}	____	________	________
3	________ mol L^{-1}	____	________	________
4	________ mol L^{-1}	____	________	________
5	________ mol L^{-1}	____	________	________

Construct a Beer's Law Plot as described in the experiment.

5. Equation of Beer's Law Line (if determined) ____________

6. Unknown Solution Absorbance ____________

7. Concentration of Unknown ____________ mol L^{-1}

Name: ______________________________

Date: __________ Section: ______________________

Questions: Experiment 13

1. A 0.85 cm cuvette is mistakenly substituted for a 0.80 cm cuvette. When the percent transmission for the solution in the "too large" cuvette is taken, will it be higher, lower, or the same as it would be in a proper sized cuvette? Explain.

2. The λ_{max} in this experiment is 615 nm, corresponding to the color orange. The copper(II) solution is certainly not orange in color. Explain.

Experiment 14 Determination of the Iron Content in a Vitamin Tablet

OBJECTIVES

- To determine the amount of iron in a vitamin tablet using visible spectroscopy.

PRIOR READING

Chapter Five: Introduction
Section 3.5: Heating Methods
Section 3.7: Filtration
Section 3.8: Solution Preparation

INTRODUCTION

Recently, there has been a growing awareness of the importance of a large number of metals and other inorganic materials in biological systems. To cite only a few examples, iron is necessary in oxygen transport and for energy storage, copper and zinc are present in many enzymes, molybdenum is important in nitrogen fixation by bacteria in plants and catalyzes various oxidation reactions, and manganese is necessary for oxygen evolution in photosynthesis.

If any of these essential elements are missing from a diet, severe illness or death may result. Lack of molybdenum, for example, prevents the oxidation of xanthine (obtained in the diet from potatoes, coffee, and tea) to uric acid, and is the primary cause of the gout. Many people take vitamin pills with "mineral additives" to ensure sufficient amounts of these necessary metals. Vitamins with iron added are quite common. The iron(II) present in vitamins is in the form of ferrous fumarate, $FeC_2H_2O_4$. It is interesting to note that iron(III) is generally useless for living systems. Many areas have soils with high iron(III) content but little or no Fe(II), and thus are nonsupportive of plant life, despite their high iron content.

In this experiment, the amount of iron in a vitamin tablet will be determined. Since iron(II) is only weakly colored (pale blue-green when uncomplexed in aqueous solution), the iron(II) will be reacted with 1,10-phenanthroline ($C_{12}H_8N_2$) prior to analysis. The iron(II) phenanthroline complex, $[Fe(phen)_3]^{2+}$ is highly colored (orange) and easy to detect. Sodium acetate ($NaCH_3COO$) is added to control the acidity of the solutions. Iron(II) in solution is easily oxidized to Fe(III), which does not form a colored complex with phenanthroline. Hydroxylaminehydrochloride, $(NH_2OH)HCl$, a reducing agent, is added to ensure that all of the dissolved iron is iron(II). The following reactions occur:

$$CH_3COO^- + H^+ \rightleftharpoons CH_3COOH$$

$$2\ NH_2OH \cdot HCl + 4\ Fe^{3+} + 5\ H_2O \rightleftharpoons N_2O + 4\ Fe^{2+} + 6\ H_3O^+ + 2\ Cl^-$$

EXPERIMENTAL PROCEDURE

- Weigh an iron-enriched vitamin tablet (such as One-A-Day Plus Iron) to the nearest mg. Place the tablet in a mortar and pestle, and crush the tablet to a fine powder. Weigh and transfer about 25 mg (± 1 mg) of the tablet to a 25 mL beaker, and add 5 mL of 0.01 M HCl.
- Place the beaker on a magnetic stirring hot plate, and with stirring, boil the solution for about 15 minutes. Allow the solution to cool to room temperature.

Depending on the type of vitamin used, much of the solid may not dissolve in the acid—this is normal. The remaining solid consists of binder and other inert ingredients. The iron will have all dissolved.

- Filter the liquid extract by suction filtration using a Hirsch funnel and water trap (Figure 3.19, p. 42), and transfer the liquid filtrate to a 10 mL volumetric flask using a Pasteur pipet. Rinse the beaker with 1 mL of deionized water. Use this to also rinse the filter cake and filter flask. **Save the rinse.** Repeat with a second 1 mL rinse.
- Transfer the combined rinses to the same 10 mL volumetric flask. Dilute to the mark with deionized water. This solution will be used as the "unknown solution" in this experiment.

Preparation of Standards

A series of standards is prepared using the following stock solutions:

A. ***0.0100 g/L Fe(II)***
B. ***1% (w/v) hydroxylamine hydrochloride***
C. ***1.00 M sodium acetate***
D. ***1% (w/v) 1,10-phenanthroline.***

The iron compound used to prepare the Fe(II) stock is ferrous ammonium sulfate hexahydrate, $Fe(NH_4)_2(SO_4)_2 \cdot 6H_2O$. The iron stock solution also contains sulfuric acid, H_2SO_4. Solution A must be pipetted exactly, as that determines the exact concentration of iron in the standard. The other reagents are added in excess, so their volumes need not be as exact.

- Make up the following four standards in 10 mL volumetric flasks (all ±0.01 mL). The blank solution is prepared in a beaker or flask.

	Vol. of Soln. A, mL	Vol. of Soln. B, mL	Vol. of Soln. C, mL	Vol. of Soln. D, mL
Standard 1:	0.500	1.00	1.00	1.00
Standard 2:	1.000	1.00	1.00	1.00
Standard 3:	1.500	1.00	1.00	1.00
Standard 4:	2.000	1.00	1.00	1.00
Blank		2.00	2.00	2.00

- Fill each volumetric flask to the mark with deionized water. Add 14 mL of deionized water to the blank solution (twice as much blank solution is prepared so the blank reading may be rechecked). Stopper and mix each solution thoroughly. Obtain the absorbance of each of the standard solutions at 508 nm. (Instructions for the operation of the Spectronic 20 and 21 are given in the Introduction to this chapter.)
- Prepare a solution of your unknown in a 10 mL volumetric flask by adding 1.000 mL of your unknown, 1.00 mL each of B, C, and D, and diluting to the mark with deionized water. Repeat for each tablet analyzed. These are the "diluted unknown solutions." Obtain the absorbance of the unknowns at 508 nm.

Calculations

- Calculate the concentration of Fe(II) in each of the standards (dilution calculation).
- Construct a Beer's Law plot from the standard data, as described in the introduction to Chapter 5.
- From the Beer's Law plot or from the Beer's Law equation for the line, determine the concentration of iron(II) in each of the diluted unknown solutions.
- Since the volume and concentration of the diluted unknown solutions are now known, the mass of iron in the diluted unknown solutions can be readily determined. This diluted unknown solution contained 1.00 mL of the original 10.00 mL (unknown solution) prepared from the tablet. Calculate the mass of Fe(II) in the 10.00 mL unknown solution.
- This unknown solution was prepared from a weighed subsample of the whole tablet. Calculate the mass of Fe(II) in the whole tablet.

Further Work

See Experiment 15 for an environmental application of this procedure.

Name: ______________________________

Date: __________ Section: ______________________________

Pre-Laboratory Questions: Experiment 14

1. Most iron(III) compounds are very weakly colored, and do not lend themselves well to visible spectrophotometric analysis. Suggest a way in which their analyzability can be improved.

2. The absorbance of iron is determined at 508 nm. From the information given in the introduction to Chapter 5, what do you think the visible absorption spectrum of $Fe(phen)_3^{2+}$ looks like? Why is this wavelength chosen?

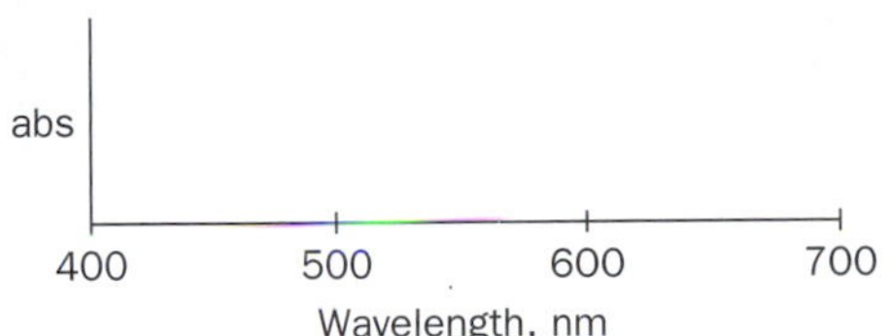

3. After graphing the absorbances of the $Fe(phen)_3^{2+}$ standards, the following equation relating absorbance (y axis) to concentration [mg Fe(II)/L, x axis] was obtained:

$$y = 310.2x + 0.007$$

The absorbance of a diluted unknown solution (volume = 10.00 mL, prepared from a vitamin tablet as in this experiment) is 0.483. What is the concentration and mass of Fe(II) in the diluted unknown solution? If this was prepared by diluting 1.00 mL of the original 10.00 mL of unknown solution, what is the mass of Fe(II) in the 10 mL of unknown solution?

Name: ______________________________

Date: __________ Section: ______________________________

Data Sheet: Experiment 14

Brand of tablet analyzed ______________________________

Advertised mass of Fe per tablet ______________________________

	Trial 1	*Trial 2*	*Trial 3*
Mass of whole tablet	______ g	______ g	______ g
Mass analyzed	______ g	______ g	______ g

Standard Solution Readings

Solution #	*Concentration*	*Absorbance*	*Corrected Absorbance*
Blank	______ mol L^{-1}	______	______
1	______ mol L^{-1}	______	______
2	______ mol L^{-1}	______	______
3	______ mol L^{-1}	______	______
4	______ mol L^{-1}	______	______

Construct a Beer's Law Plot as described in the introduction to Chapter 5.

Beer's Law equation (if determined) ______________________________

	Trial 1	*Trial 2*	*Trial 3*
Unknown absorbance	______	______	______
Corrected absorbance	______	______	______
Concentration of Dil. Unknown	______ mg L^{-1}	______ mg L^{-1}	______ mg L^{-1}
Mass of Fe(II) in Dil. Unknown	______ g	______ g	______ g
Mass of Fe(II) in Unknown	______ g	______ g	______ g
Mass of iron in whole vitamin tablet	______ g	______ g	______ g

Percent difference from advertised value:

Name: ______________________

Date: __________ Section: ______________________

Questions: Experiment 14

1. Iron is one of the biologically necessary metals. Name two biochemical functions that it serves.

2. Some plants grow in soils as high as 20% iron. If the iron is present in the form of $Fe(OH)_3$, the plants can still be iron deficient. Explain how this can be true. (Hint: consider the solubility of this compound.) How can the iron in the soil be made usable?

3. Search the internet for information on iron toxicity. What are the health effects of too much iron? The Code of Federal Regulations is available on-line. See if you can find the allowable limits for iron (and other metals) in drinking water. What does it mean when different chemical species have very different allowable limits?

Experiment 15 Determination of the Source of Iron Pollution

OBJECTIVES

- To determine the source of iron pollution in a town.

PRIOR READING

Introduction: Chapter Five
Section 3.8: Solution Preparation—Dilution

INTRODUCTION

Recently, the town of Hematite had the school system's drinking water tested because students complained that it had an odd, bitter taste. The school and the part of town downstream from the school uses water from a river as its drinking water. The water is filtered to remove particulates and bacteria. The part of town upstream uses wells, as the stream flow is too low in that part of town. Everything tested fine except that the iron content was above the recommended limit for drinking water. The townspeople have their suspicions about the source of the iron—there is an abandoned mine in the hills above town, a magnetic recording tape company, and a start-up company that produces programmable refrigerator magnets.

There was a minor problem with the Water Department's record keeping. A field chemist took water samples in bottles numbered 1 to 20. He took two samples at each of ten sample sites, mostly from streams that feed into the river. He put the samples randomly into the bottles, that is, sequentially numbered bottles are not necessarily from the same site. The bottles have no other identification, so the person analyzing the samples doesn't know which samples come from which sites.

The field chemist made a careful record of which bottle number went with which site number. Unfortunately, part of the list was lost. What remains are the site and corresponding bottle numbers for 9 samples, each from a different site. The Water Department realized, however, that all was not necessarily lost. If the samples have different enough levels of iron, the sample pairings could be reestablished, and the extra two that match each other would be the samples from the tenth site. The results could then be mapped and any other necessary sampling done to confirm results.

EXPERIMENTAL METHOD

In this experiment, the amount of iron in water samples taken around the town will be determined by the method used in Experiment 14. The iron in the samples is converted to iron(II) phenanthroline complex, $[Fe(phen)_3]^{2+}$, which is highly colored and easy to detect using a spectrophotometer. Each student will be assigned a sample to analyze in triplicate. After analyzing your sample, the results for all the samples will be collected to see if the sample pairings can be established. Then the results will be mapped and a report written as to the probable source of the iron.

Because the iron(II) phenanthroline method works in the range of 10^{-5} M (10^{-3} g Fe/L) and the range of iron levels in polluted water is often significantly higher, the samples may need to be diluted to reach the working range.

The best way to accomplish this is to do a 1:10 dilution (1.00 mL of sample diluted with deionized water to 10.00 mL in a volumetric flask). Then use 1.00 mL of that solution to prepare a colored complex (like the standard with 1.00 mL of stock). The iron stock solution is prepared with 0.045 M H_2SO_4 to help keep the iron as Fe(II). As the water samples have not been acidified, you must add 1.00 mL of 0.045 M H_2SO_4 to account for this. The absorbance of this solution will tell you if you need to dilute more or perhaps less (or perhaps even use the original sample).

EXPERIMENTAL PROCEDURE

Preparation of Standards

Your instructor will have prepared a series of stock solutions:

A. 0.0100 g/L Fe(II)
B. 1% (w/v) hydroxylamine hydrochloride
C. 1.00 M sodium acetate
D. 1% (w/v) 1,10-phenanthroline.

The iron compound used to prepare the Fe(II) stock is ferrous ammonium sulfate hexahydrate, $Fe(NH_4)_2(SO_4)_2 \cdot 6H_2O$. The iron stock solution also contains sulfuric acid, H_2SO_4. Be careful to pipet Solution A **exactly,** as that (along with the final 10.00 mL volume) determines the exact concentration of iron in the standard. The other reagents are added in excess, so their volumes need not be as exact.

- Make up the following four standards in 10 mL volumetric flasks (all ±0.01 mL). The blank solution is prepared in a beaker or flask.

	Vol. of Soln. A, mL	Vol. of Soln. B, mL	Vol. of Soln. C, mL	Vol. of Soln. D, mL
Standard 1	0.500	1.00	1.00	1.00
Standard 2	1.000	1.00	1.00	1.00
Standard 3	1.500	1.00	1.00	1.00
Standard 4	2.000	1.00	1.00	1.00
Blank	0	2.00	2.00	2.00

- Fill each volumetric flask to the mark with deionized water. Add 14 mL of deionized water to the blank solution (twice as much blank solution is prepared so the blank reading may be rechecked). Stopper and mix each solution thoroughly. Obtain the absorbance of each of the standard solutions at 508 nm.
- Record the bottle number of your sample. Rinse a 1 mL volumetric or graduated pipet with the sample solution. Pipet 1.00 mL of sample into a clean 10 mL volumetric flask. Fill to the mark with deionized water, stopper, and mix thoroughly. Label this flask "**1:10 dilution**."
- Prepare your analysis solution as follows: in a 10 mL volumetric flask, add 1.00 mL of the "**1:10 dilution**" (rinse the pipet with it first), 1 mL of 0.045 M H_2SO_4, 1 mL each of solutions B, C, and D, and dilute to the mark with deionized water. Label the flask "**analysis solution,**" and obtain its absorbance.
- **If the absorbance is in the range of 0.07 to 0.70,** you lucked out on the dilution. Prepare your "**1:10 dilution**" twice more and prepare the "**analysis solution**" from it each time, obtaining the absorbance for each trial. Discard all solutions.

  If the absorbance is below 0.07, your "**analysis solution**" is too dilute. Discard both the "**1:10 dilution**" and the "**analysis solution.**" Prepare a new "**analysis solution**" in a 10 mL volumetric flask as follows: Add 1.00 mL of the original water sample (rinse pipet first!), 1 mL of 0.045 M H_2SO_4, 1 mL each of solutions B, C, and D, and fill to the mark with deionized water. Record the absorbance at 508 nm. Perform two more trials. Discard all solutions.

If the absorbance is above 0.70, your "**1:10 dilution**" is still too concentrated. Discard the "**analysis solution.**" Prepare a more dilute solution in the following way. Rinse a 1 mL volumetric or graduated pipet with the "**1:10 dilution.**" Pipet 1.00 mL of the "**1:10 dilution**" into a clean 10 mL volumetric flask. Fill to the mark with deionized water, stopper, and mix thoroughly. Label this flask "**1:100 dilution.**" Prepare a new "**analysis sample**" by taking a clean 10 mL volumetric flask, and adding 1.00 mL of the "**1:100 dilution**" (rinse the pipet with it first), 1 mL of 0.045 M H_2SO_4, 1 mL each of solutions B, C, and D, and fill to the mark with deionized water. Obtain the absorbance of the "**analysis sample**" at 508 nm. Perform two more trials, preparing new "**1:100 dilution**" solutions, then preparing new "**analysis solution**" and obtaining the absorbances. Discard all solutions.

Calculations

- Construct a Beer's Law plot, as described in the Chapter Five Introduction. Use the Beer's Law equation obtained to calculate the concentration of iron in your analyzed solutions. Since the volume and concentration of the solutions analyzed are now known, the concentration of iron in the original water sample can be determined.
- Multiply the concentration obtained from the Beer's Law equation by the dilution factor (after all, you diluted the solution one or more times) to obtain the original concentration of iron in your original water sample. Calculate the average concentration.
- Consult your classmates to find the person whose sample has the same concentration as yours. Record their bottle number. Record the bottle numbers and the concentrations on your instructor's master list. Be sure to write legibly—transcription errors are common. Once all the results have been turned in, obtain a copy of the master list.
- Fill the results in on the attached map, and see if you can determine the source of the pollution. Write a brief report to the Hematite Town Council summarizing the problem, the analytical program, the mapped results, and your conclusions.

Name: ______________________________

Date: __________ Section: ______________________________

Pre-Laboratory Questions: Experiment 15

Base your answers to the following questions on the map that follows on page 213.

1. Suppose that the Startup Company is the pollution source. Would you expect the concentration of iron at location "I" to be higher, somewhat lower, a lot lower or the same as at location H? Explain why.

2. Suppose that the Tape Company is the pollution source. Would you expect the concentration of iron at location "D" to be higher, somewhat lower, a lot lower or the same as at location F? Explain why.

3. The analysis of one water sample required the preparation of the "**1:10 dilution**" and then the preparation of the "**1:100 dilution**" solution. The concentration in the "**analysis solution**" prepared from the "**1:100 dilution**" was found to be 2.33×10^{-3} g Fe/L. What is the dilution factor? What is the concentration of iron in the original water sample?

Name: ______________________________

Date: __________ Section: ______________________________

Data Sheet: Experiment 15

1. Zero Cell Reading ____________

2. Sample Cell Reading ____________

3. Cell-to-Cell Correction Factor ____________

4. Standard Solution Readings

Solution #	*Concentration, mol L^{-1}*	*Absorbance*	*Corrected Abs.*
1	______	______	______
2	______	______	______
3	______	______	______
4	______	______	______

Construct a Beer's Law Plot as described in the experiment.

Unknown Sample ____________.

5. Unknown solution absorbance (prepared from 1:10) ____________

6. If neccesary, absorbance of unknown prepared from original water sample ____________

If neccesary, absorbance of unknown prepared from 1:100 dilution ____________

7. Circle the dilution used: original 1:10 1:100 1:1000

Absorbances for 3 Trials (Trial 1 is one of the above readings)

	Trial 1	*Trial 2*	*Trial 3*
Absorbance	______	______	______
Corrected absorbances	______	______	______
Average absorbance		______	
Concentration of dilute solution		______ mol L^{-1}	
Concentration of original water sample		______ mol L^{-1}	

Name: ______________________________

Date: __________ Section: ______________________

Questions: Experiment 15

Write a brief report to the Hematite Town Council summarizing the problem, the analytical program, the mapped results, and your conclusions.

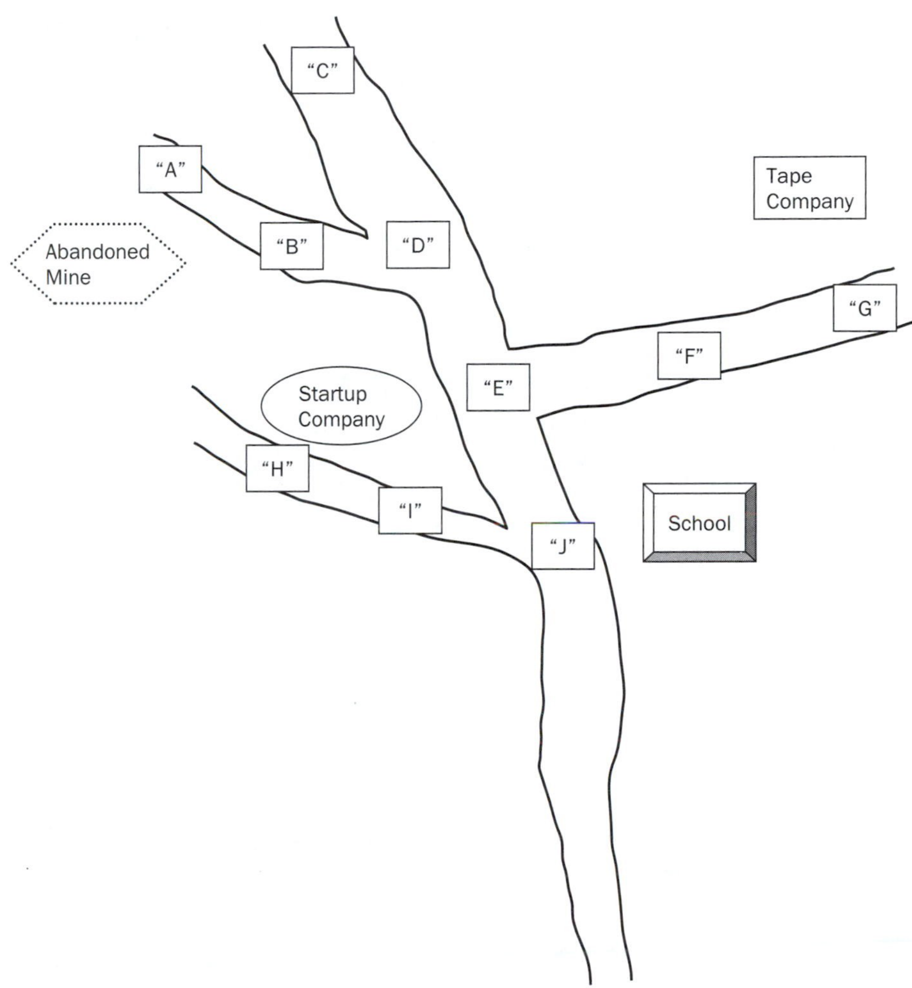

Town of Hematite—Water Sample Locations

Experiment 16 Infrared Analysis of Copper(II) Chloride–DMSO Complex

OBJECTIVES

- To synthesize a metal–DMSO (dimethylsulfoxide) complex.
- To determine the nature of bonding in a complex using IR.

PRIOR READING

Section 3.7: Filtration

Chapter 5: Introduction

INTRODUCTION

The **infrared (IR) region** occurs at longer wavelengths than the red color region of the visible spectrum. The absorptions in this region are valuable tools for determining the nature of bonding in a particular compound. The absorption spectrum can be likened to a fingerprint—no two compounds have exactly the same infrared spectrum, although similar compounds can have similar spectra. The absorbances, also called **bands,** correspond to various bond vibrations in the sample. Instead of a wavelength scale, infrared spectra use a wavenumber scale, where the wavenumber is the inverse of wavelength and the units are usually cm^{-1} (reciprocal centimeters). Bonds of particular polyatomic groups absorb at particular frequencies. For example, the double bond of the C═O group generally appears at a frequency of about 1700 cm^{-1}. The normal range covered by IR spectroscopy is from 400–4000 cm^{-1}, corresponding to wavelengths of 2500–25,000 nm. Table E16.1 lists some group frequencies for commonly encountered functional groups.

The frequency at which a particular bond will vibrate can be predicted using Hooke's law:

$$\nu = \frac{1}{(2\pi c)}\sqrt{\frac{k}{\mu}}$$

Table E16.1 Infrared Frequencies of Common Functional Groups

Bonds to Hydrogen		Moderate Mass Bonds		Bonds to Heavy Elements	
C—H	2900 cm^{-1}	B—F	1400 cm^{-1}	C—Cl	750 cm^{-1}
Si—H	2150 cm^{-1}	C—N	1100 cm^{-1}	C—Br	650 cm^{-1}
Ge—H	2100 cm^{-1}	C—O	1100 cm^{-1}	N—Br	690
N—H	3400 cm^{-1}	N—F	1070 cm^{-1}		
P—H	2300 cm^{-1}	C═C	1650 cm^{-1}	O—Cl	780
As—H	2200 cm^{-1}	C═N	1650 cm^{-1}	O—Br	710
				O—I	690
O—H	3500 cm^{-1}	C═O	1700 cm^{-1}		
S—H	2600 cm^{-1}	C≡C	2100 cm^{-1}	B—Cl	950
Se—H	2300 cm^{-1}	C≡N	2150 cm^{-1}	B—Br	800
F—H	4100 cm^{-1}	C≡O	2170 cm^{-1}	S—Cl	520
Cl—H	3000 cm^{-1}			S—Br	400
Br—H	2650 cm^{-1}			P—Cl	515
I—H	2300 cm^{-1}			P—Br	390
4000 cm^{-1}					400 cm^{-1}

where ν is the frequency, c is the speed of light, k is the force constant (related to bond strength), and μ is an averaged mass of the atoms (m_1 and m_2) making up the bond defined by the equation

$$\mu = \frac{m_1 m_2}{m_1 + m_2}$$

In this experiment, infrared spectroscopy will be used to investigate a dimethylsulfoxide (DMSO, CH_3SOCH_3) complex of copper(II) chloride. The metal forms an adduct (addition product) with DMSO:

$$CuCl_2 + 2\,(CH_3)_2S{=}O \rightarrow CuCl_2 \cdot 2\,[(CH_3)_2S{=}O]$$

DMSO enjoys some notoriety in being a possible treatment for arthritis. It is readily absorbed through the skin, and gives a garlic odor to the breath of the user. The absorption of the S=O bond in DMSO occurs at 1050 cm^{-1}.

Metals can bond to DMSO either through the oxygen or the sulfur atom. If the metal is bonded to the DMSO at the sulfur, the metal donates electron density to the SO bond, making it stronger (k increases). The frequency of the S=O absorption therefore <u>increases</u> from the usual value of 1050 cm^{-1}. If the bonding is to the oxygen of the DMSO, the metal removes electron density from the SO bond, making it weaker (k decreases). The S=O absorption therefore appears at <u>lower</u> frequency. Infrared spectroscopy can thereby tell us the nature of bonding in DMSO complexes.

EXPERIMENTAL SECTION

Part A: Preparation of $CuCl_2 \cdot 2$ DMSO

Record all data and observations on the data sheet. NOTE: DMSO is somewhat toxic. Since it is absorbed through the skin, and is also a good solvent, it can readily carry other substances with it. Be careful—do not let DMSO come in contact with your skin.

- Place about 150 mg (weighed to ±1 mg) of anhydrous copper(II) chloride in a 10 mL Erlenmeyer flask equipped with a magnetic stir bar. Add 1 mL of absolute ethanol, and stir until the copper(II) chloride dissolves.
- Slowly, add 250 μL of DMSO (automatic delivery pipet, or 5 drops from a medicine dropper). The reaction should take place almost immediately, yielding a light green precipitate. Stir the mixture for several minutes.
- Collect the product by suction filtration using a Hirsch funnel and water trap (Figure 3.19). Rinse the Erlenmeyer flask with a 500 μL portion of cold ethanol and use this to wash the product crystals in the Hirsch funnel. Repeat with a second 500 μL portion of cold ethanol.
- Dry the material on a clay tile or on filter paper. Weigh the product. Calculate a percentage yield, and determine the melting point, if desired.

Part B: Characterization of the Product

- Acquire the infrared spectrum of the copper(II) chloride–DMSO adduct (see your instructor for directions on how to operate your particular instrument), and determine the position of the S=O band.
- Compare the spectrum to that of DMSO itself. See if the S=O band moved to higher or lower wavenumber, and determine whether the DMSO is complexed at the sulfur or oxygen.

Name: ______________________

Date: __________ Section: ______________________

Pre-Laboratory Questions: Experiment 16

1. How does one determine whether a particular metal bonds to DMSO at the sulfur or at the oxygen?

2. Atoms and ions can be broadly classified into two categories: hard: Small radius, high charge; or soft: Large radius, low charge. Hard metals prefer to bond to the oxygen atom in DMSO, whereas soft metals prefer to bond to the sulfur atom. Which element (sulfur or oxygen) would you expect each metal in the following compounds to bond with in DMSO: $PtCl_2$, $SnCl_4$, $FeCl_3$, AuCl. Explain.

3. Look up the density of DMSO. Using the density equation, calculate the moles of each reagent used (see Experimental Procedure). Use the number of moles of each reagent and the stoichiometry of the reaction to show which of the two reagents is limiting. From the theoretical yield of product in moles and the molar mass of the product, calculate the theoretical yield (in grams) of the product.

Name: ______________________

Date: ________ Section: ______________________

Data Sheet: Experiment 16

Preparation of $CuCl_2 \cdot 2DMSO$

If container is not tared, report mass of container and container plus $CuCl_2$.

1. Mass of $CuCl_2$ used ____________ g

If container is not tared, report mass of container and container plus $CuCl_2 \cdot 2DMSO$.

2. Mass of $CuCl_2 \cdot 2DMSO$ recovered ____________ g

3. Percentage yield of product ____________ %

4. Melting point range ________________°C

Attach the IR spectrum, with the SO absorption marked.

5. SO absorption(s)? ________________ cm^{-1}

Is the metal bonded to S or O? Explain.

Name: ______________________________

Date: __________ Section: ______________________

Questions: Experiment 16

1. Draw the Lewis dot structure for DMSO.

2. DMSO enjoys some notoriety as being a possible treatment for arthritis. It has the interesting side effect of giving the patient "garlic breath." Most drugs are approved for human therapeutic use through so-called "double-blind" tests. Describe what a double-blind test is (see Glossary), and why DMSO cannot be evaluated in this manner.

3. Just as metals and metal ions can be classified as hard or soft (see Pre-Laboratory Question 2), anions can be classified in a like manner. Report the ionic radii of O^{2-} and S^{2-}. Which would be softer, O^{2-} or S^{2-}? Since soft metals prefer to bond at the sulfur, and hard metals at the oxygen, what simple rule (about soft/hard cations bonding preferably with soft/hard anions) does this suggest?

Chapter 6
Solution Chemistry

Experiment 17: Standardization of a NaOH Solution

Experiment 18: Determination of K_a of a Weak Acid

Experiment 19: Neutralization Power of an Antacid

Experiment 20: Buffer Solutions

Experiment 21: Redox Titration of Permanganate (Balancing Oxidation–Reduction Reactions)

Experiment 22: Solubility Product Constant and Common Ion Effect

Experiment 17 Standardization of a NaOH Solution

OBJECTIVES

- To standardize a base solution of unknown concentration.
- To do an acid–base titration.

PRIOR READING

Section 3.3: Weighing
Section 3.4: Liquid Volumes
Section 3.9: Titration

INTRODUCTION

There are many circumstances where one wishes to know the amount of a particular chemical **(analyte)** in a solution or mixture. This can be accomplished using an instrument such as a visible spectrophotometer (see Experiments 13–15). Often when solutions are involved, **volumetric analysis** can be employed using a technique known as **titration.**

In a titration, the volume of a *standard reagent solution* (called the **titrant**) required to react completely with the analyte is measured. The titrant is added to the analyte in small, controlled portions using a buret. The end of the titration, where the analyte, A, has completely reacted, is called the **equivalence point.** Titration is widely used as an analytical tool in the laboratory.

The following are requirements for a successful titration:

1. A titrant of known concentration (the standard solution)
2. A way to accurately measure the volume of titrant (a buret)
3. A signal that the end point has been reached (usually an added indicator)
4. A titration reaction must be stoichiometric, complete, and rapid

Ideally, in any titration we want to measure the equivalence point. However, what we actually measure is the **end point** of the titration. The end point is indicated by a sudden change in a physical or chemical property of the solution. In many titrations, an **indicator** is added to the solution being titrated. The indicator changes color at the end point, thereby visually indicating the end of the titration. In some cases, the color change of one of the reactants indicates the end point, while in other cases it is a change in pH, conductivity, or optical absorbance.

The end point signal, which occurs immediately after the equivalence point, tells you to stop adding titrant and to record the volume. The concentration of the titrant times the volume will equal the number of millimoles of titrant added.

The most common type of titration is an **acid–base** titration. Simple acids and bases react to form a salt and water. This reaction is called a **neutralization** reaction.

$$H^+_{(aq)} + OH^-_{(aq)} \rightleftharpoons H_2O_{(l)}$$

Here, an acid of unknown concentration is titrated with a base of known concentration (or *vice-versa*). The end point is determined by the color change of an added indicator. If a strong acid is titrated with a strong base, for example, the pH near the equivalence point will change sharply from about 5.0 to about 9.0 upon the addition of just a few drops of base. The pH at the equivalence point is 7.0. The standard indicator solution for this type of titration is **phenolphthalein** (pronounced *fee-nol-thay-leen*), which undergoes a color change from colorless in acid solution to pink in base solution at a pH of 8.3. Only a few drops of phenolphthalein are needed.

Table E17.1 Common Indicators: Color and pH Change

Indicator Name	Acid Color	Base Color	pH Range
Thymol Blue	red	yellow	1.2–2.8
Methyl Orange	red	yellow	3.2–4.4
Methyl Red	red	yellow	4.8–6.0
Litmus	red	blue	4.7–8.3
Phenolphthalein	colorless	pink	8.2–10.0
Alizarin Yellow	yellow	red	10.1–12.0

Weak acids have equivalence points in the basic region above pH 7.0 due to hydrolysis of the formed salt; weak bases have equivalence points in the acidic region below pH 7.0 for this reason. In these cases other indicators may be in order. Some common indicators are listed in Table E17.1. Various mixtures of indicators (called **universal indicators**) are also available.

Standardization of Acids and Bases

One of the difficulties in performing an accurate acid–base titration is that the concentration of the titrant must be accurately known. A solution of accurately known concentration of titrant is called a *standard solution*. Unfortunately, most commercial acids (sulfuric, nitric, hydrochloric, acetic, and phosphoric) vary to some degree from batch to batch, and must therefore have their exact concentration determined before they are used. The common strong bases (sodium and potassium hydroxide) are also problems, as the solids easily absorb CO_2 or absorb moisture (are **hygroscopic**) from the air. Thus, one doesn't really know the composition of what is being measured. It is for this reason that the common acids and bases cannot serve as standard solutions.

Compounds that can be used as **primary standards** must meet certain requirements:

1. They must exhibit high purity, typically $100.00 \pm 0.05\%$.
2. They must exhibit high stability, preferably over many years and be nonhygroscopic.
3. They should have a high equivalent weight. The equivalent weight is defined as the amount of acid that will produce 1 mole of protons (H^+) in a reaction, the amount of base that will produce 1 mole of hydroxide ions (OH^-) in a reaction or the amount of other compounds that will produce 1 mole of electrons when oxidized or reduced in a reaction.

A *standard solution* is prepared by dissolving an *accurately* weighed sample of a primary standard to make a known volume of solution. A **volumetric flask** is used for this purpose. There are several common organic acids and bases that are used. Oxalic acid ($H_2C_2O_4 \cdot 2\,H_2O$) is a common primary standard that has two acidic hydrogens **(diprotic),** and consequently reacts with two moles of sodium or potassium hydroxide:

$$H_2C_2O_{4(aq)} + 2\,NaOH_{(aq)} \rightarrow 2\,H_2O + Na_2C_2O_{4(aq)}$$

Potassium hydrogen phthalate ($KHC_8H_4O_4$, KHP) is also commonly used as a primary standard. It is somewhat hygroscopic and must be dried before use. It is monoprotic and therefore reacts with one mole of sodium or potassium hydroxide:

$$KHC_8H_4O_{4(aq)} + NaOH_{(aq)} \rightarrow H_2O + KNaC_8H_4O_4$$

Sodium or potassium hydroxide is usually standardized by titration with a solution of one of these primary standards. This accurately determines the concentration of the sodium or potassium hydroxide. Any acid solution can then be titrated with the now-standardized base solution, and its concentration determined as well. Since the primary standard solution is used to standardize all other solutions, it is imperative that an exact mass of the material and an exact volume of solution be used.

Calculations

The most common way of expressing the concentration of a solution is **molarity** (abbreviated M or c). Molarity (M) is generally expressed as mol L^{-1}, i.e., the number of moles of solute per one liter of solution. Since we are using microscale amounts (mg, mL, and μL) we may wish to express M in millimoles per millileter (mmol/mL).

To determine the molarity of the oxalic acid solution used in this experiment, one divides the number of moles of oxalic acid in the solution by the volume of the solution. Since oxalic acid is a diprotic acid, two moles of NaOH are required to neutralize each mole of oxalic acid. Since the concentration and volume of the acid are known, the number of moles of the acid can be easily calculated. Using the 2:1 base:acid stoichiometric ratio, the number of moles of added base added can be determined. Since the volume of the added base is known, the concentration of base can thereby be established.

Example: 1.00 mL of 0.09895 M oxalic acid was titrated with sodium hydroxide solution. It was found that it took exactly 198 μL of the sodium hydroxide to reach the end point. What is the concentration of the sodium hydroxide solution?

Answer: We first calculate the number of moles of oxalic acid (OA) that were added.

$$\text{mol} = (M, \text{mol L}^{-1})(V, L)$$

$$\text{mol oxalic acid (OA)} = (0.09895 \text{ mol L}^{-1})(0.00100 \text{ L}) = 9.895 \times 10^{-5} \text{ mol OA}$$

From the stoichiometry of the balanced reaction equation, we calculate the mol of NaOH:

$$9.895 \times 10^{-5} \text{ mol OA (2 mol NaOH/1 mol OA)} = 1.979 \times 10^{-4} \text{ mol NaOH}$$

We used 198 μL of the sodium hydroxide:

$$V = 198\ \mu\text{L}\ (1\ \text{L}/1.00 \times 10^{6}\ \mu\text{L}) = 1.98 \times 10^{-4}\ \text{L NaOH}$$

$$\text{M NaOH} = \text{mol/L} = (1.979 \times 10^{-4} \text{ mol NaOH})/(1.98 \times 10^{-4} \text{ L NaOH})$$

$$\text{M NaOH} = 1.00 \text{ M}$$

EXPERIMENTAL PROCEDURE

Optional Procedure A: Microscale

- Place about 120 mg of oxalic acid dihydrate in a 10 mL volumetric flask. Record the mass to ±1 mg or better. Add about 5 mL of distilled water and shake the flask to dissolve all of the acid. Now, fill the flask with distilled water to the mark. Calculate the molarity of this solution and record the value in the data table. This solution will serve as your primary standard. For more accurate results, more than one standard solution should be used, and the results compared.

- Obtain 25 mL of approximately 0.1 M NaOH solution to be standardized; this solution will be used in several subsequent experiments, and should be kept in a tightly stoppered polyethylene bottle.
- Obtain a 2.0 mL microburet, and rinse it twice with the sodium hydroxide solution, making sure that the solution wets the entire inner surface. Refill the microburet with the sodium hydroxide solution, and note the level on the data sheet.
- Pipet (using an automatic delivery pipet or a second microburet) exactly 100 μL (0.100 mL) of the oxalic acid solution into a 10 mL Erlenmeyer flask. Add about 2 mL of distilled water, and then add 1 small drop of phenolphthalein indicator solution. **NOTE: No more than one drop of indicator solution (delivered with a fine-tipped pipet) should be used, as the accuracy of your results could be affected severely, since the indicator is acidic.** Place a white piece of paper under the flask for color contrast.
- Dropwise, add the sodium hydroxide solution from the buret to the oxalic acid solution in the flask. Swirl after each drop is added. The end point is reached when the pink color of phenolphthalein indicator persists for at least 30 sec after addition of a drop of NaOH solution.
- Note the final buret level on the data sheet. Calculate the concentration of the NaOH solution.
- Repeat the procedure two more times using volumes of 110 and 120 μL of the oxalic acid solution, or until your results are precise to ±1%. Calculate the molarity of the sodium hydroxide solution.

Optional Further Study, A

- Weigh out 10–20 mg (±1.0 mg) of a solid, unknown acid. Record the weight on the data sheet, as well as the unknown number. Dissolve it in about 2 mL of water, in a 10 mL Erlenmeyer flask. Add one small drop of phenolphthalein indicator solution.
- Titrate the unknown solid acid solution with your standardized sodium hydroxide solution. Determine the molecular weight of the unknown acid, assuming it has one acidic hydrogen per molecule.

Optional Procedure B: Macroscale

Since the materials used in this experiment are relatively inexpensive and safe to use, this experiment can be efficiently run at a larger scale, if desired. The following procedure should be followed.

- Place about 1.2 g of oxalic acid dihydrate (±1 mg) in a 100 mL volumetric flask. Record the mass to ±1 mg or better. Add about 50 mL of distilled water and shake the flask to dissolve all of the acid. Fill the flask with distilled water to the mark. Calculate the molarity of this solution and record the value in the data table. This solution will serve as your primary standard. For more accurate results, more than one standard solution should be used, and the results compared.
- Obtain 400 mL of approximately 0.1 M NaOH solution to be standardized; this solution will be used in several subsequent experiments, and should be kept in a tightly stoppered polyethylene bottle.
- Obtain a 50 mL buret, and rinse it twice with the sodium hydroxide solution, making sure that the solution wets the entire inner surface. Refill the buret with the sodium hydroxide solution, and note the level on the data sheet.

- Pipet exactly 20.00 mL of the oxalic acid solution into a 125 mL Erlenmeyer flask. Add about 20 mL of distilled water, and then add 2–3 drops of phenolphthalein solution. Place a white piece of paper under the flask for color contrast.
- Add the sodium hydroxide solution from the buret, fairly rapidly at first, with swirling, until the pink color of the indicator begins to persist. At this point, add the NaOH solution dropwise. Swirl after each drop is added. The end point is reached when the pink color of phenolphthalein indicator persists for at least 30 sec.
- Note the final buret level on the data sheet. Calculate the concentration of the NaOH solution.
- Repeat the procedure with 18.00 and 19.00 mL portions of oxalic acid solution, or until your results are precise to $\pm 1\%$.

Optional Further Study, B

- Weigh out 0.20–0.3 g (± 1.0 mg) of a solid, unknown acid. Record the weight on the data sheet, as well as the unknown number. Dissolve it in about 100 mL of water in a 250 mL Erlenmeyer flask. Add 2–3 drops of phenolphthalein indicator solution.
- Titrate this unknown solid acid solution with your standardized sodium hydroxide. Determine the molecular weight of the unknown acid, assuming it has one acidic hydrogen per molecule.

Name: ______________________________

Date: __________ Section: ______________________________

Pre-Laboratory Questions: Experiment 17

1. Why are NaOH and HCl not used as primary standards for acid–base titrations? What properties should a primary standard possess?

2. When 1.05 mL of an unknown concentration solution of NaOH is titrated with standardized 0.100 M oxalic acid ($H_2C_2O_4$), 295 μL are needed to reach the end point. What is the molarity of the unknown?

3. The NaOH solution in Question 2 is used to standardize an unknown solution of HCl. It requires 134 μL of the NaOH solution to titrate 1.00 mL of the HCl to the end point. What is the molarity of the HCl?

Name: ______________________

Date: ________ Section: ______________________

Data Sheet: Experiment 17

Mass of oxalic acid ______________ g

Molarity of oxalic acid standard ______________ mol L^{-1}

Standardization of Sodium Hydroxide

	Trial 1	*Trial 2*	*Trial 3*
1. Volume of Oxalic Acid Titrated	_____ mL	_____ mL	_____ mL
2. Initial Buret Reading	_____ mL	_____ mL	_____ mL
3. Final Buret Reading	_____ mL	_____ mL	_____ mL
4. Volume of NaOH added	_____ mL	_____ mL	_____ mL
5. Calculate the molarity of the NaOH	_____ mol L^{-1}	_____ mol L^{-1}	_____ mol L^{-1}

Average Molarity ______________ mol L^{-1}

Optional Further Study

Unknown Number ____________

1. Mass of Unknown Solid Acid ______________ g

2. Initial Volume of NaOH ______________ mL

3. Final Volume of NaOH ______________ mL

4. Moles of Base Used ______________ mL

5. Moles of Unknown Acid ______________ mol

6. Molecular Weight of Unknown Acid ______________ mol L^{-1}

Name: ____________________

Date: ________ Section: ____________________

Questions: Experiment 17

1. Describe how you could prepare a 0.3 M oxalic acid solution using a 10 mL volumetric flask.

2. 255 mg of oxalic acid dihydrate are dissolved in 10.00 mL of water. What is the molarity of the resulting solution?

3. Potassium hydrogen phthalate is used to standardize a solution of KOH. The KHP solution was made up by dissolving 155 mg of KHP to a final volume of 10 mL. Titration of 1.00 mL of this KHP solution with KOH solution required 0.78 mL to reach the end point. What is the molarity of the KOH solution?

Experiment 18 Determination of K_a for a Weak Acid

OBJECTIVES

- To determine the dissociation constant, K_a, for a weak acid by titration with a standardized base.
- To learn the use of a pH meter.

PRIOR READING

Section 3.3: Weighing
Section 3.4: Liquid Volumes
Section 3.9: Titration

INTRODUCTION

Strong Acids and Bases

When a strong acid (such as HCl) or base (such as NaOH) is dissolved in water, it completely dissociates into its constituent ions:

$$HCl_{(aq)} + H_2O \rightarrow H_3O^+_{(aq)} + Cl^-_{(aq)}$$

$$NaOH_{(aq)} \rightarrow Na^+_{(aq)} + OH^-_{(aq)}$$

Since the reaction is complete, the resulting concentration of H_3O^+ is the same as the original concentration of HCl, and the resulting concentration of OH^- is the same as the original concentration of NaOH. Table E18.1 lists some of the more common strong acids and bases.

The dissociation of water can also produce H_3O^+ and OH^-:

$$H_2O + H_2O \rightleftharpoons H_3O^+ + OH^-$$

However, this is an equilibrium reaction—it does not proceed to completion. In fact, very little water actually dissociates. This amount can be expressed as a dissociation constant for water, K_w:

$$K_w = [H_3O^+][OH^-] = 1 \times 10^{-14} \text{ (at 25°C)}$$

The value of K_w varies slightly with temperature, but can be treated as constant for most purposes. Thus, if the concentration of H_3O^+ is known, the concentration of OH^- can be calculated, and *vice versa*.

The concentrations of H_3O^+ and OH^- can vary over a wide range. It is therefore convenient to express concentrations and dissociation constants on a logarithmic scale, commonly called the pH scale [pH is from the French *puissance d'Hydrogen*, the power of hydrogen]:

$$pH = -\log[H_3O^+]$$

$$pOH = -\log[OH^-]$$

$$pK_w = -\log K_w = 14$$

Table E18.1 Common Strong Acids and Bases

Acids:	HNO_3	H_2SO_4*	HCl	HBr	HI	$HClO_4$
Bases:	NaOH	KOH				

*Strong only for the first dissociation.

Since the reactions take place in water, the pH and the pOH must be related to each other by the dissociation constant of water:

$$pK_w = 14 = pH + pOH$$

Thus, if any of the quantities $[H_3O^+]$, $[OH^-]$, pH, or pOH are known, the other three can be easily calculated.

Weak Acids and Bases

A weak acid or base is one that does not completely dissociate. At equilibrium, some of the undissociated acid or base is still present. The extent of dissociation of any weak acid is given by the acid dissociation constant, K_a or for a weak base, K_b. For the dissociation equilibrium of a weak acid,

$$\underset{\text{weak acid}}{HA} + \underset{\text{water}}{H_2O} \rightleftharpoons \underset{\text{hydronium ion}}{H_3O^+} + \underset{\text{conjugate base}}{A^-}$$

the equilibrium expression may be written

$$K_a = \frac{[H_3O^+][A^-]}{[HA]}$$

where the square brackets represent the concentrations at equilibrium. Note that in a simple dissociation of a weak acid in deionized water,

$$[H_3O^+] = [A^-]$$

since they are produced together in the same stoichiometric ratio. Similar expressions can be written for bases, involving $[OH^-]$.

Percent Ionization of Weak Acids and Bases

One way of measuring the relative strength of weak acids or bases is by measuring their *percent ionization* in aqueous solutions. For example, a 0.10 M solution of acetic acid (CH_3COOH) is ionized to the extent of 1.3%, whereas a 0.10 M solution of nitrous acid (HNO_2) is ionized 6.5%. Thus, on a relative basis, nitrous acid produces a higher concentration of H_3O^+ ions and is the stronger acid of the two.

How does one calculate the percent ionization of an acid? Let us consider 0.10 M acetic acid. We know the ionization reaction:

$$HC_2H_3O_2 + H_2O \rightleftharpoons H_3O^+ + C_2H_3O_2^-$$

and that the ionization constant, $K_a = 1.8 \times 10^{-5}$ (see Table E18.2). The equilibrium equation can be written as

$$K_a = \frac{[H_3O^+][C_2H_3O_2^-]}{[HC_2H_3O_2]}$$

We know the initial concentration of acetic acid (0.10 M) and that some of it (call this x) dissociated (ionized). At equilibrium, then, $[HC_2H_3O_2] = 0.1 - x$. For every acetic acid that dissociated, one hydronium and one acetate ion formed. Thus, $[H_3O^+] = [C_2H_3O_2^-] = x$. Substituting into the equilibrium expression,

$$K_a = 1.80 \times 10^{-5} = (x)(x)/(0.10 - x)$$

Table E18.2 Properties of Common Weak Acids

Acid	Formula	Mol. Wt.	K_a	pK_a
Acetic	$HC_2H_3O_2$	60.1	1.8×10^{-5}	4.75
Ascorbic	$HC_6H_7O_6$	176.1	6.8×10^{-5}	4.17
Benzoic	$HC_7H_5O_2$	122.1	6.5×10^{-5}	4.19
Propionic	$HC_3H_5O_2$	74.1	1.3×10^{-5}	4.87
Sodium bisulfate	$NaHSO_4 \cdot H_2O$	138.0 (FW)	1.2×10^{-2}	1.92
Sodium bisulfite	$NaHSO_3$	104.1 (FW)	6.3×10^{-8}	7.20

This may be solved using the quadratic equation. Alternatively, we can recognize that since the acid is weak, x will be small compared to 0.10, and the equation thereby simplifies to:

$$K_a = 1.80 \times 10^{-5} = (x)(x)/(0.10)$$

$$x^2 = 1.80 \times 10^{-6}$$

$$x = 1.34 \times 10^{-3}$$

which is the amount of acetic acid ionized. The % ionization is then

$$\%\ \text{ionization} = 100[(\text{amount ionized})/(\text{original amount})]$$

$$\%\ \text{ionization} = 100(1.34 \times 10^{-3}/0.10) = 1.34\%$$

Weak base ionizations can be determined in like manner.

Calculation of pH Values

If the equilibrium constant, K, for a weak acid or base dissociation is known, the pH can be calculated directly from the initial concentration of the weak acid or base. For example, consider the dissociation of a 1.00 M solution of acetic acid. We know the K_a value to be 1.8×10^{-5}. Using the equilibrium expression derived above, we obtain

$$K_a = 1.8 \times 10^{-5} = (x)(x)/(1.00 - x)$$

This may be solved using the quadratic equation. Alternatively, we can recognize that since the acid is weak, x will be small compared to 1.00, and the equation thereby simplifies to:

$$K_a = 1.80 \times 10^{-5} = (x)(x)/(1.00)$$

$$x^2 = 1.80 \times 10^{-5}$$

$$x = 4.24 \times 10^{-3} = [H_3O^+]$$

$$pH = -\log[H_3O^+] = -\log(4.24 \times 10^{-3})$$

$$pH = 2.37$$

Titration of a Weak Acid—Determination of K_a

Titration curves (see Fig. E18.1) are plots of pH versus the volume of standarized acid or base added in a titration. These curves are valuable because they show the point where equivalent amounts of weak acid (HA) and strong

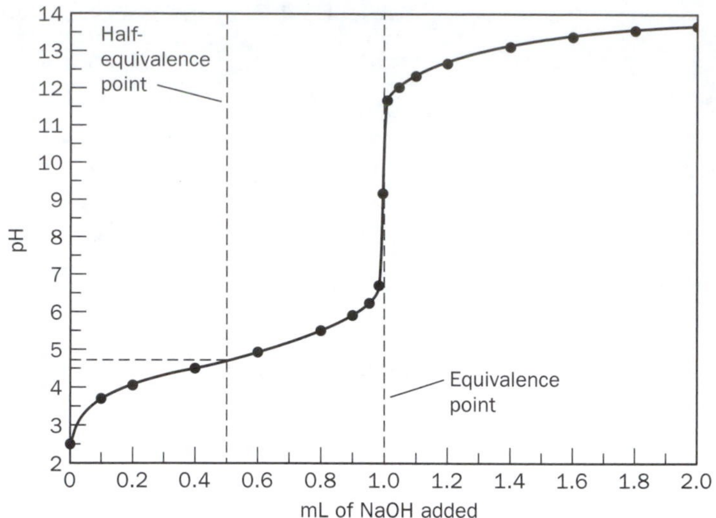

Figure E18.1 *Titration curve for titration of 1.0 mL of 1 M $HC_2H_3O_2$ with 1 M NaOH.*

base titrant (OH^-) have been added [or weak base (BOH) and strong acid titrant (H_3O^+)]. This is termed the **equivalence point,** and is seen on the curve where the rapid rise of the pH value occurs (see Fig. E18.1). In this experiment, we are dealing with a weak acid in which we wish to determine the K_a value. In cases of this type, if the volume of base required to reach the *equivalence point* of the acid is known, the dissociation constant K_a for the acid may be determined in the following way:

We define the **half-equivalence point** as when the amount of base necessary to titrate *half of the acid* present has been added. Letting x be the amount of weak acid initially present, the amount of acid present at the half-equivalence point will be $0.5x$. Since half the acid will have reacted at this point, there will be $0.5x$ of the conjugate base present as well. Thus, $[HA] = [A^-] = 0.5x$. Substituting into the equilibrium expression,

$$K_a = \frac{[H_3O^+][0.5x]}{[0.5x]} = [H_3O^+]$$

or

$$pK_a = pH$$

Thus, the pH at the half-equivalence point is equal to the pK_a of the acid. This is shown graphically for the titration of 1.0 mL of 1.0 M acetic acid with 1.0 M NaOH in Figure E18.1.

In the graph, the number of mL of NaOH added is plotted on the x axis, and the experimental pH on the y axis. The equivalence point is easily determined as the center of the "fast rising" portion of the graph. It therefore required 1.0 mL of 1.0 M NaOH to reach the equivalence point. The half-equivalence point is therefore at 0.5 mL of NaOH added, and a vertical line is drawn there. The pH is read at this amount of NaOH added. To obtain this value, a horizontal line is constructed where the half-equivalence point line meets the titration curve.

We can therefore read the pH value as about 4.75. This is equal to the pK_a value. The K_a is obtained by changing the sign, and taking the inverse log of 4.75. Thus, $K_a = 10^{-4.75} = 1.8 \times 10^{-5}$, which is the correct value for acetic acid.

Table E18.2 lists molecular weights (formula weights for salts), K_a and pK_a values for several common weak acids.

EXPERIMENTAL PROCEDURE

Optional Procedure A: Microscale

- Weigh about 20 mg (±1 mg) of an unknown acid, and place it in a 30 mL beaker containing a magnetic stir bar set atop a magnetic stirrer. Add 15 mL of water, and stir to dissolve the acid. Add 1 drop of phenolphthalein indicator.
- Obtain a calibrated pH meter (Figure E18.2), and immerse the electrodes into the acid solution, making sure that the electrodes are not low enough to come into contact with the stir bar. Clamp a microburet filled with the standardized NaOH prepared in Experiment 17 to titrate the acid solution. If Experiment 17 was not done, use a 0.100 M solution of NaOH.
- Add the NaOH solution dropwise, stopping every two drops to allow the solution to reach equilibrium for 1–2 minutes, and record the pH.
- When the pH changes by more than 0.2 pH units during the addition, the equivalence point is near. At this point, stop after every drop, allow 1 minute for equilibration, and take a reading.
- After the equivalence point is passed, resume stopping every two drops until at least an additional 0.5 mL of NaOH solution have been added.
- If time allows, repeat with a second sample of the same unknown acid.

Optional Procedure B: Macroscale

- Weigh about 2.5 g (±1 mg) of an unknown acid, and place it in a 250 mL beaker containing a magnetic stir bar set atop a magnetic stirrer. Add 50 mL of water, and stir to dissolve the acid. Add 2–3 drops of phenolphthalein indicator.
- Obtain a calibrated pH meter, and immerse the electrodes into the acid solution, making sure that the electrodes are not low enough to come into contact with the stir bar. Clamp a 50 mL buret filled with standardized NaOH, prepared in Experiment 17, to titrate the acid solution. If Experiment 17 was not done, use a 0.100 M solution of NaOH.
- Add the NaOH solution in one mL increments, allow the solution to reach equilibrium for 2–3 minutes, and record the pH.

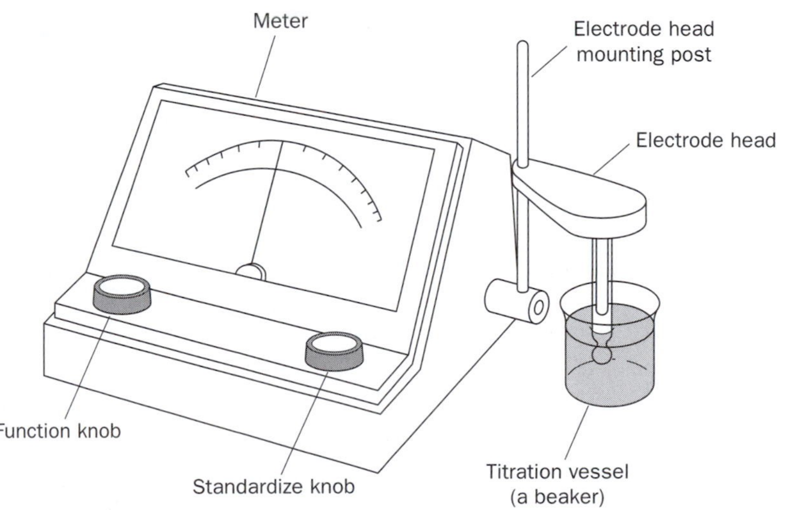

Figure E18.2 *A pH meter.*

- When the pH changes by more than 0.2 pH units during the addition, the equivalence point is near. At this point, stop after every five drops, allow 1 minute for equilibration, and take a reading.
- After the equivalence point is passed, resume stopping every mL until at least an additional 10 mL of NaOH solution have been added.
- If time allows, repeat with a second sample of the same unknown acid.

Treatment of Data

- Plot the titration data, and determine the equivalence point, half-equivalence point, pK_a and K_a of your unknown.
- From the number of mL of NaOH needed to reach the equivalence point, calculate the number of mmol of NaOH added, and thus the number of mmol of H^+ in your sample.

Determine the molecular weight (or formula weight) of the acid, assuming only one acidic hydrogen is present.

Name: ______________________________

Date: __________ Section: ______________________________

Pre-Laboratory Questions: Experiment 18

1. Calculate the pH of a 0.35 M solution of acetic acid.

2. A particular solution of ammonia ($K_b = 1.8 \times 10^{-5}$) has a pH of 8.3. What is the concentration of ammonia in this solution?

$$NH_3 + H_2O \rightleftharpoons NH^+_{4(aq)} + OH^-_{(aq)}$$

3. The following data were obtained for the titration of 500 μL of 1.0 M acetic acid with 1.0 M NaOH:

Volume NaOH, μL	0	100	200	300	400	440	480	490
pH	2.87	4.14	4.57	4.92	5.35	5.61	6.13	6.44

Volume NaOH, μL	498	500	502	510	520	560	600	700	800
pH	7.14	8.72	10.30	11.00	11.29	11.75	11.96	12.22	12.36

On a sheet of graph paper, plot the pH curve, and determine (a) the equivalence point, and (b) the pK_a of acetic acid based on these data.

Name: ______________________________

Date: __________ Section: ______________________________

Data Sheet: Experiment 18

Unknown Number: ______________________________

1. Mass of Unknown Acid ____________ g

2. Concentration of Standardized NaOH ____________ mol L^{-1}

3. Titration Data

	Vol. NaOH, mL	*pH*		*Vol. NaOH, mL*	*pH*
1.	_____	_____	21.	_____	_____
2.	_____	_____	22.	_____	_____
3.	_____	_____	23.	_____	_____
4.	_____	_____	24.	_____	_____
5.	_____	_____	25.	_____	_____
6.	_____	_____	26.	_____	_____
7.	_____	_____	27.	_____	_____
8.	_____	_____	28.	_____	_____
9.	_____	_____	29.	_____	_____
10.	_____	_____	30.	_____	_____
11.	_____	_____	31.	_____	_____
12.	_____	_____	32.	_____	_____
13.	_____	_____	33.	_____	_____
14.	_____	_____	34.	_____	_____
15.	_____	_____	35.	_____	_____
16.	_____	_____	36.	_____	_____
17.	_____	_____	37.	_____	_____
18.	_____	_____	38.	_____	_____
19.	_____	_____	39.	_____	_____
20.	_____	_____	40.	_____	_____

Name: ____________________

Date: __________ Section: ____________________

Data Sheet: Experiment 18, p. 2

4. Construct a pH plot of your data on a sheet of graph paper.

5. How many mL of NaOH were needed to reach the equivalence point?

6. Determine the K_a of the unknown acid.

7. Determine the molecular weight (or formula weight) of the unknown acid (assuming one acidic hydrogen per molecule).

Name: ____________________

Date: ________ Section: ____________________

Questions: Experiment 18

1. Calculate the percent ionization in a 0.1 M solution of hydrofluoric acid ($K_a = 3.53 \times 10^{-4}$).

2. For any polyprotic acid, verify that the concentration of the dianion is always approximately equal to K_2, regardless of the concentration. (For example, the concentration of HPO_4^{2-} for a solution of phosphoric acid is 6.23×10^{-8} M). Explain why this is true.

3. The % ionization of a weak acid always declines in a solution that contains the salt of the anion, relative to a solution of the acid alone. For example, the % dissociation of acetic acid is less in a solution that contains sodium acetate than if the acetic acid were alone. Verify this for a 0.5 M solution of acetic acid and 0.5 M each acetic acid/acetate, and explain why.

Experiment 19 Neutralization Power of an Antacid

OBJECTIVES

- To determine (and test the advertised claims to) the neutralization power of an antacid.
- To perform an acid–base back-titration.

PRIOR READING

Section 3.3: Weighing
Section 3.4: Liquid Volumes
Section 3.9: Titration

INTRODUCTION

The stomach contains parietal cells that secrete hydrochloric acid at a concentration of about 0.155 M. This secretion of HCl increases when food enters the stomach, to aid in the digestion of your food. If you drink or eat too much, your digestive system may generate too much acid. This can lead to heartburn or indigestion. We often take an antacid to neutralize this excess stomach acid.

There are many over-the-counter products available to treat excess stomach acidity. Some of the better-known trade names are Rolaids, Maalox, and Tums. Most are *solid bases*, which have the capacity to neutralize stomach acid. There is great similarity between the commercial products, as the majority contain the carbonate (CO_3^{2-}) ion as the active agent. The carbonate ion reacts with the hydrochloric acid in your stomach to form carbonic acid, which decomposes into carbon dioxide and water. Shown here is the reaction of acid with $CaCO_3$, the main ingredient of Tums.

$$CaCO_{3(s)} + H^+_{(aq)} \rightleftharpoons Ca^{2+}_{(aq)} + HCO^-_{3(aq)}$$

$$HCO^-_{3(aq)} + H^+_{(aq)} \rightarrow H_2O_{(l)} + CO_{2(g)}$$

Notice that CO_2 is generated and is the cause of the burp that often is expelled.

In addition to the base present, many antacids also contain other ingredients such as sucrose, starch, talc, mineral oil, natural and artificial flavors, adipic acid, and color additives. The antacids differ in the cations present. Some contain sodium carbonate, some calcium carbonate, and some magnesium carbonate. In some cases, OH^- is the active species such as in $Mg(OH)_2$. Table E19.1 gives a list of some commercial brands of antacid

Many people are concerned with minimizing sodium in their diet, and in contrast, adding calcium. While sodium is necessary in the diet, an excess increases water retention and leads to high blood pressure, possibly resulting in stroke. Sodium is present in most processed foods to a very high degree,

Table E19.1 Common Antacids

Antacid Name	Main Component
Rolaids	$CaCO_3$
Tums	$CaCO_3$
Maalox	$Mg(OH)_2 + Al(OH)_3$
Milk of Magnesia	$Mg(OH)_2$
Alka-Seltzer II	$NaHCO_3 + KHCO_3$
Gaviscon	$Al(OH)_3$
Amphojel	$Al(OH)_3$

and it is therefore present in most peoples' diets in larger amounts than necessary. Calcium, on the other hand, is beneficial for formation of strong bones and teeth, and is generally not found to any great degree in processed foods. Overconsumption of antacids containing magnesium can lead to magnesium poisoning. The choice of an appropriate antacid is not necessarily a trivial decision.

In this experiment, the acid-neutralizing power of one or more antacids will be determined. The antacid will be allowed to react with an *excess* of acid. The unreacted acid will then be **back-titrated** with standardized NaOH, to a phenolphthalein end point.

Back-Titration

In several instances, the direct titration of an analyte with a reagent is not feasible. For example, in this experiment, the titration of an antacid material that may have low solubility in acid is not practical because you would have to wait for complete reaction to take place after each addition of acid. To overcome this problem, we add an excess of the acid (over the amount of the antacid used), allow the reaction to reach completion, and then **titrate the excess acid** with a standard solution of base. The amount of NaOH needed is equal to the amount of excess acid. Subtracting the excess acid from the amount of acid that you added gives the amount of acid the antacid neutralized.

$$\text{Acid neutralized} = (\text{Acid added}) - (\text{Excess acid titrated})$$

Here, the system has gone from being basic, *past the equivalence point* to the acid side (excess acid), and then *back to the equivalence point* (neutralization). The final titration to the equivalence point is called a back-titration.

One critical aspect of the procedure you are to follow is the removal of CO_2 gas from the solution by boiling. CO_2 gas dissolved in the system (from the carbonate in the antacid or from the water used) is converted to carbonic acid, H_2CO_3, which can react with the NaOH. Thus, the boiling step to remove this gas is essential so the titration results will only determine the amount of *excess* HCl you added to the reaction mixture.

Neutralization of Stomach Acid

Most advertising claims for antacids refer to the weight of stomach acid (approximately 0.15 M HCl) that a given weight of antacid can neutralize. "Rolaids consumes twenty times its weight in stomach acid" is an example of such a claim. The truth of these claims will be tested in this experiment.

EXPERIMENTAL PROCEDURE

Optional Procedure A: Microscale

- Weigh a whole antacid tablet to ±1 mg and record this weight in the data table. Then take three tablets and crush them using a clean mortar and pestle.
- Weigh and place about 25 mg (to ±1 mg) of the finely crushed antacid in a 50 mL Erlenmeyer flask, and add 20 mL of water. Assist the dissolution of the antacid by stirring with a magnetic stir bar on a magnetic hot plate, a glass rod, or by swirling the flask. If all of the tablet does not dissolve after 3–5 minutes, continue to the next step.
- Add one drop of phenolphthalein solution from a fine capillary dropper. The solution may turn light pink at this point, due to the presence of the basic antacid. Using a graduated 2 mL pipet, add 2.00 mL (±0.01 mL) of stomach acid (standardized HCl of approximately 0.15 M), and stir well until all remaining solids have dissolved. The solution should be acidic at this point, and therefore colorless.

- Add a boiling stone (or use a magnetic stir bar and magnetic stirrer) and heat the solution to boiling for about 5 minutes. A fair amount of bubbling should occur, as CO_2 gas is generated and evolves from the solution. If the solution turns pink during this boiling period, cool the flask to room temperature and add an additional 0.50 mL of acid. Be sure to record the **total** amount of HCl that was added.
- Cool the flask to room temperature by immersing it in an ice–water bath or under a stream of cold running water. Using a microburet, titrate the remaining acid with standardized NaOH (see Experiment 17) until the first permanent trace of pink color develops. **Note: Be careful. It may only require a few drops.** If Experiment 17 was not done, use standardized 0.100 M NaOH.
- Perform a second run on the same antacid. Repeat the procedure for the other antacid(s), if time permits.

Optional Procedure B: Macroscale

- Weigh a whole antacid tablet to ±1 mg and record this weight in the data table. Then take three tablets and crush them using a clean mortar and pestle.
- Weigh and place about 250 mg (to ±1 mg) of the finely crushed antacid in a 125 mL Erlenmeyer flask, and add 50 mL of water. Assist the dissolution of the antacid by stirring with a magnetic stir bar on a magnetic hot plate, glass rod, or by swirling the flask. If all of the tablet does not dissolve after 3–5 minutes, continue to the next step.
- Add 2–3 drops of phenolphthalein solution. The solution may turn light pink at this point, due to the presence of the basic antacid. From a 25 mL pipet, add 20.00 mL of stomach acid (approximately 0.15 M HCl), and stir well until all remaining solids have dissolved. The solution should be acidic at this point, and therefore colorless.
- Add a boiling stone (or use a magnetic stir bar and magnetic stirrer) and heat the solution to boiling for about 5 minutes. A fair amount of bubbling should occur, as CO_2 gas is generated and evolves from the solution. If the solution turns pink during this boiling period, cool the flask to room temperature and add an additional 1.00 mL of acid. Be sure to record the **total** amount of HCl that was added.
- Cool the flask to room temperature by immersing it in an ice/water bath or under a stream of cold running water. Using a 50 mL buret, titrate the remaining acid with the standardized NaOH prepared in Experiment 17, until the first permanent trace of pink color develops. **Note: Be careful. It may only require a few drops.** If Experiment 17 was not done, use standardized 0.100 M NaOH.
- Perform a second run on the same antacid. Repeat the procedure for the other antacid(s), if time permits.

Optional Further Study

- Use a pH meter to verify that phenolphthalein changes color at the equivalence point (where the pH rises rapidly). For many weak bases (and some of these solutions will contain weak bases at the equivalence point), phenolphthalein is not an accurate indicator of the equivalence point. $Mg(OH)_2$, $Ca(OH)_2$ and $Al(OH)_3$ are weak bases.

Name: ______________________________

Date: __________ Section: ______________________________

Pre-Laboratory Questions: Experiment 19

1. Assuming that stomach acid is 0.150 M HCl (ρ = 1.00 g mL^{-1}), how much mass of acid would a 500 mg antacid tablet composed of Na_2CO_3 neutralize?

2. What is the purpose of heating the acid solution with the antacid to boiling? Why is this a critical step in the procedure?

3. a. Identify four different cations that make up available antacids.

b. Identify the three common anions that make up antacids.

Name: ______________________________

Date: __________ Section: ______________________

Data Sheet: Experiment 19

Concentration of HCl ____ mol L^{-1} Concentration of NaOH ____ mol L^{-1}

	Trial 1 Antacid 1	*Trial 2 Antacid 1*	*Trial 1 Antacid 2*	*Trial 2 Antacid 2*
1. Name of antacid	______________		______________	
2. Mass of whole tablet	_____ g	_____ g	_____ g	_____ g
3. Mass of antacid used	_____ g	_____ g	_____ g	_____ g
4. Volume of HCl added	_____ mL	_____ mL	_____ mL	_____ mL
5. Initial volume, NaOH	_____ mL	_____ mL	_____ mL	_____ mL
6. Final volume, NaOH	_____ mL	_____ mL	_____ mL	_____ mL
7. Volume of NaOH added	_____ mL	_____ mL	_____ mL	_____ mL
8. Moles of HCl added	_____ mol	_____ mol	_____ mol	_____ mol
9. Moles of NaOH used	_____ mol	_____ mol	_____ mol	_____ mol
10. Moles neutr. by antacid	_____ mol	_____ mol	_____ mol	_____ mol
11. Neutralizing power of the antacid (mL standard acid neutralized per gram antacid)	____ mL g^{-1}	____ mL g^{-1}	____ mL g^{-1}	____ mL g^{-1}
12. Average neutralizing power of antacid	_____ mL g^{-1}	_____ mL g^{-1}		

Name: ______________________________

Date: __________ Section: ______________________________

Questions: Experiment 19

1. List two possible sources of error in this experiment. For each indicate what effect it would have on the end result.

2. Give a plausible reason for why *three* antacid tablets were ground to obtain the sample for analysis.

3. The active ingredient in Philips Milk of Magnesia is $Mg(OH)_2$, in Amphojel it is $Al(OH)_3$, and in Alka-Seltzer it is $NaHCO_3$. Write balanced equations showing how each neutralizes stomach acid.

Experiment 20 Buffer Solutions

OBJECTIVES

- To learn how to prepare buffer solutions.
- To investigate the effect of added acid and base on buffer solutions.
- To investigate the effect of dilution on buffer capacity.

PRIOR READING

Section 3.3: Weighing
Section 3.4: Liquid Volumes
Section 3.9: Titration

INTRODUCTION

A buffer is a solution that resists change in pH when small amounts of acid or base are added, or when it is diluted. It contains appreciable quantities of both a weak acid and its conjugate base (or a weak base and its conjugate acid). In the **Brønsted-Lowry theory** of acids and bases, an acid is a proton donor and a base is a proton acceptor. When an acid gives up a proton, it becomes a base because it can now accept a proton. Similarly, when a base accepts a proton, it becomes an acid. Such reactant-product couples are called **conjugate acid–base pairs.**

Commonly used buffer solutions are the acetic acid/acetate buffer:

$$HC_2H_3O_2 + H_2O \rightleftharpoons C_2H_3O_2^- + H_3O^+ \tag{1}$$

$$C_2H_3O_2^- + H_2O \rightleftharpoons HC_2H_3O_2 + OH^- \tag{2}$$

and the ammonia/ammonium buffer:

$$NH_4^+ + H_2O \rightleftharpoons NH_3 + H_3O^+ \tag{3}$$

$$NH_3 + H_2O \rightleftharpoons NH_4^+ + OH^- \tag{4}$$

When acetic acid reacts with water, some of the molecules lose H^+ to form acetate and hydronium ions (reaction 1). Although acetate is formed and is a weak base, **the solution is acidic due to the formation of H_3O^+.** When acetate ion reacts with water, it gains a proton and an acetic acid molecule is formed along with a hydroxide ion, OH^- (reaction 2). Although an acetic acid molecule is formed, **the solution is basic due to the formation of OH^-.** Similar reactions occur in the ammonia/ammonium system (reactions 3–4). Note that while many acids have the word "acid" in their name, some (such as the ammonium ion) do not.

pH of a Buffer Solution

The pH of a buffer solution can be determined from the dissociation constant of the weak acid or base from which it is made, and by the ratio of conjugate acid and base present. Consider the generic weak acid, HA, and its conjugate base, A^-.

$$HA + H_2O \rightleftharpoons A^- + H_3O^+ \quad K_a = 1.0 \times 10^{-5} \tag{5}$$

The equilibrium constant, K_a, is called the acid dissociation constant and the expression for this acid is

$$K = 1.0 \times 10^{-5} = \frac{[H_3O^+][A^-]}{[HA]}$$

or, taking the negative log (p) of both sides,

$$pK_a = 5.0 = -\log [H_3O^+] - \log ([A^-]/[HA])$$

Since pH = $-\log [H_3O^+]$, solving for the pH:

$$pH = 5.0 + \log ([A^-]/[HA])$$

Since the HA and A^- are both present in the same solution, the term $[A^-]/[HA]$ may be simplified to (moles of A^-)/(moles of HA). It is usually easier to calculate the change in the number of moles of HA and A^- in a buffer than to calculate the change in their concentrations.

From the equation above, we can easily see that for a solution containing equal numbers of moles of HA and A^-, the pH will be 5.0. If more HA is present than A^-, the term log ($[A^-]/[HA]$) will be less than zero and the pH will be less than 5.0. Since more conjugate acid is present, it seems reasonable that the solution should be more acidic than pK_a. Similarly, if more A^- is present than HA, the term log ($[A^-]/[HA]$) will be greater than zero, and the pH will be greater than 5 (more basic than pK_a).

Addition of Acid or Base to a Buffer

What is the effect when a small amount of acid or base is added to a buffer? If H_3O^+ is added, it will react with A^- to form HA, shifting reaction (5) to the left. This results in a slight increase in the number of moles of HA, and a slight decrease in the number of moles of A^-. The ratio ($[A^-]/[HA]$) is therefore slightly smaller, and the pH will be slightly lower. This is logical—when adding acid, the solution should become more acidic (lower pH).

Similarly, if OH^- is added, it will react with HA, forming A^-, shifting reaction (5) to the right. This results in a slight decrease in the number of moles of HA and a slight increase in the number of moles of A^-. The ratio ($[A^-]/[HA]$) is therefore slightly larger and the pH will be slightly higher. This is also logical—when adding base, the solution should become more basic (higher pH).

Buffers are able to withstand changes in the pH when a small amount of acid or base are added, a small amount being up to approximately half of the acid or base originally present. Suppose a solution has 0.10 moles each of HA and A^-. If half of the A^- is used up by adding 0.050 moles of strong acid, then we will have 0.15 moles of HA and 0.050 moles of A^-. The mole ratio has changed from 1.0 to 0.33, and the log of the mole ratio has changed from 0.00 to −0.48. Thus the pH will be lowered by 0.48 pH units, from 5.00 to 4.52. This is a reasonably small change in the pH. If significantly more than half is used, however, the buffering capacity of the solution will be exceeded, and the pH will change substantially. Similarly, if 0.050 moles of strong base had been added to the buffer solution, the pH would have risen by 0.48 to 5.48.

Preparation of a Buffer

Buffer solutions can be prepared at any moderate value of pH that is desired. Suppose we wish to prepare a buffer with a pH of 4.80 using an acid with a pK_a of 5.00. Using HA and A^- in equal molar quantities, the pH would be 5.00. In order to achieve the lower pH, we need to prepare a buffer which has more HA than A^-. We can calculate the required mole ratio of A^- to HA by substituting the desired pH into the buffer equation:

$$pH = pK_a + \log([A^-]/[HA])$$

$$4.80 = 5.0 + \log([A^-]/[HA])$$

$$[A^-]/[HA] = 0.63$$

This tells us that the ratio of the base to its conjugate acid should be 0.63:1 in order to achieve the desired pH. Buffer solutions can also be prepared by partially reacting a weak acid (or weak base) with a strong base (or strong acid) until the desired ratio of conjugates is present. Regardless of how two buffers are prepared, if the amounts of conjugate acid and base are the same, the buffers should have the same pH and the same behavior with added strong acid or strong base.

Behavior of Buffers with Added Acid or Base

In the previous example, suppose that the total number of moles of HA and A^- was 0.20. We can solve for the number of moles of HA and A^- by solving the simultaneous equations:

$$[HA] + [A^-] = 0.20 \qquad \text{(since the total number of moles is 0.20)}$$

$$[A^-]/[HA] = 0.63 \qquad \text{(from above)}$$

Rearranging the second equation gives $[A^-] = 0.63\,[HA]$. Substituting this into the first equation gives

$$[HA] + 0.63\,[HA] = 0.20$$

$$1.63\,[HA] = 0.20$$

$$[HA] = 0.123$$

Substituting the value of [HA] into the first equation again gives $[A^-] = 0.077$.

What happens if we add 0.050 moles of a strong acid or of a strong base? We can assume that small amounts of strong acid or base will react to completion. If 0.050 moles of a strong acid is added, it will react with the A^- to form HA. Thus, the amount of acid will increase by 0.050 moles (to 0.173 moles) and the amount of the base A^- will decrease by 0.050 moles (to 0.027 moles). The pH will change to 4.20, a more acidic value. Note that the change is larger here than in the previous example, as relatively less base was present in the buffer to start with.

If 0.050 moles of a strong base is added, it will react with the HA to form A^-. Thus, the amount of base will increase by 0.050 moles (to 0.127 moles) and the amount of acid will decrease by 0.050 moles (to 0.073 moles). The pH will then be 5.24, a more basic value. Note that the change is smaller here than in the previous example, as relatively more acid was present in the buffer to start with.

This buffer is better at controlling pH when base is added, not because the pH is acidic, but because there is more conjugate acid than base present in the original buffer. We can now define the term **buffer capacity.** Buffer capacity is the amount (in moles) of strong acid or strong base required to change the pH of a buffer by 1.00 pH units. The value of 1.00 is arbitrary, but reasonable, as it means the ratio ($[A^-]/[HA]$) has changed by a factor of 10. If it changes by more than that, the buffer's pH starts to change more rapidly as more acid or base is added and the solution is no longer a good buffer.

Dilution

How are buffer solutions different from simple acid or base solutions? If a small amount of base is added to an acid solution, the pH changes relatively little (except near the equivalence point). The same is true of adding a small amount of acid to base. In this characteristic, acid or base solutions are similar to buffers. However, if a strong acid or base solution is diluted, the pH changes dramatically. What happens to a buffer when it is diluted? Consider the example using the generic acid HA once more.

If a quantity of buffer is diluted (say by a factor of 2), the concentration of HA and A^- both decrease by a factor of 2, but the ratio (moles of A^-)/(moles of HA) remains unchanged. **Thus, the pH of a buffer does not change upon dilution.** The capacity of the buffer to resist pH change also does not change upon dilution, since the number of moles of acid and base present have not changed. Note, however, that 1.0 mL of the diluted buffer will have a smaller buffer capacity than 1.0 mL of the undiluted buffer, since it has half the number of moles of acid and base present in it.

EXPERIMENTAL PROCEDURE

Standardization and Use of the pH Meter

If the pH meter has not already been standardized, follow your laboratory instructor's directions, and standardize the pH meter using pH 7 and 10 buffers. All readings should be taken with the buffer in a 16 × 100 mm test tube.

Preparations

- Before inserting the electrode into a new solution, rinse it well using a wash bottle of distilled water. Blot the electrode dry but do not rub it. Rinse the test tube with distilled water and shake as much of the distilled water out as possible each time before pipetting into it.
- Obtain 10 mL of each of the following three buffers in stoppered 25 mL Erlenmeyer flasks:

  1. Buffer A containing 0.020 mmol $NH_{3(aq)}$ and 0.040 mmol NH_4^+ per mL.
  2. Buffer B containing 0.030 mmol $NH_{3(aq)}$ and 0.030 mmol NH_4^+ per mL.
  3. Buffer C containing 0.040 mmol $NH_{3(aq)}$ and 0.020 mmol NH_4^+ per mL.

Part A: Determination of Buffer Capacity

You will add strong acid dropwise to a buffer and monitor the pH until the pH changes by 1.0 pH units. This is repeated with a new sample of buffer and adding strong base.

- Obtain a 2 mL pipet, and rinse it with a small amount of Buffer A. Fill the pipet with Buffer A. Pipet 2.00 mL (±0.01 mL) of Buffer A into a 16 × 100 mm test tube, and record the pH of the solution. Touch the electrode against the side of the test tube as you remove it, so that any drops of solution will remain in the test tube. **Do not rinse the electrode.**

- Obtain a 2 mL graduated pipet, and rinse it with a small amount of 0.100 M HCl. Fill the pipet to the 2.00 mL mark with the 0.100 M HCl. Add 1 drop of 0.100 M HCl from the graduated pipet to a 16 $\times$ 100 mm test tube, and tap the test tube to mix the solution. Reinsert the electrode. Raise and lower the tube two or three times, using the electrode to further mix the solution, and note the pH (you need not record this). Repeat this procedure adding 1–2 drops of 0.100 M HCl at a time until the pH has changed by 1.0 (get as close as you can to a change of 1.0). Record the pH and the volume of 0.100 M HCl used.
- Discard the solution, rinse the test tube and electrode, and repeat with another 2.00 mL of Buffer A, this time adding 0.100 M NaOH dropwise until the pH increases by 1.0. Use a second 2 mL graduated pipet for the 0.100 M NaOH.
- Repeat these steps for Buffers B and C.

Part B: The Effect of Dilution/Moles on Buffer Capacity

You will examine the buffer capacity of diluted buffer by (1) using the same volume of buffer as Part A and adding an equal volume of deionized water, and (2) by using half as much buffer as Part A and diluting to the same degree. This examines whether dilution affects buffer capacity or whether it is the number of moles of buffer constituents present that is the governing factor. If dilution is the key, both of these will have lower buffer capacities than Part A because they are more dilute. If moles are the key, then the dilute buffers with the same moles as Part A should have the same buffer capacity as A and only the buffers with fewer moles will have lower buffer capacities.

Same Number of Moles, Half the Concentration of Part A

- Rinse the 2 mL buffer pipet with a small amount of Buffer A and then fill it with the buffer. Pipet 2.00 mL ($\pm$0.01 mL) of Buffer A into the test tube. Rinse a second pipet with distilled water, fill it, and pipet 2.00 mL of distilled water into the same test tube. Tap the tube to mix the solution. Insert the electrode as before, raising and lowering the tube two or three times, using the electrode to further mix the solution. Record the pH of the solution.
- Touch the electrode against the side of the test tube as you remove it so that any drops of solution will remain in the test tube; do not rinse the electrode. Using the same HCl pipet as in Part A, add 1–2 drops of 0.100 M HCl at a time, monitoring the pH until the pH has changed by 1.0 (get as close as you can to a change of 1.0). Record the pH and the volume of 0.100 M HCl used.
- Discard the solution, rinse the test tube and electrodes, and repeat with 2.00 mL of Buffer A and 2.00 mL distilled water. Using the same NaOH pipet as Part A, add 1–2 drops of 0.100 M NaOH at a time, monitoring the pH until the pH has changed by 1.0 (get as close as you can to a change of 1.0). Record the pH and the volume of 0.100 M NaOH used.

Half the Number of Moles, Half the Concentration of Part A

- Rinse the 2 mL buffer pipet with a small amount of Buffer A and then fill it with the buffer. Pipet 1.00 mL ($\pm$0.01 mL) of Buffer A into the test tube. Rinse a second pipet with distilled water, and pipet 1.00 mL of distilled water into the same test tube. Tap the tube to mix the solution. Insert the electrode as before, raising and lowering the tube two or three times, using the electrode to further mix the solution. Record the pH of the solution.
- Touch the electrode against the side of the test tube as you remove it so that any drops of solution will remain in the test tube; do not rinse the

electrode. Using the same HCl pipet as in Part A, add 1–2 drops of 0.100 M HCl at a time, monitoring the pH until the pH has changed by 1.0 (get as close as you can to a change of 1.0). Record the pH and the volume of 0.100 M HCl used.

- Discard the solution, rinse the test tube and electrodes, and repeat with 1.00 mL of Buffer A and 1.00 mL distilled water. Using the same NaOH pipet as in Part A, add 1–2 drops of 0.100 M NaOH at a time, monitoring the pH until the pH has changed by 1.0 (get as close as you can to a change of 1.0). Record the pH and the volume of 0.100 M NaOH used.

Part C: The Effect of Preparation Method on Buffer pH and Buffer Capacity

A buffer is prepared by two different methods and the buffer capacities compared.

Method 1

Prepare a buffer by starting with conjugate base and adding strong acid.

- Rinse a 2 mL pipet with 0.100 M $NH_{3(aq)}$ and then fill it with the solution. Pipet 1.20 mL ($\pm$0.01 mL) of this solution into a clean test tube. Using the same HCl pipet as in Part A, pipet 0.80 mL of 0.100 M HCl into the same test tube. Tap the tube to mix the solution. Insert the electrode, raising and lowering the tube two or three times, using the electrode to further mix the solution. Record the pH of the solution.
- Touch the electrode against the side of the test tube as you remove it so that any drops of solution will remain in the test tube; do not rinse the electrode. Using the same HCl pipet, add 1–2 drops of 0.100 M HCl at a time, monitoring the pH until the pH has changed by 1.0 (get as close as you can to a change of 1.0). Record the pH and the volume of 0.100 M HCl used.
- Discard the solution, rinse and repeat with 1.20 mL of 0.100 M $NH_{3(aq)}$ and 0.80 mL of 0.100 M HCl. Using the same NaOH pipet as in Part A, add 1–2 drops of 0.100 M NaOH at a time, monitoring the pH until the pH has changed by 1.0 (get as close as you can to a change of 1.0). Record the pH and the volume of 0.100 M NaOH used.

Method 2

Prepare a buffer by starting with conjugate base and adding conjugate acid.

- Rinse a 2 mL pipet with 0.040 M $NH_{3(aq)}$ and then fill it with the solution. Pipet 1.00 mL ($\pm$0.01 mL) into a clean test tube. Rinse a second 2 mL pipet with 0.080 M $NH_4Cl_{(aq)}$. Pipet 1.00 mL into the same test tube. Repeat the measurements and additions of acid and base of Method 1 above.

Note: All undiluted buffers in the experiment are designed to give the same total number of moles of conjugate acid and base [moles HA + moles A^- = constant, although the ratio A^-/HA changes]. The diluted buffers have exactly one-half as many moles of each.

CALCULATIONS

The most informative way to examine the data generated from Parts A and B of this experiment is to plot the number of mmol of acid and base (y axis) required to change the pH by 1.0 versus the mole ratio of conjugate acid to conjugate base (x axis) for the 3 buffers.

- Calculate the initial mole ratio of conjugate acid to conjugate base in each of the three buffer solutions.
- Calculate the mmol of acid and base required to change the pH by one unit. Note: (mmol/mL is the same as mol/L).

$$\text{mmol} = (M, \text{mmol/mL})(V, \text{mL})$$

- Construct two graphs from the data:
  Graph 1 summarizes the data for adding strong acid to the three buffers. There will be three lines on the graph. Use a different color for each line.

  Line 1 will be for the addition of 0.100 M HCl in Part A. Plot the mmol of acid used for each buffer. Draw a point-to-point line.

  Line 2 will be for the addition of 0.100 M HCl in Part B (1). Plot the mmol of acid used for each buffer. Draw a point-to-point line.

  Line 3 will be for the addition of 0.100 M HCl in Part B (2). Plot the mmol of acid used for each buffer. Draw a point-to-point line.

  Graph 2 is the same as graph 1, except it should summarize the data for adding strong base to the buffers.

Name: ______________________________

Date: __________ Section: ______________________

Pre-Laboratory Questions: Experiment 20

1. A buffer solution contains 0.500 M acetic acid ($K_a = 1.8 \times 10^{-5}$) and 0.500 M sodium acetate. What is the pH of the buffer solution?

2. Calculate the moles of acetic acid and sodium acetate in 2.00 mL of the buffer in Question 1. 0.50 mL of 0.500 M NaOH is then added. Calculate the new amounts (moles) of acetic acid and sodium acetate. What is the pH of the resulting solution?

3. 0.50 mL of 0.500 M HCl is added to a new 2.00 mL sample of the buffer in Question 1. Calculate the new amounts (moles) of acetic acid and sodium acetate. What is the pH of the resulting solution?

Name: ______________________________

Date: __________ Section: ______________________

Data Sheet: Experiment 20

Part A. Undiluted Buffers

	Buffer A	*Buffer B*	*Buffer C*
Add HCl			
pH of original buffer solutions	_____	_____	_____
Volume of 0.100 M HCl to change pH by 1.0	_____ mL	_____ mL	_____ mL
Final pH	_____	_____	_____
Add NaOH			
pH of original buffer solutions	_____	_____	_____
Volume of 0.100 M NaOH to change pH by 1.0	_____ mL	_____ mL	_____ mL
Final pH	_____	_____	_____

Part B. Diluted Buffers

	Buffer A	*Buffer B*	*Buffer C*
Add HCl, same # moles as Part A			
pH of original buffer solutions	_____	_____	_____
Volume of 0.100 M HCl to change pH by 1.0	_____ mL	_____ mL	_____ mL
Final pH	_____	_____	_____
Add NaOH, same # moles as Part A			
pH of original buffer solutions	_____	_____	_____
Volume of 0.100 M NaOH to change pH by 1.0	_____ mL	_____ mL	_____ mL
Final pH	_____	_____	_____

Name: ______________________

Date: ________ Section: ______________________

Data Sheet, Experiment 20, p. 2

Part B. Diluted Buffers

	Buffer A	*Buffer B*	*Buffer C*
Add HCl, half # moles as Part A			
pH of original buffer solutions:	_____	_____	_____
Volume of 0.100 M HCl to change pH by 1.0	_____ mL	_____ mL	_____ mL
Final pH	_____	_____	_____
Add NaOH, half # moles as Part A			
pH of original buffer solutions	_____	_____	_____
Volume of 0.100 M NaOH to change pH by 1.0	_____ mL	_____ mL	_____ mL
Final pH	_____	_____	_____

Part C. Preparation of Buffers

	Method 1	*Method 2*
Add HCl		
pH of original buffer solutions	_____	_____
Volume of 0.100 M HCl to change pH by 1.0	_____ mL	_____ mL
Final pH	_____	_____
Add NaOH		
pH of original buffer solutions:	_____	_____
Volume of 0.100 M NaOH to change pH by 1.0	_____ mL	_____ mL
Final pH	_____	_____

	Buffer A	*Buffer B*	*Buffer C*
Mole ratio of conjugate acid to conjugate base, $(NH_4^+):(NH_3)$	_______	_______	_______

Construct Graphs 1 and 2 as described under Experimental Procedure on a piece of graph paper.

Name: ______________________________

Date: __________ Section: ______________________________

Questions: Experiment 20

1. Comment on the graphs prepared from your data. Discuss the buffer capacities within any one line on your graphs. As the mole ratio of A^- to HA increases, is there more conjugate acid or base present initially? Which buffer should have the highest capacity (largest amount, mmol, to change the pH by one unit) for the addition of base? Addition of acid? Is this what you observed?

2. Compare the buffer capacity of:

a. The different buffers when adding strong acid

b. The different buffers when adding strong base

c. The full strength buffer versus the diluted buffer (Part B) with same number of moles of buffer as Part A.

d. The full strength buffer versus the diluted buffer (Part B) with half the number of moles of buffer as Part A.

3. Calculate the number of moles of conjugate acid and base present in each buffer in Part C (remember that in Method 1, the acid reacts with the ammonia to produce ammonium ions). Discuss the following. Are the numbers of moles the same? Do they have the same initial pH? Do they have similar buffer capacities for acid and base? Is there a significant difference between the two?

Experiment 21 Redox Titration of Permanganate (Balancing Oxidation–Reduction Reactions)

OBJECTIVES

- To learn to balance oxidation–reduction reactions
- To perform an oxidation–reduction titration

PRIOR READING

Section 3.3: Weighing
Section 3.4: Liquid Volumes
Section 3.9: Titration

INTRODUCTION

Many elements possess more than one oxidation state. These include representatives of both the **main group elements** (the s and p blocks, such as N, S, and P) and the **transition elements** (the d and f blocks, such as Fe, Co, Cr). Manganese, for example, is known in the 0, II, III, IV, V, VI, and VII oxidation states. In most cases for transition metals, each oxidation state exhibits a characteristic color.

In this experiment, the more accessible oxidation states of manganese will be investigated by reducing the Mn(VII) in permanganate ion (MnO_4^-) in a series of redox titrations. In a reduction, the material being reduced gains electrons, and consequently has its oxidation state lowered. In acid solution, for example, Mn(VII) is reduced to Mn(II) (manganous ion) through the gain of five electrons, as shown in the following unbalanced equation:

$$MnO_4^- + 5\,e^- + H^+ \rightleftharpoons Mn^{2+}$$

These equations are called **half-reactions** because only one species is changing oxidation state, requiring electrons in the balanced equation. In neutral solution, Mn(VII) is reduced to MnO_2 [Mn(IV), manganese dioxide], and in basic solution, Mn(VII) is reduced to Mn(VI) (MnO_4^{2-}, manganate ion):

$$MnO_4^- + 3\,e^- \rightleftharpoons MnO_2$$

$$MnO_4^- + e^- + OH^- \rightleftharpoons MnO_4^{2-}$$

Since the electrons gained by the Mn(VII) must come from somewhere, for every material being reduced, some species must simultaneously be **oxidized.** In an oxidation, the material being oxidized loses electrons, and consequently has its oxidation state increased. In this experiment, for example, it is the bisulfite ion (HSO_3^-) that is oxidized to sulfate ion (SO_4^{2-}):

$$HSO_3^- \rightleftharpoons SO_4^{2-} + 2\,e^-$$

The oxidation state of sulfur in the bisulfite is +4, whereas in sulfate, the oxidation state of sulfur is +6. Thus, the sulfur lost two electrons. This half reaction is the same in acid, base, or neutral solution.

Balancing Redox Reactions in Acidic Solution

In acidic solutions, hydrogen ion and water are present in addition to the materials undergoing oxidation and reduction. Thus, H^+ and H_2O can be used to balance the half reactions. Consider the redox reaction of the dichromate ion ($Cr_2O_7^{2-}$) with iodide (I^-) in acid solution. The two unbalanced half reactions are

$$Cr_2O_7^{2-} \rightleftharpoons Cr^{3+}$$

$$I^- \rightleftharpoons I_2$$

Since there are two chromium atoms on the left and only one on the right, one mole of $Cr_2O_7^{2-}$ must produce two moles of Cr^{3+} (chromic ion). Since there is one iodine ion on the left and there are two atoms on the right, two moles of iodide must produce one mole of molecular iodine (I_2). The half-reactions are then

$$Cr_2O_7^{2-} \rightleftharpoons 2\ Cr^{3+}$$

$$2\ I^- \rightleftharpoons I_2$$

To balance the oxygen atoms, note that the only source of oxygen in solution is H_2O. We must therefore add one water for every missing oxygen.

$$Cr_2O_7^{2-} \rightleftharpoons 2\ Cr^{3+} + 7\ H_2O$$

$$2\ I^- \rightleftharpoons I_2$$

We can balance this by adding 14 H^+ to the left side, to balance the hydrogens.

$$Cr_2O_7^{2-} + 14\ H^+ \rightleftharpoons 2\ Cr^{3+} + 7\ H_2O$$

$$2\ I^- \rightleftharpoons I_2$$

Finally, the total charge on each side must be balanced by adding electrons to the side with the more positive (less negative) charge. In the chromium half-reaction, the left side has a charge of +12 and the right side +6, so six electrons must be added to the left side, and similarly, in the iodine half-reaction, two electrons must be added to the right side.

$$Cr_2O_7^{2-} + 14\ H^+ + 6\ e^- \rightleftharpoons 2\ Cr^{3+} + 7\ H_2O$$

$$2\ I^- \rightleftharpoons I_2 + 2\ e^-$$

Note that the electrons appear on the left side of one half-reaction, and on the right side of the other. This must always be the case, as one reaction is a reduction and the other must be an oxidation.

In order to obtain the total reaction, the two half-reactions must be added up in such a way that the electrons cancel out. Thus, the iodine half-reaction must be multiplied by three:

$$Cr_2O_7^{2-} + 14\ H^+ + 6\ e^- \rightleftharpoons 2\ Cr^{3+} + 7\ H_2O$$

$$6\ I^- \rightleftharpoons 3\ I_2 + 6\ e^-$$

The two half-reactions may now be added, giving the net redox reaction:

$$Cr_2O_7^{2-} + 6\ I^- + 14\ H^+ + \rightleftharpoons 2\ Cr^{3+} + 3\ I_2 + 7\ H_2O$$

Balancing Redox Reactions in Base

Perhaps the simplest way to balance a reaction in base is to first balance it in acid, then, for every H^+ in the reaction, add that many OH^- to each side. On one side, the H^+ and OH^- become H_2O, leaving only OH^- in the reaction on the other side.

Alternatively, OH^- and H_2O are used to balance the oxidation and reduction half reactions. The chromium and iodine atoms are balanced in the same way as above in acid solution:

$$Cr_2O_7^{2-} \rightleftharpoons 2\,Cr^{3+}$$

$$2\,I^- \rightleftharpoons I_2$$

Balancing oxygen in basic solution is different than in acid. For each missing oxygen, two hydroxide atoms are added (one will end up as H_2O).

$$Cr_2O_7^{2-} \rightleftharpoons 2\,Cr^{3+} + 14\,OH^-$$

$$2\,I^- \rightleftharpoons I_2$$

To balance the added hydrogens, we add a number of waters equal to half the number of hydroxides to the other side. Since 14 hydroxide ions were added to the right side, we must now add 7 water molecules to the left side.

$$Cr_2O_7^{2-} + 7\,H_2O \rightleftharpoons 2\,Cr^{3+} + 14\,OH^-$$

$$2\,I^- \rightleftharpoons I_2$$

The total charge must be balanced as before. The number of electrons added in a half-reaction will always be the same, regardless of whether it is balanced in acid or base solution.

$$Cr_2O_7^{2-} + 7\,H_2O + 6\,e^- \rightleftharpoons 2\,Cr^{3+} + 14\,OH^-$$

$$2\,I^- \rightleftharpoons I_2 + 2\,e^-$$

In order to obtain the total reaction, the two half-reactions must be added up in such a way as the electrons cancel out, as before.

$$Cr_2O_7^{2-} + 7\,H_2O + 6\,e^- \rightleftharpoons 2\,Cr^{3+} + 14\,OH^-$$

$$6\,I^- \rightleftharpoons 3\,I_2 + 6\,e^-$$

The two half-reactions may now be added, giving the net redox reaction:

$$Cr_2O_7^{2-} + 6\,I^- + 7\,H_2O \rightleftharpoons 2\,Cr^{3+} + 3\,I_2 + 14\,OH^-$$

Reactions carried out in neutral solution may be balanced by either the acid or base method.

EXPERIMENTAL PROCEDURE

Record all observations and measurements on the data sheets.

- Using a graduated or volumetric pipet, place 1.00 mL (±0.01 mL) of 0.0100 M $KMnO_4$ (potassium permanganate) solution into each of three 10 mL Erlenmeyer flasks. Label these flasks A, B, and C. **NOTE: Additional water may be added to the permanganate solutions in order to increase the volume for visual clarity, if desired.**

Acid Solution Titration

- Add 1.0 mL of 1 M H_2SO_4 (sulfuric acid) to flask A. Fill a microburet with 0.0200 M $NaHSO_3$ (sodium bisulfite) solution, and titrate the permanganate solution dropwise. The purple color of the solution will disappear suddenly at the endpoint, so the $NaHSO_3$ must be added slowly. The manganese is being reduced from the dark purple MnO_4^- to the almost colorless (pale pink) Mn^{2+}. Record the volume of $NaHSO_3$ used on the data sheet.

Neutral Solution Titration

- Refill the microburet with 0.0200 M $NaHSO_3$ solution and titrate the $KMnO_4$ solution in flask B. The purple color of permanganate will disappear at the endpoint, as a brown suspension of MnO_2 is formed. Record the volume of $NaHSO_3$ used on the data sheet.

Base Solution Titration

- Add 1.0 mL of 1 M NaOH to the permanganate solution in flask C. Refill the microburet with the $NaHSO_3$ solution and titrate to the dark green-colored end point of the manganate ion, MnO_4^{2-}. This endpoint is difficult to hit exactly, due to the dark colors of the reactant and product. Look for the last of the purple color. One way of reaching this end point more accurately is to calculate the amount of $NaHSO_3$ solution needed to provide one mole of electrons to the permanganate solution, as is readily seen from the balanced equation for this reaction.

Further Study

- In hot acidic solution, permanganate reacts quantitatively with oxalate $(C_2O_4)^{2-}$, oxidizing it to CO_2 gas, while the permanganate is reduced to Mn^{2+}. Using the acid titration conditions boil dissolved, weighed samples of oxalate-containing compounds (or unknowns) and titrate while hot with 0.0200 M $KMnO_4$ solution to determine the mass percent oxalate in the compound.

Name: ______________________________

Date: __________ Section: ______________________________

Pre-Laboratory Questions: Experiment 21

1. Balance the following oxidation-reduction reactions:

a. $Mn^{2+} + BiO_3^- \rightleftharpoons MnO_4^- + Bi^{3+}$ (acid)

b. $NO_3^- + I_2 \rightleftharpoons NO_2 + IO_3^-$ (acid)

c. $ClO_3^- + S^{2-} \rightleftharpoons Cl^- + S^0$ (base)

d. $Mo(OH)_3 + HO_2^- \rightleftharpoons MoO_4^{2-}$ (base)

2. Do the volumes of H_2SO_4 and NaOH used in the experiment need to be accurately measured? Explain.

3. If 542 μL of 0.0402 M $KMnO_4$ are required to titrate 200.0 μL of NaI solution to the end point in acid, what was the concentration of iodide ion in the original solution? The products are Mn(II) and I_2^0. Write the balanced redox equations in acid.

4. Balance the three redox reactions in this experiment by combining each of the three permanganate half-reactions on p. 273 with the bisulfite half-reaction. One should be in acid and one should be in base as indicated.

Name: ______________________________

Date: __________ Section: ______________________________

Data Sheet: Experiment 21

Acid Solution Titration

1. Volume of 0.0100 M $KMnO_4$ used: ______________ mL

2. Volume of 0.0200 M $NaHSO_3$ used: ______________ mL

3. Describe mixture at end point.

4. Calculate moles of $NaHSO_3$ used and moles of $KMnO_4$ used. What is the ratio? Does this correspond to the ratio predicted by the **balanced** oxidation–reduction reaction?

Neutral Solution Titration

1. Volume of 0.0100 M $KMnO_4$ used: ______________ mL

2. Volume of 0.0200 M $NaHSO_3$ used: ______________ mL

3. Describe mixture at end point.

4. Calculate moles of $NaHSO_3$ used and moles of $KMnO_4$ used. What is the ratio? Does this correspond to the ratio predicted by the **balanced** oxidation–reduction reaction?

Base Solution Titration

1. Volume of 0.0100 M $KMnO_4$ used: ______________ mL

2. Volume of 0.0200 M $NaHSO_3$ used: ______________ mL

3. Describe mixture at end point.

4. Calculate moles of $NaHSO_3$ used and moles of $KMnO_4$ used. What is the ratio? Does this correspond to the ratio predicted by the **balanced** oxidation–reduction reaction?

Name: ______________________________

Date: __________ Section: ______________________

Questions: Experiment 21

1. Write the electronic configurations for Mn, Mn^{2+}, Mn^{4+}, Mn^{6+}, and Mn^{7+}.

2. Sodium bismuthate ($NaBiO_3$) oxidizes $MnCl_2$ to MnO_4^- and is consequently reduced to $Bi(OH)_3$. How many grams of $MnCl_2$ will be oxidized by 5.0 g of $NaBiO_3$?

3. In each of the titrations performed in this experiment, the end point was apparent due to a color change in the material being oxidized or reduced. Suppose that none of the materials involved in the oxidation–reduction reaction are colored. Suggest a way one could determine the endpoint in such a case. (Hint: Think about acid–base titrations.)

Experiment 22 The Solubility Product Constant and Common Ion Effect

OBJECTIVES

- To determine the solubility constant of a salt.
- To determine the effect on solubility of adding a common ion.

PRIOR READING

Section 3.3: Weighing
Section 3.4: Liquid Volumes
Section 3.9: Titration

INTRODUCTION

For the general reaction

$$aA + bB \rightleftharpoons cC + dD$$

the form of the equilibrium constant expression is:

$$K_{eq} = \frac{[C]^c\,[D]^d}{[A]^a\,[B]^b}$$

For solutions, concentrations are usually expressed in terms of molarity. Pure solid and liquid components do not appear in the equilibrium expression as their concentrations (densities) do not change, and are therefore combined with the K.

For the dissolution of a slightly soluble salt, such as silver chloride, AgCl, or calcium hydroxide, $Ca(OH)_2$,

$$AgCl_{(s)} \rightleftharpoons Ag^+_{(aq)} + Cl^-_{(aq)}$$

$$Ca(OH)_{2(s)} \rightleftharpoons Ca^{2+}_{(aq)} + 2\ OH^-_{(aq)}$$

the equilibrium expression is referred to as a **solubility product constant,** K_{sp}, which is a constant for a given solvent at constant temperature. It can be affected by temperature, solvent and the presence of other dissolved species. Since these compounds are solids, the equilibrium expressions are (K_{sp} values may be found in Appendix G):

$$K_{sp} = [Ag^+][Cl^-] = 1.77 \times 10^{-10}$$

$$K_{sp} = [Ca^{2+}][OH^-]^2 = 4.68 \times 10^{-6}$$

If K_{sp} is known, the solubility (mol L^{-1} or g/mL^{-1}) of the compound can be calculated. In the case of AgCl, the concentrations of Ag^+ and Cl^- will be equal. Letting both equal x, we can solve for $[Ag^+]$.

$$K_{sp} = [Ag^+][Cl^-] = 1.77 \times 10^{-10}$$

$$(x)(x) = 1.77 \times 10^{-10}$$

$$(x) = 1.33 \times 10^{-5}\ M$$

The solubility, S (mol L^{-1}), is equal to the $[Ag^+]$, since one mole of salt dissolves for every mole of Ag^+ that is produced. Thus,

$$S(\text{AgCl}) = 1.33 \times 10^{-5}\ \text{M}$$

For $Ca(OH)_2$, the concentration of OH^- will be twice that of Ca^{2+}. Letting $x = [Ca^{2+}]$,

$$K_{sp} = [\text{Ca}^{2+}][\text{OH}^-]^2 = 4.68 \times 10^{-6}$$

$$4.68 \times 10^{-6} = (x)(2x)^2 = 4x^3$$

$$x = 1.05 \times 10^{-2}\ \text{M}$$

The solubility, S (mol L^{-1}), is equal to the $[Ca^{2+}]$, since one mole of salt dissolves for every mole of Ca^{2+} that is produced. Thus,

$$S[\text{Ca(OH)}_2] = 1.05 \times 10^{-2}\ \text{M}$$

If a mixture is made from solutions containing the component ions, the K_{sp} equation allows determination of whether precipitation will occur. The concentrations, after accounting for dilution, are used to calculate the ion product according the K_{sp} expression. **If the resulting value is less than or equal to *K_{sp}*, *no precipitation will occur. If the ion product is greater than K_{sp}, precipitation occurs (the reaction shifts to the left), lowering the concentrations, until the ion product is once again equal to K_{sp}.***

In this experiment, the solubility product constant of the organic salt, potassium hydrogen tartrate (KHT) will be determined. KHT has the formula $KHC_4H_4O_6$, where one of the acidic hydrogens of tartaric acid has been replaced by a potassium ion. When KHT dissolves in water, the salt dissociates:

$$\text{KHT}_{(s)} \rightleftharpoons \text{K}^+_{(aq)} + \text{HT}^-_{(aq)}$$

The hydrogen tartrate ion, HT^-, does not dissociate to any appreciable extent. The solubility product expression for the dissociation can therefore be written as:

$$K_{sp} = [\text{K}^+][\text{HT}^-]$$

To calculate the K_{sp} of potassium hydrogen tartrate, the equilibrium concentrations of K^+ and HT^- must be determined for a saturated solution. If a saturated solution of KHT is prepared in deionized water, the equilibrium concentrations of K^+ and HT^- will be identical due to the stoichiometry of the dissociation. Thus,

$$K_{sp} = [\text{HT}^-]^2$$

The equilibrium concentration of HT^-, a weak monoprotic acid, can easily be determined by acid-base titration, using phenolphthalein as the indicator.

$$\text{HT}^-_{(aq)} + \text{OH}^-_{(aq)} \rightarrow \text{H}_2\text{O}_{(l)} + \text{T}^{2-}_{(aq)}$$

The saturated solution must be filtered to separate it from the excess solid KHT, which would dissolve during the titration and also be titrated, leading to an erroneously high [HT^-]. At the endpoint of the titration, moles OH^- added is equal to initial moles HT^- in the solution titrated. For solutions, moles = M V, thus:

$$[HT^-] = (M_{NaOH})(V_{NaOH})/(V_{HT^-})$$

Concentrations of reactants and products at equilibrium can be affected by a number of factors. One of these is the presence of a common ion (called the **common ion effect**). In the case of AgCl discussed earlier, if additional chloride ion (say, from NaCl) is added to a saturated solution of AgCl, the equilibrium will shift to the left, and AgCl will precipitate. This illustrates **Le Chatelier's principle,** which states that when a system at equilibrium is put under a stress that changes one or more concentrations, it will shift to alleviate that stress. The K_{sp} equation also indicates that if solid AgCl is added to a solution that already contains a source of Cl^-, much less AgCl will dissolve than would dissolve in deionized water. The already high concentration of Cl^- means that only a small concentration of Ag^+ is necessary to satisfy K_{sp}.

The common ion effect will be investigated by determining the solubility of KHT in solutions containing K^+ from another source (dissolved KNO_3). Le Chatelier's principle predicts that KHT will be less soluble in solutions containing other potassium salts.

EXPERIMENTAL PROCEDURE

Optional Procedure A: Microscale Procedure

Preparation of Saturated Solutions of KHT in Deionized Water and a KNO_3 Solution

- Place two clean 25 mL Erlenmeyer flasks and two filter funnels in the oven to dry. Weigh out approximately 70 mg (±5 mg) of finely powdered KHT into a third clean 25 mL Erlenmeyer flask. Add 10 mL of deionized water (this is the KHT–water solution), a magnetic stir bar, and stir continuously for at least 20 minutes using a magnetic stirrer. Place several layers of towelling between the flask and the stirrer to prevent heating.
- Prepare a second solution of 70 mg (±5 mg) of finely powdered KHT into a fourth clean 25 mL Erlenmeyer flask. Add 10 mL of the assigned concentration of KNO_3 (this is the KHT–KNO_3 solution).
- When the KHT–water solution has 5 minutes left to stir, remove a flask and funnel from the oven, and allow them to cool to room temperature. Place a dry piece of filter paper in the funnel. When the KHT–water solution has stirred the remaining 5 minutes, remove it from the stirrer.
- Rinse the stir bar with deionized water at the sink, add it to the KHT–KNO_3 flask and place that on the magnetic stirrer to stir for 20 minutes.
- Filter the KHT–water solution through the dry filter into the dry flask. Do not rinse the filter paper. Stopper the flask, label, and set aside. Discard the filter paper and any solid KHT left in the original flask.
- Similarly, filter the KHT–KNO_3 solution through the second oven-dried funnel with dry paper into the second oven-dried flask after it has stirred for 20 minutes, stopper, label, and set aside.

Solubility Product Constant of KHT

- Rinse and fill a clean 2 mL microburet with the filtered KHT–water solution. Rinse and fill a second clean microburet with standardized 0.0400 M NaOH solution. Record the initial volumes of the two burets to ±0.001 mL.

- Release about 500 μL of the KHT solution into a clean 25 mL Erlenmeyer flask. Add 10 mL of deionized water and 1 drop of phenolphthalein indicator (fine tipped Pasteur pipet). Titrate dropwise to the first appearance of pink with the NaOH solution. Record the final volumes of both solutions to ± 0.001 mL.
- Repeat the process twice more using different volumes of the KHT–water solution. Dispose of the solutions according to your laboratory instructor's directions.

Solubility of KHT in a KNO_3 Solution

- Drain any remaining KHT–water solution from its buret, rinse the buret once with deionized water, then rinse and fill with the filtered KHT–KNO_3 solution.
- Repeat the titration procedure above (3 trials). Dispose of solutions according to your laboratory instructor's directions. **NOTE: This will require less NaOH solution to titrate if the common ion effect is real, as there will be less dissolved KHT. Use a somewhat larger volume of KHT solution (up to 2500 μL–consult your lab instructor).**

Optional Procedure B: Macroscale Procedure

Preparation of Saturated Solutions of KHT in Deionized Water and a KNO_3 Solution

- Place two clean 250 mL Erlenmeyer flasks and two filter funnels in the oven to dry. Weigh out approximately 0.7 g (±0.05 g) of finely powdered KHT into a third clean 250 mL Erlenmeyer flask. Add 100 mL of deionized water (this is the KHT–water solution), a magnetic stir bar, and stir continuously for at least 20 minutes using a magnetic stirrer. Place several layers of towelling between the flask and the stirrer to prevent heating.
- Prepare a second solution of 0.7 g (±0.05 g) of finely powdered KHT into a fourth clean 250 mL Erlenmeyer flask. Add 100 mL of the assigned concentration of KNO_3 (this is the KHT–KNO_3 solution).
- When the KHT–water solution has 5 minutes left to stir, remove a flask and funnel from the oven, and allow them to cool to room temperature. Place a dry piece of filter paper in the funnel. When the KHT–water solution has stirred for the remaining 5 minutes, remove it from the stirrer.
- Rinse the stir bar with deionized water at the sink, add it to the KHT–KNO_3 flask and place that on the magnetic stirrer to stir for 20 minutes. Filter the KHT–water solution through the dry filter into the dry flask. Do not rinse the filter paper. Stopper the flask, label, and set aside. Discard the filter paper and any solid KHT left in the original flask.
- Similarly, filter the KHT–KNO_3 solution through the second oven-dried funnel with dry paper into the second oven-dried flask after it has stirred for 20 minutes, stopper, label, and set aside.

Solubility Product Constant of KHT

- Rinse and fill a clean 25 mL buret with the filtered KHT–water solution. Rinse and fill a second clean buret with standardized 0.0400 M NaOH solution. Record the initial volumes of the two burets to ±0.01 mL.
- Release about 12 mL of the KHT solution into a clean 250 mL Erlenmeyer flask. Add 100 mL of deionized water and 2–3 drops of phenolphthalein indicator. Titrate dropwise to the first appearance of pink with the NaOH solution. Record the final volumes of both solutions to ±0.01 mL.

- Repeat the process twice more using different volumes of the KHT–water solution. Dispose of the solutions according to your laboratory instructor's directions.

Solubility of KHT in a KNO_3 Solution

- Drain any remaining KHT–water solution from its buret, rinse the buret once with deionized water, then rinse and fill with the filtered KHT–KNO_3 solution.
- Repeat the titration procedure above (3 trials). Dispose of solutions according to your laboratory instructor's directions. **NOTE: This will require less NaOH solution to titrate if the common ion effect is real, as there will be less dissolved KHT. Use a somewhat larger volume of KHT solution (up to 30 mL—consult lab instructor).**

Calculations

- Calculate the concentration of HT^- for each run in the deionized water and in your KNO_3 solution. Determine the average concentration of HT^- in each. Report these values to your laboratory instructor.
- Calculate K_{sp} from the deionized water data.
- Summarize the class's results by preparing a plot of $[HT^-]$ on the y axis versus $[KNO_3]$ on the x axis, including your average $[HT^-]$ in deionized water (where $[KNO_3] = 0.00$ M).

Further Study

- The solubility of KHT in the KNO_3 solutions is also affected by the $[NO_3^-]$. Although the solubility decreases as $[KNO_3]$ increases, the K_{sp} calculated is not constant. The presence of extra ions makes the KHT a little more soluble (not as big a decrease in solubility as we would expect), so K_{sp} seems to increase with $[KNO_3]$. A correction can be calculated for the concentration by determining the solubility of KHT in a solution of the same concentration of ions of a related salt but with no common ion.
- Prepare a solution of KHT in the same concentration of $NaNO_3$ and titrate it to determine the $[HT^-]$. This should be greater than the concentration of HT^- in deionized water. The correction factor for the concentration of HT^-, γ (gamma), is

$$\gamma = [HT^-]_{H_2O}/[HT^-]_{NaNO_3}$$

- There will be a different γ for each concentration. To see if K_{sp} is fairly constant, calculate a corrected K_{sp} for each concentration using the corresponding γ and the $[HT^-]$ in KNO_3 as calculated previously and the equation:

$$\text{corrected } K_{sp} = \gamma[HT^-][K^+]_{total}$$

$$\text{where } [K^+]_{total} = [K^+]_{KNO_3} + [K^+]_{KHT} = [KNO_3] + [HT^-]$$

Name: ______________________________

Date: __________ Section: ______________________________

Pre-Laboratory Questions: Experiment 22

1. A mixture is made from 1.0 mL of 0.10 M $Mg(NO_3)_2$ and 3.0 mL of 0.010 M KF. Will MgF_2 precipitate? $K_{sp} = 3.7 \times 10^{-8}$ for MgF_2. MgF_2 completely dissociates into Mg^{2+} and F^- ions. Don't omit the change in concentration due to dilution (mixing).

2. The K_{sp} of $PbBr_2$ is 2.1×10^{-6} at 25°C. Calculate the solubility of $PbBr_2$ in mg mL^{-1} of solution.

3. Why must the receiving flask, funnel and filter paper be dry when filtering the saturated solutions? Why don't the KHT buret and the titration flask have to be dry before use?

Name: ______________________________

Date: __________ Section: ______________________________

Data Sheet: Experiment 22

Concentration of NaOH: ______________________________ mol L^{-1}

Solubility Constant Data, Deionized Water

	Trial 1	*Trial 2*	*Trial 3*
1. Initial volume, KHT	_____ mL	_____ mL	_____ mL
2. Final volume, KHT	_____ mL	_____ mL	_____ mL
3. Initial volume, NaOH	_____ mL	_____ mL	_____ mL
4. Final volume, NaOH	_____ mL	_____ mL	_____ mL

Calculation of K_{sp} in Deionized Water

5. Volume of KHT added	_____ mL	_____ mL	_____ mL
6. Volume of NaOH added	_____ mL	_____ mL	_____ mL
7. $[HT^-]$	_____ mol L^{-1}	_____ mol L^{-1}	_____ mol L^{-1}

8. Calculate the value of K_{sp} for each trial, and determine the average value.

Name: ____________________

Date: ________ Section: ____________________

Data Sheet: Experiment 22, p. 2

Common Ion Effect

Concentration of KNO_3: _____mol L^{-1} Concentration of NaOH:____mol L^{-1}

	Trial 1	*Trial 2*	*Trial 3*
1. Initial volume, KHT	_____ mL	_____ mL	_____ mL
2. Final volume, KHT	_____ mL	_____ mL	_____ mL
3. Initial volume, NaOH	_____ mL	_____ mL	_____ mL
4. Final volume, NaOH	_____ mL	_____ mL	_____ mL

Calculation of Solubility of KHT in KNO_3

5. Volume of KHT added	_____ mL	_____ mL	_____ mL
6. Volume of NaOH added	_____ mL	_____ mL	_____ mL
7. $[HT^-]$	_____ mol L^{-1}	_____ mol L^{-1}	_____ mol L^{-1}

8. Calculate the average solubility of KHT in the presence of the common ion from KNO_3. Obtain values from your classmates for their solutions, and construct a graph of the solubility of KHT (*y* axis) versus $[KNO_3]$ (*x* axis). What does this graph show? Does this verify the common ion effect?

Name: ______________________________

Date: __________ Section: ______________________________

Questions: Experiment 22

1. The Environmental Protection Agency recommends that Pb^{2+} have a concentration not exceeding 1.00×10^{-7} M in drinking water. What concentration of NaCl could be used to remove the lead from the water to bring it to a safe level? K_{sp} for $PbCl_2$ is 1.6×10^{-5}. What would the drawbacks of doing this be?

2. Could Pb^{2+} be separated from Ba^{2+} by precipitation of the insoluble fluorides from a solution that, initially, has $[Pb^{2+}] = [Ba^{2+}] = [F^-] = 0.0100$ M? Briefly explain why or why not. If one precipitates, which is it? K_{sp} $PbF_2 = 8.0 \times 10^{-8}$, $BaF_2 = 1.7 \times 10^{-6}$. As an extra challenge, what % of each precipitates?

3. Calculate the total $[K^+]$ in each solution of KHT in KNO_3. Calculate K_{sp} for each solution. Are they approximately constant? Is this in conflict with your graph of solubility versus $[KNO_3]$? Discuss this, considering what is in the KHT/KNO_3 solutions versus KHT/deionized water solutions and what factors can affect K_{sp}.

Chapter 7

Physical Chemistry

Experiment 23: Determination of Heats of Reaction

Experiment 24: Hess's Law: Calculation of a Heat of Reaction

Experiment 25: Colligative Properties: Freezing Point Depression

Experiment 26: Kinetics: The Iodine Clock Reaction

Experiment 23 Determination of the Heat of Reaction

OBJECTIVES

- To determine the stoichiometry of acid–base reactions.
- To determine the enthalpy of acid–base reactions.

PRIOR READING

Section 2.6: Graphing of Data
Section 3.3: Weighing
Section 3.4: Liquid Volumes
Section 3.9: Titration

INTRODUCTION

Thermodynamics is the study of heat in motion. When a physical or chemical change occurs, it is usually accompanied by a change in the heat content **(enthalpy)** of the materials in question. Suppose you are heating one liter of water from room temperature to its boiling point at one atmosphere pressure. The product (water at 100°C) has a higher energy content than the starting material (water at 25°C). In order to accomplish this change, the **first law of thermodynamics** states that energy must be conserved, that is, the difference in energy must be supplied. This change in heat energy is usually expressed as a change in **enthalpy, ΔH.**

$$\Delta H = H_{\text{final}} - H_{\text{initial}}$$

Here, ΔH would represent the amount of heat that was needed to accomplish the heating, H_{final} would be the heat content of the hot water, and H_{initial} would be the heat content of the water at room temperature.

Enthalpy is defined as the heat content of a material under a particular set of conditions, called a **state.** Since there is only one value of enthalpy for any given state, enthalpy is one of a number of thermodynamic variables called **state functions.** Most of the time, we are not interested in any particular value of enthalpy. Instead, we are interested in the change in the enthalpy, ΔH, in going from one state to another, or in going from reactants to products. This is because there is no way of directly measuring an enthalpy under a given set of conditions—we can only measure how much it changes.

We must therefore define a starting temperature and pressure, and choose an arbitrary value for the enthalpy of that state. The state chosen to define enthalpy is 298 K and 1 atm pressure, at which the enthalpy of formation of any element **in its normal state** (e.g., oxygen as a diatomic gas, iron as a solid metal, bromine as a diatomic liquid, etc.) is defined as zero. Furthermore, for ions, the enthalpy of formation of the H^+ ion at a concentration of 1 M is defined as zero. All other ion enthalpies are calculated as changes from this standard state.

In a chemical reaction, the products usually have a different energy content than the reactants, even if they are at the same temperature and pressure. Stable materials generally have low (negative) enthalpies, while less stable materials have high (more positive) enthalpies. Unstable materials will often react spontaneously, or with little activation energy, to form more stable products. Such reactions are said to be **enthalpy driven,** that is, they proceed because ΔH for the reaction is so negative.

It is somewhat difficult to measure the amount of heat transferred in a chemical reaction. To make an accurate measurement, it is important that any heat change occurring in the system be isolated from the surroundings.

A device used to measure enthalpy changes is called a **calorimeter,** of which the most familiar type is a Thermos bottle. Due to its insulating vacuum jacket, the Thermos bottle does not exchange heat with its surroundings, and any liquid in it retains its temperature (hot or cold). If the calorimeter is perfectly insulated, any change in the temperature of its interior must be due to a chemical change taking place. In order for a calorimetric measurement to be accurate, the reaction taking place must be fast, complete, and lead to only one set of products.

Most often, the calorimeter is filled with a known quantity of water (or some other liquid of known heat capacity). The reaction then takes place, and the calorimeter liquid heats up or cools off, depending on whether the reaction releases heat (is **exothermic**) or absorbs heat (is **endothermic**). By measuring the temperature change, the heat of reaction may be calculated via the equation

$$\Delta H = m\, C_p\, \Delta T$$

where m is the number of grams of liquid, C_p is the **gram heat capacity** of the liquid (4.184 J g^{-1} K^{-1} for water), and ΔT is the temperature change (in either °C or K).

The heat capacity is defined as the amount of heat necessary to raise the temperature of a quantity of material by one degree. The two most familiar units of heat capacity in the United States are the **BTU** (British Thermal Unit), defined as the amount of heat necessary to raise the temperature of one pound of water by 1°F and the **calorie,** defined as the amount of heat necessary to raise the temperature of one gram of water by 1°C.

In an ideal system, all the heat produced by the reaction would be taken up by the solution. In reality, some of the heat produced is lost, heating up the apparatus (the test tube, thermometer, and Styrofoam block). This must be taken into account in the calculations. We therefore need to know how much energy is required to raise the temperature of the apparatus by 1°C. This can be determined by performing a reaction that releases a known amount of heat into the solution (and thereby the apparatus). By measuring the ΔT of the solution, the amount of heat taken up by the solution can be calculated. The remainder must have been absorbed by the apparatus or lost to the surroundings.

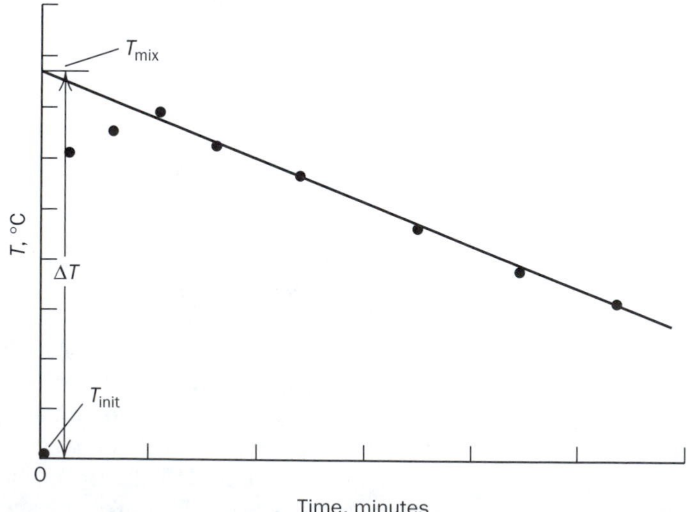

Figure E23.1 *Extrapolation of* ΔT.

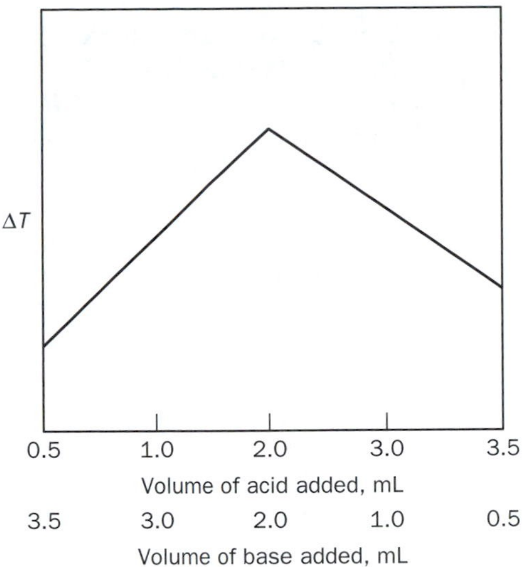

Figure E23.2 *Reaction stoichiometry plot.*

The simplest such "reaction" is to mix equal quantities of hot and cold water. If the system absorbs no heat, the equilibrium temperature of the mixture would be midway between the hot and cold temperatures. In reality, the apparatus will absorb some heat. It is initially at the same temperature as the cold water, and absorbs some heat from the hot water when it is added. Therefore, the equilibrium temperature will be somewhat less than halfway between the hot and cold water temperatures.

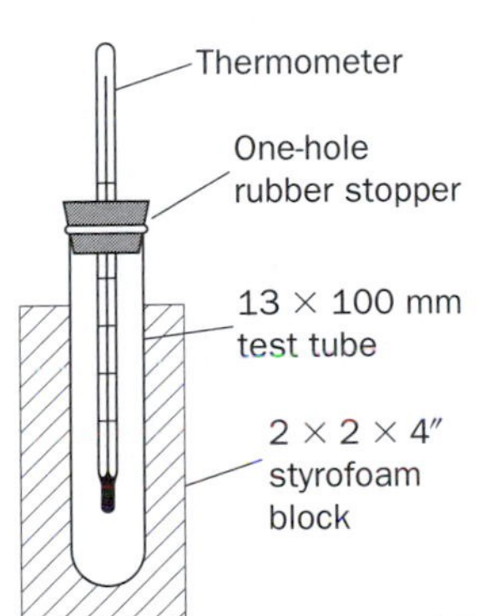

Figure E23.3 *Apparatus for measuring heat of reation.*

In this experiment, the Job's Plot method (also employed in Experiment 6) will be used to determine the stoichiometry and enthalpy of neutralization of two acid–base reactions. Varying amounts of an acid will be reacted with varying amounts of base. The temperature of the reaction (*y* axis) will be plotted versus time (*x* axis) in each case. A straight line is drawn through the relatively linear part of the data after the temperature change has peaked and begins to return to room temperature. Generally, there are a few points above or below the post-mixing line, which should be ignored in drawing the line, as they represent incomplete reaction or incomplete mixing (see Figure E23.1).

The line is extended back to the time of mixing (0 minutes) to extrapolate the temperature change that would have been observed had an instantaneous reaction taken place. The temperature at which the straight line intersects the *y* axis is the maximum T for the reaction. The quantity ΔT is the difference between this temperature and the initial temperature. A graph is then made of ΔT (*y* axis) versus volume of acid and base added (*x* axis). The maximum of this plot corresponds to the reaction stoichiometry (see Figure E23.2). The enthalpy of neutralization (corresponding to the heat of the reaction) is calculated using this value of ΔT.

EXPERIMENTAL PROCEDURE

Preparation of Calorimeter

- Obtain a 2″ × 2″ × 4″ block of Styrofoam, and drill a hole in it to hold a 13 × 100 mm test tube. Loosely insert a one-hole rubber stopper, fitted with a thermometer, in the test tube. The stopper should be briefly removed when adding chemicals to the tube, and then immediately replaced. This apparatus is shown in Figure E23.3.

The Styrofoam block cools off slowly. This tends to add heat to later runs. To avoid this, bring the block back to room temperature by inserting a test tube of room temperature water for several minutes after each run.

If two thermometers are used in this experiment, check them against each other by inserting them both in room temperature water and comparing the readings. Adjust the recorded temperature of the hot water for any difference.

If only one thermometer is used, it should be inserted into a beaker of hot water, then wiped dry before inserting it into the hot water used in the experiment. Similarly, it should be inserted in a beaker of room temperature water, then wiped dry before inserting it in the room temperature water used in this experiment. This prevents the loss or gain of heat in the next solution. Since we are only using small volumes, a significant amount of heat can be transferred to or from the thermometer if these steps are not followed.

Part A: Determining the Calorimeter Constant

- Pipet 2.00 mL (±0.01 mL) of room temperature water into a 13 × 100 mm test tube and place it in the Styrofoam block. Insert the stopper-thermometer assembly.
- Obtain 400–500 mL of 45–50°C water in a 600 mL beaker. Pipet 2.00 mL of this hot water into a second test tube and place it in the hot water bath for several minutes. Warm the thermometer in the hot water bath, wipe it dry, and then insert it into the hot water in the test tube. Record the temperature of the hot water once it stabilizes. Similarly, record the temperature of the room temperature water in the Styrofoam block test tube.
- Remove the stopper momentarily, tilt the Styrofoam block and quickly pour the hot water in the test tube into the room temperature water, noting the time of mixing. Immediately re-insert the stopper-thermometer assembly. Record the temperature to ±0.1°C every 30 sec for 5 minutes. Discard the contents, bring the Styrofoam block back to room temperature, and repeat the run.

Part B: Job's Law Titration of HCl with NaOH

NOTE: ***The HCl and NaOH used in the experiment should be allowed to stand at room temperature for at least one hour.***

- Trial 1: Place 0.50 mL (±0.01 mL) of 1.00 M standardized HCl in the test tube using a graduated pipet. Place the test tube in the Styrofoam block, and insert the stopper-thermometer assembly. Record the initial temperature to ±0.1°C. Removing the stopper momentarily, add 3.50 mL of 1.00 M standardized NaOH all at once. Immediately replace the stopper-thermometer assembly. Swirling the mixture, take temperature readings every 10 sec until the temperature begins to fall, and every 30 sec thereafter for at least 3 minutes.

- Repeat the procedure using the following quantities of acid and base:

	Trial			
	2	3	4	5
Volume of acid, mL	1.00	2.00	3.00	3.50
Volume of base, mL	3.00	2.00	1.00	0.50

Note that the total volume is the same in each case.

Part C: Job's Law Titration of H_2SO_4 with NaOH

- Repeat the procedure in Part B, except using 0.500 mL of 1.00 M standardized H_2SO_4, and 3.50 mL of 1.00 M standardized NaOH in Trial #1.
- Repeat the procedure using the following quantities of acid and base:

	Trial			
	2	3	4	5
Volume of acid, mL	1.00	2.00	3.00	3.50
Volume of base, mL	3.00	2.00	1.00	0.50

Calculations

For all trials, calculate ΔT for each reaction by plotting temperature (y axis) versus time (x axis) and extrapolating back to time zero as shown in Figure E23.1 to find $T_{mixture}$ and ΔT. For all solutions, C_p is 4.184 J g^{-1} °C^{-1}. The density of the solutions can be taken to be 1.00 g mL^{-1}.

To Determine the Calorimeter Constant

- ΔT_{cold} is the ΔT of the cold water (and the apparatus):

$$\Delta T_{cold} = T_{mixture} - T_{original\ of\ cold\ water}, °C$$

ΔT_{hot} is the ΔT of the hot water:

$$\Delta T_{hot} = T_{mixture} - T_{original\ of\ the\ hot\ water}, °C$$

(Note that ΔT_{cold} will be positive and ΔT_{hot} will be negative)
C_{cal} is the calorimeter constant.

- Since this is a well-insulated system, the heat lost by the hot water plus that gained by the cold water and apparatus should add up to zero.

$$\text{heat lost by hot water} = m_{hot}\, C_p\, \Delta T_{hot}$$

$$\text{heat absorbed by cold water} = m_{cold}\, C_p\, \Delta T_{cold}$$

$$\text{heat absorbed by apparatus} = C_{cal}\, \Delta T_{cold}$$

where m_{hot} and m_{cold} are the masses of the hot and cold water, respectively.

Thus,

$$m_{hot}\, C_p\, \Delta T_{hot} + m_{cold}\, C_p\, \Delta T_{cold} + C_{cal}\, \Delta T_{cold} = 0$$

Solve this equation for C_{cal}.

To Determine the Heat of Reaction, ΔH

- Make a plot of ΔT (y axis) versus volume of acid/base added (x axis) as shown in Figure E23.2 for the reactions involving HCl. The maximum temperature change will occur when the reactants are in the stoichiometric ratio. Note that this may not occur at one of your data points. Use the point of intersection of the two lines to find the maximum ΔT and the corresponding stoichiometric ratio. Since the concentrations and volumes of the acid and base are known, calculate the number of moles of each and determine the stoichiometric ratio.
- Calculate the heat of reaction using the equation

$$-\Delta H = mC_p\Delta T + C_{cal}\Delta T \qquad \text{units: J}$$

where m is the total mass of the acid and base (both are changing temperature) and ΔT is the maximum ΔT from the graph.
- The **molar enthalpy, $\Delta H°$,** is equal to the ΔH obtained above, divided by the number of moles of acid.

$$\Delta H° = (\Delta H)/(\text{moles of acid}) \qquad \text{units: J mol}^{-1}$$

- Repeat this procedure for the sulfuric acid reactions.

Name: ______________________________

Date: __________ Section: ______________________

Pre-Laboratory Questions: Experiment 23

1. If concentrated acid had been used instead of dilute acid in this experiment, the heat of dilution of the acid would have to be taken into account. Explain.

2. In most thermodynamic calculations, the heat capacity at constant pressure, C_p, is used rather than the heat capacity at constant volume, C_v. Explain why.

3. When 530 mg of a compound was dissolved in a calorimeter containing 5.0 mL of water, the temperature dropped from 25.5 to 22.4°C. The heat capacity of the calorimeter is 55 J/°C. Calculate the mass heat of solution (J g^{-1}) of the compound. What % of the heat is absorbed by the calorimeter?

Name: ______________________________

Date: __________ Section: ______________________

Data Sheet: Experiment 23

Part A: Calorimeter Constant

Initial temperature of room temp. water in calorimeter __________ °C

Initial temperature of hot water in hot water bath __________ °C

First Trial		*Second Trial*	
Time, sec	*Temperature, °C*	*Time, sec*	*Temperature, °C*

Calculations

$T_{mixture}$ (extrapolated from graph) __________ °C

ΔT_{hot} __________ °C

ΔT_{cold} __________ °C

Heat from hot water __________ J

Heat to cold water __________ J

Calorimeter constant __________ $J\ °C^{-1}$

Name: ______________________________

Date: __________ Section: ______________________________

Data Sheet: Experiment 23, p. 2

1. 0.5 mL HCl and 3.5 mL NaOH

Time, sec	____	____	____	____	____
Temperature, °C	____	____	____	____	____
Time, sec	____	____	____	____	____
Temperature, °C	____	____	____	____	____

2. 1.0 mL HCl and 3.0 mL NaOH

Time, sec	____	____	____	____	____
Temperature, °C	____	____	____	____	____
Time, sec	____	____	____	____	____
Temperature, °C	____	____	____	____	____

3. 2.0 mL HCl and 2.0 mL NaOH

Time, sec	____	____	____	____	____
Temperature, °C	____	____	____	____	____
Time, sec	____	____	____	____	____
Temperature, °C	____	____	____	____	____

4. 3.0 mL HCl and 1.0 mL NaOH

Time, sec	____	____	____	____	____
Temperature, °C	____	____	____	____	____
Time, sec	____	____	____	____	____
Temperature, °C	____	____	____	____	____

5. 3.5 mL HCl and 0.5 mL NaOH

Time, sec	____	____	____	____	____
Temperature, °C	____	____	____	____	____
Time, sec	____	____	____	____	____
Temperature, °C	____	____	____	____	____

Name: ______________________________

Date: __________ Section: ______________________________

Data Sheet: Experiment 23, p. 3

Reaction with Sulfuric Acid

1. 0.5 mL H_2SO_4 and 3.5 mL NaOH

Time, sec	____	____	____	____	____
Temperature, °C	____	____	____	____	____
Time, sec	____	____	____	____	____
Temperature, °C	____	____	____	____	____

2. 1.0 mL H_2SO_4 and 3.0 mL NaOH

Time, sec	____	____	____	____	____
Temperature, °C	____	____	____	____	____
Time, sec	____	____	____	____	____
Temperature, °C	____	____	____	____	____

3. 2.0 mL H_2SO_4 and 2.0 mL NaOH

Time, sec	____	____	____	____	____
Temperature, °C	____	____	____	____	____
Time, sec	____	____	____	____	____
Temperature, °C	____	____	____	____	____

4. 3.0 mL H_2SO_4 and 1.0 mL NaOH

Time, sec	____	____	____	____	____
Temperature, °C	____	____	____	____	____
Time, sec	____	____	____	____	____
Temperature, °C	____	____	____	____	____

5. 3.5 mL H_2SO_4 and 0.5 mL NaOH

Time, sec	____	____	____	____	____
Temperature, °C	____	____	____	____	____
Time, sec	____	____	____	____	____
Temperature, °C	____	____	____	____	____

Name: ______________________________

Date: __________ Section: ______________________

Data Sheet: Experiment 23, p. 4

6. For each of the above reactions, construct a plot of temperature (y axis) versus time (x axis) and determine ΔT by extrapolation. Attach the plots to the data sheets.

7. Construct a plot of ΔT (y axis) versus volume of acid/base mixed (x axis) for each of the two acids. What is the mole stoichiometry for each of the two acids?

8. Determine the molar heat of neutralization for each of the acids. Assume that the density of the solutions is 1.00 g mL^{-1}, and that $C_p = 4.184$ J mL^{-1} $°C^{-1}$.

9. Compare the calculated $\Delta H°$ values for HCl and H_2SO_4. How do you think they <u>should</u> compare?

Name: ____________________

Date: ________ Section: ____________________

Questions: Experiment 23

1. After a reaction, a calorimeter containing 1.0 mL of aqueous solution experienced a temperature rise of 8.3°C. What was the heat of the reaction? If the amount of reactant corresponded to 0.10 mmol, what is the molar heat of reaction? Assume $C_{cal} = 2.5$ J/°C.

2. The value of $\Delta H°$ is -52.0 kJ mol^{-1} for the reaction

$$NH_3 \text{ (aq)} + HCl \text{ (aq)} \rightarrow NH_4Cl \text{ (aq)}$$

Estimate the increase in temperature that will occur when 1.0 mL of 0.5 M NH_3 in a Styrofoam calorimeter is neutralized by 3.0 mL of a solution containing excess HCl. Assume $C_{cal} = 1.8$ J/°C.

3. Could the same stoichiometry have been obtained if the experiment had been performed with a constant amount of acid and a variable amount of base? Explain. Assume that the calorimeter heat capacity does not vary with volume.

EXPERIMENT 24 Hess's Law: Calculation of a Heat of Reaction

OBJECTIVES

- To determine the heat of reaction of a chemical process.
- To use Hess's Law to indirectly calculate an enthalpy from experimental data.

PRIOR READING

Section 3.3: Weighing
Section 3.4: Liquid Volumes
Section 3.9: Titration
Experiment 23: Introduction

INTRODUCTION

It is difficult or impossible to measure the heat of reaction (ΔH) for many chemical processes. This is due to a number of factors:

a. The reaction of interest may occur along with other side reactions. How would we know how much heat to assign to each reaction?
b. The reaction of interest may be very slow. This would allow some heat to escape, no matter how well the system was insulated.
c. The reaction of interest may only take place at an inconvenient pressure or temperature.
d. The reaction of interest may involve dangerous reagents or products.

In such cases (and many others), it is convenient to calculate the heat of reaction indirectly, using **Hess's Law.**

Enthalpy (H) is defined as the heat content of a given material under a given set of conditions, called a **state.** Since there is only one value of enthalpy for any given state, the enthalpy is one of a number of thermodynamic quantities called **state functions.** Most of the time, we are not interested in any particular value of enthalpy. Instead, we are interested in the change in the enthalpy, ΔH, in going from one state to another, or in going from reactants to products. This is because there is no way of directly measuring an enthalpy under a given set of conditions—we can only measure how much it changes. We must therefore define a starting temperature and pressure, and choose an arbitrary value for the enthalpy of that state. The state chosen to define enthalpy is 298 K and 1 atm pressure, at which the enthalpy of formation of any element **in its normal state** (e.g., oxygen as a diatomic gas, iron as a solid metal, bromine as a diatomic liquid, etc.) is defined as zero. Furthermore, for ions, the enthalpy of formation of the H^+ ion at a concentration of 1 M is defined as zero. All other ion enthalpies are calculated as changes from this standard state.

Hess's Law states that if reactions A and B can be added up to give reaction C, then

$$\Delta H_C = \Delta H_A + \Delta H_B$$

that is, the heats of reaction can also be added up to give the heat of the summed reaction.

In this experiment, the heat of formation of MgO will be determined, using Hess's law. Magnesium metal can be dissolved in acid, according to the equation

$$Mg_{(s)} + 2\ H^+_{(aq)} \rightarrow Mg^{2+}_{(aq)} + H_{2(g)}$$

By definition, the heats of formation of Mg and H_2 are zero, since they are elements in their standard states. The heat of formation is also zero for 1 M H^+, by definition. Since the ΔH for any reaction is simply the heats of the products minus that of the reactants, the heat of formation of Mg^{2+} is simply the heat of this reaction (call it ΔH_A).

$$\Delta H_A = \Delta H\,(\text{products}) - \Delta H\,(\text{reactants})$$

$$\Delta H_A = \Delta H_f\,(Mg^{2+}) + \Delta H_f\,(H_2) - \Delta H_f\,(Mg) - 2\,\Delta H_f\,(1M\ H^+)$$

$$\Delta H_A = \Delta H_f\,(Mg^{2+}) + 0 - 0 - 2(0)$$

$$\Delta H_A = \Delta H_f\,(Mg^{2+})$$

If hydrogen and oxygen gas are put into a calorimeter (called a bomb calorimeter) and ignited, the heat of formation of water can be directly measured.

$$H_{2(g)} + 1/2\ O_{2(g)} \rightarrow H_2O_{(l)}$$

Since hydrogen and oxygen are also elements in their standard states, the heat of formation of water is equal to the heat of this reaction as well (call it ΔH_B):

$$\Delta H_B = \Delta H_f\,(H_2O) - \Delta H_f\,(H_2) - 1/2\ \Delta H_f\,(O_2)$$

$$\Delta H_B = \Delta H_f\,(H_2O) - 0 - 1/2\ (0)$$

$$\Delta H_B = \Delta H_f\,(H_2O)$$

This reaction is too dangerous to perform in a general chemistry laboratory, and we will use the literature value of ΔH_B (the heat of formation of liquid water), -285.8 kJ/mol. The negative sign indicates that heat is given off when water is formed, i.e., the reaction is **exothermic.**

In a similar fashion, the magnesium oxide, MgO, can be dissolved in acid, to yield magnesium ions:

$$MgO + 2\ H^+ \rightarrow Mg^{2+} + H_2O$$

The heat of this reaction (ΔH_C), is once again the heat of the products less that of the reactants:

$$\Delta H_C = \Delta H_f\,(Mg^{2+}) + \Delta H_f\,(H_2O) - \Delta H_f\,(MgO) - 2\ \Delta H_f\,(H^+)$$

Here, the heats of the other reactants and products are not zero, and the heat of formation of MgO must be experimentally determined.

In this particular case, there are two ways to solve for the heat of formation of MgO. The simpler way is to note that we already know ΔH_f for Mg^{2+} (this was ΔH_A), and for H_2O (this was ΔH_B). We can substitute them into the equation for ΔH_C, and solve for ΔH_f MgO. Thus,

$$\Delta H_C = \Delta H_f\,(Mg^{2+}) + \Delta H_f\,(H_2O) - \Delta H_f\,(MgO) - 2\ \Delta H_f\,(1M\ H^+)$$

$$\Delta H_C = \Delta H_A + \Delta H_B - \Delta H_f\,(MgO) - 2(0)$$

$$\Delta H_f\,(MgO) = \Delta H_A + \Delta H_B - \Delta H_C$$

The more general way to solve for ΔH_f MgO is to sum up reactions to give the desired reaction of the formation of MgO from the elements in their standard states:

$$Mg_{(s)} + 1/2\ O_{2(g)} \rightarrow MgO_{(s)}$$

In this case, the following three reactions may be summed:

$$Mg_{(s)} + 2\ H^+_{(aq)} \rightarrow Mg^{2+}_{(aq)} + H_{2(g)} \qquad \Delta H_A$$

$$H_{2(g)} + 1/2\ O_{2(g)} \rightarrow H_2O_{(l)} \qquad \Delta H_B$$

$$Mg^{2+}_{(aq)} + H_2O_{(l)} \rightarrow MgO_{(s)} + 2\ H^+_{(aq)} \qquad -\Delta H_C$$

$$Mg_{(s)} + 1/2\ O_{2(g)} \rightarrow MgO_{(s)}$$

Note that since the third reaction is the reverse of the reaction we had above, we had to change the sign of ΔH_C.

When the three reactions are added up, their sum is the desired reaction for the formation of MgO. Hess's Law tells us that the heat of this reaction is the sum of the three heats:

$$\Delta H_f\ (MgO) = \Delta H_A + \Delta H_B - \Delta H_C$$

which is, of course, precisely the same result as we obtained before.

For calculating molar enthalpies, a record of the temperature is plotted versus time (see Experiment 23). A straight line is drawn through the relatively linear part of the data after the temperature change has peaked and begins to return to room temperature. Generally there are a few points above or below the post-mixing line, which should be ignored in drawing the line, as they represent incomplete reaction or incomplete mixing (see Figure E24.1). The line is extended to the time of mixing (0 minutes) to extrapolate the temperature change that would have been observed had an instantaneous reaction taken place. The temperature at which the post-mixing line intersects the time of mixing is $T_{mixture}$ for the reaction. ΔT is obtained from this plot.

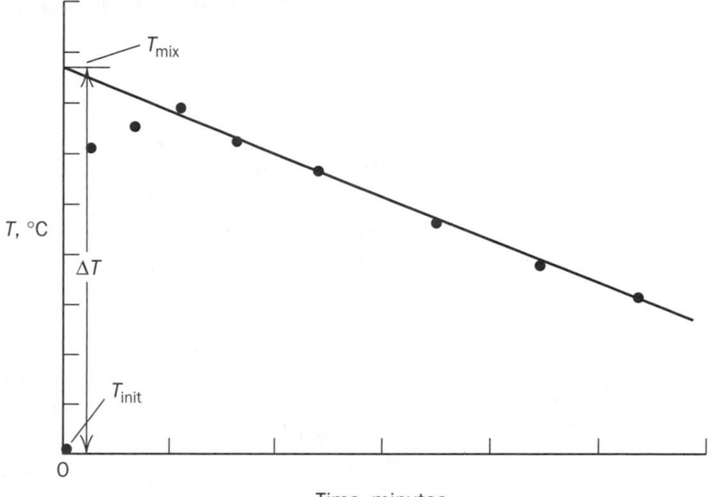

Figure E24.1 *Extrapolation of ΔT.*

In an ideal system, all the heat produced by the reaction would be taken up by the solution. In reality, some of the heat produced is lost, heating up the apparatus (the test tube, thermometer, and Styrofoam block). This should be taken into account in the calculations.

We therefore need to know how much energy is required to raise the temperature of the apparatus by 1°C. This can be determined by performing a reaction that releases a known amount of heat into the solution (and thereby, the apparatus). By measuring the ΔT of the solution, the amount of heat taken up by the solution can be calculated. The remainder must have been absorbed by the apparatus or lost to the surroundings.

The simplest "reaction" is to mix equal quantities of hot and cold water. If the system absorbs no heat, the equilibrium temperature of the mixture would be midway between the hot and cold temperatures. In reality, the apparatus absorbs some heat. It is initially at the same temperature as the cold water, and absorbs some heat from the hot water when it is added. Therefore, the equilibrium temperature will be somewhat less than halfway between the hot and cold water temperatures.

EXPERIMENTAL PROCEDURE

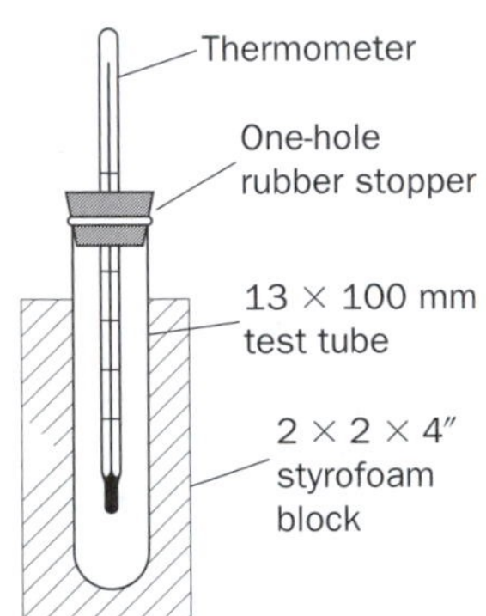

Figure E24.2 *Apparatus for measuring heat of reation.*

Preparation of Calorimeter

- Obtain a 2 × 2 × 4″ block of Styrofoam, and drill a hole in it to hold a 13 × 100 mm test tube. Insert a one-hole rubber stopper, fitted with a thermometer, in the tube. The stopper should be briefly removed when adding chemicals to the tube, and then immediately replaced. The apparatus is shown in Figure E24.2.

The Styrofoam block cools off slowly. This tends to add heat to later runs. To avoid this, bring the block back to room temperature by inserting a test tube of room temperature water for several minutes after each run.

If two thermometers are used in this experiment, check them against each other by inserting them both in room temperature water and comparing the readings. Adjust the recorded temperature of the hot water for any difference.

If only one thermometer is used, it should be inserted into a beaker of hot water, then wiped dry before inserting it into the hot water used in the experiment. Similarly, it should be inserted in a beaker of room temperature water, then wiped dry before inserting it in the room temperature water used in this experiment. This prevents the loss or gain of heat in the next solution. Since we are only using small volumes, a significant amount of heat can be transferred to or from the thermometer if these steps are not followed.

Part A: Determining the Calorimeter Constant

- Pipet 2.00 mL (±0.01 mL) of room temperature water into a 13 × 100 mm test tube and place it in the Styrofoam block. Insert the stopper-thermometer assembly.
- Obtain 400–500 mL of 45–50°C water in a 600 mL beaker. Pipet 2.00 mL of this hot water into a second test tube and place it in the hot water bath for several minutes. Warm the thermometer in the hot water bath, wipe it

dry, and then insert it into the hot water in the test tube. Record the temperature of the hot water once it stabilizes. Similarly, record the temperature of the room temperature water in the Styrofoam block test tube.

- Remove the stopper momentarily, tilt the Styrofoam block and quickly pour the hot water in the test tube into the room temperature water, noting the time of mixing. Immediately re-insert the stopper-thermometer assembly. Record the temperature to ±0.1°C every 30 sec for 5 minutes. Discard the contents, bring the Styrofoam block back to room temperature and repeat the run.

Part B: Determining the Heat of Formation of Mg^{2+}

- Add 2.00 mL (±0.01 mL) of standardized 1.00 M HCl to the test tube, and place it in the Styrofoam block. Insert the stopper-thermometer assembly. After waiting 15 minutes for temperature equilibration, measure the temperature of the solution.
- Weigh 8–10 mg (±1 mg) of sanded magnesium ribbon, remove the stopper, and add the magnesium to the acid solution. Immediately, loosely re-stopper the tube. Swirling the mixture, take temperature readings every 15 sec until the temperature begins to fall, and every 30 sec thereafter for at least 3 minutes.
- Complete the data table showing your results. Collect the solution in a waste flask. Bring the block back to room temperature, and if time permits, perform a duplicate trial.

Part C: Heat of Reaction of MgO with Acid

- Add 2.00 mL (±0.01 mL) of standardized 1.0 M HCl to the test tube, and place it in the Styrofoam block. Insert the stopper-thermometer assembly. After waiting 15 minutes for temperature equilibration, measure the temperature of the solution.
- Weigh about 35 mg (±1 mg) of magnesium oxide, remove the stopper, and add this to the acid solution. Immediately, loosely stopper the tube. **Make sure that all the magnesium oxide is in the acid solution. You may need to tip the test tube and block so that the acid rinses the walls of the test tube.** Swirling the mixture, take temperature readings every 15 sec until the temperature begins to fall, and every 30 sec thereafter for at least 3 minutes.
- Complete the data table showing your results. Bring the Styrofoam block back to room temperature, and if time permits, perform a duplicate run. Collect the solution in a waste flask.

Calculations

For all runs, determine ΔT for each reaction by plotting temperature (y axis) versus time (x axis) and extrapolating back to time zero as shown in Figure E24.1 to find $T_{mixture}$ and ΔT. For all solutions, Cp is 4.184 J g^{-1} °C^{-1} and the density is 1.00 g $mL^{-}1$.

To Determine the Calorimeter Constant:

- ΔT_{cold} is the ΔT of the cold water (and the apparatus):

$$\Delta T_{cold} = T_{mixture} - T_{original\ of\ cold\ water,}\ °C$$

ΔT_{hot} is the ΔT of the hot water:

$$\Delta T_{hot} = T_{mixture} - T_{\text{original of the hot water,}} \ ^{\circ}C$$

(Note that ΔT_{cold} will be positive and ΔT_{hot} will be negative.)
C_{cal} is the calorimeter constant.

- Since this is an well-insulated system, the heat lost by the hot water plus that gained by the cold water and apparatus should add up to zero.

$$\text{heat lost by hot water} = m_{hot}\, C_p\, \Delta T_{hot}$$

$$\text{heat absorbed by cold water} = m_{cold}\, C_p\, \Delta T_{cold}$$

$$\text{heat absorbed by apparatus} = C_{cal}\, \Delta T_{cold}$$

where m_{hot} and m_{cold} are the masses of the hot and cold water, respectively. Thus,

$$m_{hot}\, C_p\, \Delta T_{hot} + m_{cold}\, C_p\, \Delta T_{cold} + C_{cal}\, \Delta T_{cold} = 0$$

Solve this equation for C_{cal}.

To Determine the Molar Enthalpies, ΔH°_A and ΔH°_C

- Determine the enthalpy of reaction for each Mg and MgO trial using the equation

$$-\Delta H = m_{acid}\, C_p\, \Delta T + C_{cal}\, \Delta T$$

- The molar enthalpy is calculated from:

$$\Delta H^{\circ} = (\Delta H)/(\text{moles of limiting reagent})$$

- Using the literature value for ΔH°_B and the calculated values for ΔH°_A and ΔH°_C, calculate the molar enthalpy of formation, ΔH°_f of MgO.

Name: ______________________________

Date: __________ Section: ______________________________

Pre-Laboratory Questions: Experiment 24

1. Calculate the heat of formation for carbon monoxide

$$C(graphite) + 0.5\ O_{2(g)} \rightarrow CO_{(g)}$$

from the following data:

$$CO_{(g)} + 0.5\ O_{2(g)} \rightarrow CO_{2(g)} \qquad \Delta H^\circ = -283.0\ kJ$$

$$C(graphite) + O_{2(g)} \rightarrow CO_{2(g)} \qquad \Delta H^\circ = -393.5\ kJ$$

2. Assuming that you used 10.0 mg of Mg and 30.0 mg of MgO in this experiment, calculate the number of moles of reagents in each reaction and determine whether the acid or the magnesium (Part B) or magnesium oxide (Part C) is the limiting reagent in each part.

3. Why should the test tube be as dry as possible at the beginning of each trial?

Name: ____________________

Date: ________ Section: ____________________

Data Sheet: Experiment 24

Part A: Calorimeter Constant

Initial temperature of room temp. water in calorimeter ________ °C

Initial temperature of hot water in hot water bath ________ °C

First Trial		*Second Trial*	
Time, sec	*Temperature, °C*	*Time, sec*	*Temperature, °C*
________	________	________	________
________	________	________	________
________	________	________	________
________	________	________	________
________	________	________	________
________	________	________	________
________	________	________	________
________	________	________	________
________	________	________	________
________	________	________	________
________	________	________	________
________	________	________	________

Calculations

$T_{mixture}$ (extrapolated from graph) ________ °C

ΔT_{hot} ________ °C

ΔT_{cold} ________ °C

Heat from hot water ________ J

Heat to cold water ________ J

Calorimeter constant ________ $J\ °C^{-1}$

Name: ____________________

Date: ________ Section: ____________________

Data Sheet: Experiment 24, p. 2

Part B: Heat of Formation of Mg^{2+}

Initial temperature of 1.0M HCl in calorimeter ________ °C

Mass of magnesium ________ mg

First Trial		*Second Trial*	
Time, sec	*Temperature, °C*	*Time, sec*	*Temperature, °C*
________	________	________	________
________	________	________	________
________	________	________	________
________	________	________	________
________	________	________	________
________	________	________	________
________	________	________	________
________	________	________	________
________	________	________	________
________	________	________	________
________	________	________	________
________	________	________	________

Calculations

$T_{mixture}$ (extrapolated from graph) ________ °C

ΔT ________

Heat of reaction ________ kJ

Moles of magnesium ________

Molar heat of reaction, ΔH_A ________ kJ mol^{-1}

Name: ______________________________

Date: __________ Section: ______________________

Data Sheet: Experiment 24, p. 3

Part C: Heat of Reaction of MgO with Acid

Initial temperature of 1.0 M HCl in calorimeter ____________ °C

Mass of MgO ____________ mg

First Trial		*Second Trial*	
Time, sec	*Temperature, °C*	*Time, sec*	*Temperature, °C*
________	________	________	________
________	________	________	________
________	________	________	________
________	________	________	________
________	________	________	________
________	________	________	________
________	________	________	________
________	________	________	________
________	________	________	________
________	________	________	________
________	________	________	________
________	________	________	________

Calculations

$T_{mixture}$ (extrapolated from graph) ____________ °C

ΔT ____________

Heat of reaction ____________ kJ

Moles of MgO ____________

Molar heat of reaction, ΔH_C ____________ kJ mol^{-1}

Heat of Formation of MgO = $\Delta H_A + \Delta H_B - \Delta H_C$ ____________ kJ mol^{-1}

Name: ______________________________

Date: __________ Section: ______________________________

Questions: Experiment 24

1. In this experiment, we have accounted for the heat capacity of the calorimeter in calculating the enthalpy of reaction. What effect would ignoring this have on the calculation?

2. Hess's Law is especially convenient when one wishes to obtain the heat of a reaction that cannot be performed in the laboratory. From the following data, calculate the enthalpy change for the transformation of graphite into diamond.

$$\text{C (graphite)} + O_2 \rightarrow CO_2 \qquad \Delta H^\circ = -390.5 \text{ kJ}$$

$$\text{C (diamond)} + O_2 \rightarrow CO_2 \qquad \Delta H^\circ = -393.5 \text{ kJ}$$

3. In theory, could one transform graphite into diamond by application of the ΔH difference between the two (calculated above)? What other factors must be taken into account?

Experiment 25 Colligative Properties: Freezing Point Depression

OBJECTIVES

- To determine a freezing point.
- To determine a molal freezing point depression constant.

PRIOR READING

Section 3.8: Solution Preparation

INTRODUCTION

When a pure solid is dissolved in a solvent, several physical properties of that solvent change in a way that depends only on the relative amounts of the solute and solvent present. Such properties are called **colligative properties.** There are four common colligative properties: boiling point elevation, freezing point depression, vapor pressure reduction, and osmotic pressure. The most familiar of these is the freezing point depression. If salt is placed on an icy roadway, the freezing point of the water will be lowered, and the ice will melt, depending only on the amount of salt that has been added.

Pure water freezes at 0.00°C. If 1.00 mole of nonvolatile dissolved particles, molecules, or ions are added to 1.00 kg of water, the freezing point of water drops from 0.00 to −1.86°C. This drop is called the **freezing point depression constant** for water. Other liquids have different constants. The relationship between the concentration of solute particles and the freezing point depression is expressed as

$$\Delta T_{fp} = k_{fp}\, m$$

where ΔT_{fp} is the actual change in the freezing point, k_{fp} is the freezing point depression constant, and m is the **molality** of the solution. Molality (m) is defined as the number of moles of solute in one kilogram of solvent. Thus, the freezing point depression constant for water is $-1.86°C/m$.

Different solutes affect the freezing point depression in different ways. One mole of a molecular material (for example, sugar) will lower the freezing point of 1.00 kg of water by 1.86°C. One mole of NaCl, on the other hand, will lower the freezing point of 1.00 kg of water by about 3.72°C, twice as much. This is because NaCl will dissociate in water to form Na^+ and Cl^- ions, and therefore, twice as many particles are present in solution.

In Part A of this experiment, the freezing point of a pure solvent, cyclohexane, will be determined. In Part B, a molecular solid, naphthalene, will be added to the cyclohexane, and the freezing point of the mixture will be measured. A plot of temperature (y axis) versus time (x axis) is used for this purpose. As shown in Figure E25.1a, the temperature will level off, reaching a constant value (the freezing point temperature), as the solution freezes. (In some cases, the solution may temporarily cool below the freezing point before crystallization takes place. Such solutions are said to be supercooled, as shown in the Figure E25.1b.) Then as crystallization begins, the temperature jumps back up to the true freezing point temperature (due to the release of heat during crystallization). This complication is avoided by gentle stirring.

Since the molality of the solution is known, by calculating the freezing point lowering, the freezing point depression constant, k_{fp} can be determined. This can be used to determine the molar mass of an unknown by determining how much it lowers the freezing point of cyclohexane.

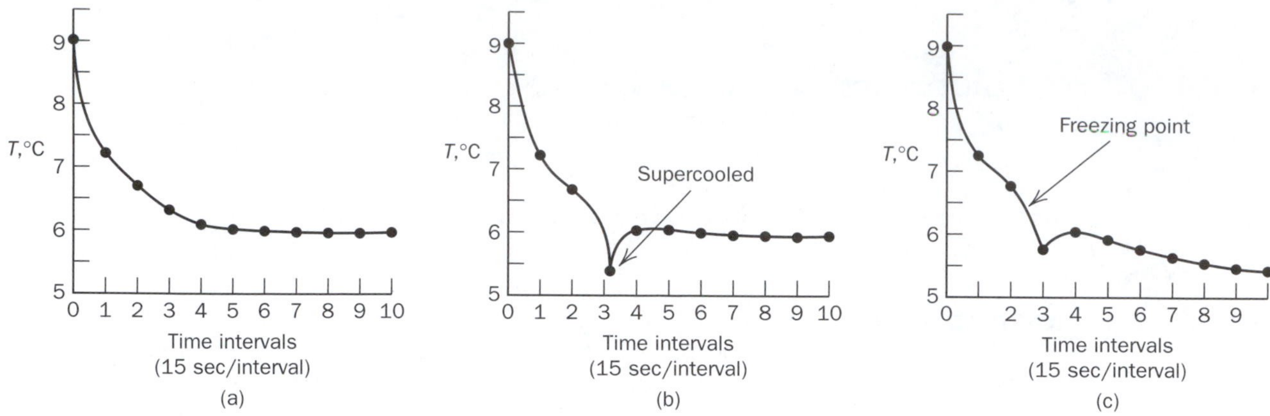

Figure E25.1 *Cooling curves of a pure solvent (a, b) and of a solution (c).*

EXPERIMENTAL PROCEDURE

> **NOTE: *This experiment is most conveniently done in partners. Cyclohexane is extremely flammable. No flames are allowed during this experiment. Record all measurements and observations on the data sheet.***

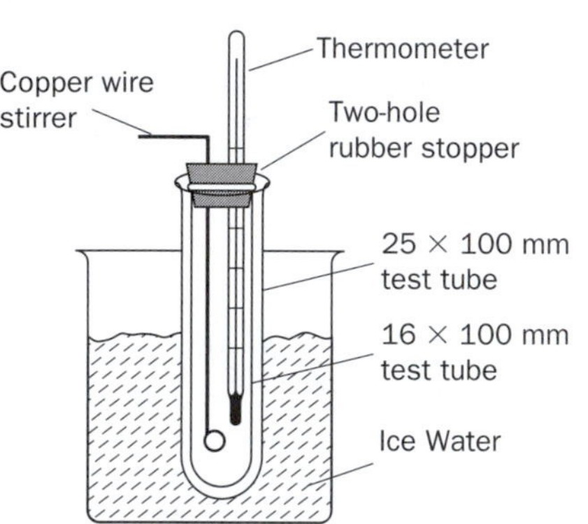

Figure E25.2 *Apparatus for freezing point depression measurement.*

Part A: Freezing Point of Pure Cyclohexane

- Prepare a freezing bath mixture of ice and water in a 250 mL beaker. Obtain a 16 × 100 mm test tube. Pipet 5.00 mL (±0.01 mL) of cyclohexane (density = 0.779 g/mL) in the test tube. Insert a stopper fitted with a thermometer and copper-wire stirrer in the test tube (See Figure E25.2).
- Obtain a 25 × 100 mm test tube and put a little water in it. Place the test tube containing the cyclohexane in the larger test tube and place the whole assembly in the ice/water bath. While one partner gently and constantly agitates the solution, the other should take temperature readings every 15 sec until the cyclohexane is completely solid.
- Melt the cyclohexane, and repeat the procedure a second time. The second test tube with water slows the rate of cooling, ensuring more accurate results. If the temperature decrease is slow, take readings less frequently.

Part B: Freezing Point Depression of Cyclohexane Solution

- Weigh approximately 60 mg of naphthalene (MW = 128 g/mol) to the nearest mg. Remove the stopper assembly (avoiding loss of adhering liquid) from the test tube just far enough to add the sample of naphthalene to the cyclohexane. If necessary, warm the solution in order to allow the naphthalene to completely dissolve.
- Repeat the previous procedure to determine the freezing point of this solution. Melt the mixture and discard it according to instructions. Rinse the test tube, thermometer and stirrer with 1–2 mL of fresh cyclohexane and discard. Remove as much of the cyclohexane as possible.

Part C: Molecular Weight Determination of an Unknown

- Pipet a fresh 5.00 mL (±0.01 mL) portion of cyclohexane into the test tube. Add 60 mg of an unknown, weighed to the nearest mg, and insert this test tube into the larger one as before. Determine the freezing point of the resulting solution.
- Melt and repeat the determination. Melt the mixture and discard it according to instructions.

Calculations

- Plot cooling curves for all trials of Parts A, B, and C as shown in Figure E25.1. The curve for part A should look like that of E25.1a, while the curves for B and C should look like that in E25.1c. As a solution freezes, it is the solvent that is freezing. The remaining liquid becomes more concentrated in solute and thus has a lower freezing point. The temperature continues to drop during freezing, reflecting this, but not as quickly, thus there is a break in the cooling curve.
- To obtain the true freezing point, draw straight lines through the two portions of the cooling curve (see Fig. E25.1c). The temperature where these two lines intersect is the true freezing point temperature. Determine the freezing points for all trials.
- Calculate the change in the freezing point, ΔT_{fp} for Part B. Calculate the molality of the naphthalene/cyclohexane solution (mol of naphthalene/kg of cyclohexane, and determine the freezing point depression constant, k_{fp}, °C/m. Calculate the average of the two trials.
- For the unknown, use the equation in the introduction, plus the average k_{fp} value determined from the cyclohexane/naphthalene data to calculate the molality, m (moles of solute per kg of solvent). Since the number of kg of solvent used is known, the number of moles of solute can be determined. Since the mass of the solute is known, the molar mass can be calculated.

Name: ______________________________

Date: __________ Section: ______________________________

Pre-Laboratory Questions: Experiment 25

1. If a solution contains 0.517 g naphthalene (MW = 128 g/mol) dissolved in 24.3 g of cyclohexane, calculate the molality of the solution, *m*.

2. Naphthalene has a molal freezing point constant of −6.90°C/*m* and a normal freezing point of 80.20°C. Calculate the freezing point of a solution prepared by dissolving 0.0108 mol biphenyl (MW = 154.2 g/mol, a molecular solid) in 210.5 g of naphthalene.

3. In the wintertime, it is common to sprinkle salt ($CaCl_2$) on icy roads and driveways. Why does one do this?

Name: ______________________________

Date: __________ Section: ______________________________

Data Sheet: Experiment 25

Part A: Freezing Point of Pure Cyclohexane

1. Volume of cyclohexane ________ mL Mass of cyclohexane ________ g

2. Freezing Point Data

Trial 1		*Trial 2*	
Time, sec	*Temperature, °C*	*Time, sec*	*Temperature, °C*
________	________	________	________
________	________	________	________
________	________	________	________
________	________	________	________
________	________	________	________
________	________	________	________
________	________	________	________
________	________	________	________
________	________	________	________
________	________	________	________
________	________	________	________
________	________	________	________
________	________	________	________
________	________	________	________
________	________	________	________
________	________	________	________
________	________	________	________

Graph T versus time for each trial and report the freezing temperature of cyclohexane:

Trial 1 __________ °C Trial 2 __________ °C

Name: ____________________

Date: ________ Section: ____________________

Data Sheet: Experiment 25, p. 2

Part B: Molal Freezing Point Constant for Cyclohexane

1. Mass of naphthalene ____________ g

2. Freezing Point Data

Trial 1		*Trial 2*	
Time, sec	*Temperature, °C*	*Time, sec*	*Temperature, °C*
______	______	______	______
______	______	______	______
______	______	______	______
______	______	______	______
______	______	______	______
______	______	______	______
______	______	______	______
______	______	______	______
______	______	______	______
______	______	______	______
______	______	______	______
______	______	______	______
______	______	______	______
______	______	______	______
______	______	______	______

Graph T versus time for each trial and report the freezing temperature of the mixture:

Trial 1 ____________ °C Trial 2 ____________ °C

Calculate the molality of naphthalene/cyclohexane solution: _____ mol kg^{-1}

Calculate the molal freezing point constant of cyclohexane, °C/m for each trial:

Trial 1 ____________ °C/m Trial 2 ____________ °C/m

Average ____________ °C/m

Name: ______________________________

Date: __________ Section: ______________________________

Data Sheet: Experiment 25, p. 3

Part C: Molecular Mass Determination of Unknown

1. Mass of unknown __________ g

2. Volume of cyclohexane _________ mL Mass of cyclohexane ________ g

3. Freezing Point Data

Trial 1		*Trial 2*	
Time, sec	*Temperature, °C*	*Time, sec*	*Temperature, °C*

Graph *T* versus time for each trial and report the freezing temperature of the mixture:

Trial 1 ______________ °C Trial 2 ______________ °C

Calculate the molality of naphthalene/cyclohexane solution _____ mol kg^{-1}

Calculate the molar mass of the unknown

Trial 1 ______________ g mol^{-1} Trial 2 ______________ g mol^{-1}

Average ______________ g mol^{-1}

Name: ______________________________

Date: __________ Section: ______________________________

Questions: Experiment 25

1. The normal freezing point of water is 0.00°C. What is the freezing point of a solution containing 200.0 mg of ethylene glycol (MW = 62.07 g/mol, molecular compound) dissolved in 1.0 g of water? Ethylene glycol is the main ingredient in permanent antifreeze.

2. A student removes the thermometer and stirring wire in Part B of this experiment to add the naphthalene, and wipes them off, losing 0.50 mL of cyclohexane. What percent error will this introduce if the original 5.00 mL is used in the calculations? Explain.

3. Solvents 'R Us checks the purity of its solvents before selling them by determining the freezing point of each lot. Pure carbon tetrachloride, CCl_4, freezes at −23.00°C and has a k_{fp} of −29.8°C/*m*. The latest lot has a freezing point of −23.31°C. What is the molality of impurities in the solvent? If the impurity is largely hexachloroethane, C_2Cl_6, how many grams of this impurity are there per kg of CCl_4? Calculate the mass % purity:

Mass % purity = 100[(1000g solvent)/(1000. g solvent + g impurities per kg solvent)]

Experiment 26 Kinetics: The Iodine Clock Reaction

OBJECTIVES

- To determine the kinetic parameters (rate, order, rate constant) of a reaction.
- To examine the effect of a catalyst on the rate of reaction.

PRIOR READING

Section 2.4: Graphing of Data
Section 3.4: Liquid Volumes

INTRODUCTION

In a kinetic study, one investigates the rate at which a reaction occurs and how that rate is affected by changes in the temperature, pressure, concentration, and the presence of a catalyst. Once these factors are understood, the reaction conditions may be optimized to speed up or slow down a reaction as desired.

In this experiment, the reaction between the iodide ion, I^-, and the peroxydisulfate ion, $S_2O_8^{2-}$, is studied:

$$3\ I^-_{(aq)} + S_2O_8^{2-}{}_{(aq)} \rightarrow I^-_{3(aq)} + 2\ SO_4^{2-}{}_{(aq)}$$

The reaction kinetics will be investigated using the **method of initial rates.** The amount of time that it takes for a certain fixed amount of the reactants to react is measured. By seeing how this time interval changes under various conditions of concentration and temperature, the rate law can be determined.

For reactions occurring in solution, the rate depends upon a number of factors, the most important being the concentrations of the reactants and the **rate constant, *k*.** The rate constant, in turn, depends on the reaction being studied, but for any particular reaction, varies only with temperature. The rate is calculated as a change in concentration over a time interval ($\Delta[X]/\Delta t$) and has the units mol L^{-1} sec^{-1}. During each of several trials under different conditions, we will determine the length of time (Δt) required for the concentration of one of the reactants ($S_2O_8^{2-}$) to change by a certain fixed amount ($\Delta[X]$). This is similar to timing runners in a race. By measuring the amount of time it takes each runner to run 1000 meters, we can calculate their rates of speed [Δ(distance)/Δ(time)], units of meters sec^{-1}.

Determining the Order of Reaction

The kinetics of any reaction can be approximated using the general rate law:

$$\text{Rate} = k[\text{reactant } 1]^a[\text{reactant } 2]^b[\text{reactant } 3]^c \ldots$$

where *a*, *b*, and *c* are called the **orders of reaction** with respect to reactant 1, reactant 2, and reactant 3, respectively. In the case of this experiment's reaction of iodide and peroxydisulfate ions, the general rate law becomes:

$$\text{Rate} = k[I^-]^a[S_2O_8^{2-}]^b$$

This rate law is fairly complex, having five variables (the two concentrations, the orders of reaction *a* and *b*, and the rate constant, *k*). We now perform two experimental trials at the same temperature. The first trial is run with some particular $[I^-]$ and some particular $[S_2O_8^{2-}]$, and the time that it takes for the reaction to occur (the rate) is measured. In the second trial, we change the concentration of only one of the reactants (I^- here), for example by halving it, and measure the rate again.

For the first trial, the rate equation is

$$R_1 = k[I^-]_1^a[S_2O_8]^b$$

and for the second trial:

$$R_2 = k[I^-]_2^a[S_2O_8]^b$$

We know that the $[I^-]$ in run 1 is twice that of run 2 (i.e., $[I^-]_1 = 2\,[I^-]_2$). Recalling that $[S_2O_8^{2-}]$ is the same in both runs, we can substitute $2\,[I^-]_2$ for $[I^-]_1$ in the first rate equation, obtaining

$$R_1 = k(2\,[I^-]_2)^a[S_2O_8]^b$$

$$R_2 = k[I^-]_2^a[S_2O_8]^b$$

If we then divide the first rate equation by the second, the k and the $[S_2O_8^{2-}]^b$ terms cancel out, and we get the following expression:

$$\frac{R_1}{R_2} = (2)^a$$

Since both R_1 and R_2 have been measured, this equation has only one unknown and can be solved for ***a***, the order of the reaction with respect to $[I^-]$.

$$a = \frac{\ln (R_1/R_2)}{\ln 2}$$

The order **b** can be calculated in a similar manner, by comparing the rates of two trials where only the $[S_2O_8^{2-}]$ changes.

Determining the Rate Constant

Once the orders of reaction ***a*** and ***b*** are known, the original rate equation has only one unknown—the rate constant, k, which can then be solved for. Remember that the rate constant for any particular reaction depends only on the temperature. This dependency may be studied by performing similar trials (with the same initial concentrations of all reactants) at several different temperatures. The relationship between k and temperature has the form

$$k = Ae^{-E_a/RT}$$

This equation is known as the **Arrhenius equation,** where R is the gas law constant, T is temperature in K, E_a is the **activation energy** of the reaction, and A is the Arrhenius proportionality factor (unique to each reaction). The activation energy is the minimum amount of energy required for a collision between reactants to result in products.

A more useful form of the Arrhenius equation can be obtained by taking the ln of both sides and comparing it to the equation of a straight line:

$$\ln k = \ln A - \frac{E_a}{RT}$$

Since we are interested in how k varies with T, we can rearrange this to become

$$\ln k = -\frac{E_a}{R}\left(\frac{1}{T}\right) + \ln A$$

$$y = \quad m \quad x + b$$

Thus, a plot of ln k (y axis) versus $1/T$ (x axis) gives a straight line with a slope equal to $-E_a/R$ and a y intercept equal to ln A.

In this experiment, we will determine the rate law for the reaction between iodide and peroxydisulfate. We will also examine the effect of a catalyst on the rate of the reaction. Many metal cations act as catalysts in reactions because they can exist in solution in two or more different oxidation states. They can easily gain an electron from one reactant and pass it on to another. The effect of the catalyst will be calculated by comparing trials with identical concentrations and temperatures, with and without the catalyst. Catalysts are very important in industrial applications, as they can greatly speed up a process or allow it to run at lower temperatures or pressures.

The main difficulty in performing a kinetics experiment is to devise a way to measure the rate of the reaction, i.e., a way to signal when the reaction has progressed to a certain point. If this can be done accurately, then $\Delta[S_2O_8^{2-}]$ will be constant. In that case, all we need do is measure the time it takes for each trial to reach that point. We can make the reaction end point obvious by adding a combination of thiosulfate ($S_2O_3^{2-}$) and starch as an indicator.

When thiosulfate and starch are present, the following three reactions occur:

$$3\, I^-_{(aq)} + S_2O_8^{2-}{}_{(aq)} \rightarrow I^-_{3\,(aq)} + 2\, SO_4^{2-}{}_{(aq)} \tag{1}$$

$$I^-_{3\,(aq)} + 2\, S_2O_3^{2-}{}_{(aq)} \rightarrow 3\, I^-_{(aq)} + S_4O_6^{2-}{}_{(aq)} \tag{2}$$

$$\text{starch}_{(aq)} + I^-_{3\,(aq)} \rightarrow \text{starch} - I_3^- \text{ complex} \tag{3}$$

Reaction 2 occurs as long as thiosulfate is present in solution. Thiosulfate reacts with the triiodide ions (I_3^-) produced in reaction 1, converting them back to iodide ion. When the thiosulfate is used up, triiodide will start to accumulate, and reaction 3 occurs, producing a **dark blue** starch–triiodide complex. Thus, two colorless solutions will be mixed together, and after some time interval (15 sec to 4 minutes, depending on concentration and temperature), there will be a sharp change in color from colorless to dark blue, indicating that $[S_2O_8^{2-}]$ has reached the desired level. Since the same amount of thiosulfate is added to each run, the same amount of peroxydisulfate is used up in each run, making $\Delta[S_2O_8^{2-}]$ constant for all runs. Since rate = $\Delta[S_2O_8^{2-}]/\Delta$time, if two runs have similar times, their rates are similar. Shorter times indicate faster rates.

EXPERIMENTAL PROCEDURE

> **NOTE: *This experiment is most conveniently performed with partners. The concentrations in the tables are stock concentrations. Calculate the real concentrations of I^- and $S_2O_8^{2-}$, taking into account the dilution factor.***

Table E26.1 Reagent Volumes for Determining the Order with Respect to Iodide

	Reagents Used (mL, ±0.01 mL)					
	0.2%	0.012 M	0.20 M	0.20 M	0.20 M	0.20 M
Trial	Starch	$Na_2S_2O_3$	KI	KNO_3	$(NH_4)_2S_2O_8$	$(NH_4)_2SO_4$
1	0.10	0.20	0.80	0	0.40	0.40
2	0.10	0.20	0.40	0.40	0.40	0.40
3	0.10	0.20	0.20	0.60	0.40	0.40
4	0.10	0.20	0.10	0.70	0.40	0.40

Part A: Determining, the Order of Reaction with Respect to Iodide

Refer to Table E26.1 for the amounts of reagents to be used for each trial. The first four reagents (starch, $Na_2S_2O_3$, KI, and KNO_3) should be pipetted into a 13 × 100 mm test tube (along with a magnetic spin vane, if desired). The other two reagents ($(NH_4)_2S_2O_8$, $(NH_4)_2SO_4$) should be pipetted into a second 13 × 100 mm test tube.

- Pipet the required amounts (from Table E26.1) of the various reagents for Trial 1 into the two 13 × 100 mm test tubes. Pour the contents of the second test tube into the first, and note the time of mixing to the nearest second. Pour the solution back and forth between the test tubes three times to thoroughly mix the two solutions. When the colorless mixture turns dark blue, note the elapsed time.
- Discard the contents of the tube in a proper waste container. Repeat Trial 1, and measure the reaction time once again. Times for the duplicate trials should be within 10% of each other, or else they should be repeated.
- Repeat this procedure for Trials 2–4.

Part B: Determining the Order of the Reaction with Respect to Peroxydisulfate

Refer to Table E26.2 for the amounts of reagents to be used for each trial.

- Follow the same procedure for Trials 5–8, performing each trial twice, or until the reaction times are within 10% of each other.

Table E26.2 Reagent Volumes for Determining the Order with Respect to Peroxydisulfate

	Reagents Used (mL, ±0.01 mL)					
	0.2%	0.012 M	0.20 M	0.20 M	0.20 M	0.20 M
Trial	Starch	$Na_2S_2O_3$	KI	KNO_3	$(NH_4)_2S_2O_8$	$(NH_4)_2SO_4$
5	0.10	0.20	0.40	0.40	0.80	0.00
6	0.10	0.20	0.40	0.40	0.40	0.40
7	0.10	0.20	0.40	0.40	0.20	0.60
8	0.10	0.20	0.40	0.40	0.10	0.70

Part C: Determining the Activation Energy of the Reaction (Trials 9–12)

> ***Refer to Trial 7 for the amounts of reagents to be used for each temperature run.***

- Follow the same procedure for Trials 9–12. Do Trial 9 at about 10° below room temperature (in a 600 mL beaker of cold water), Trial 10 at room temperature (already done in Trial 7), and Trials 11 and 12 at 10° and 20° above room temperature (in a 250 mL beaker of warm water).
- Insert the two test tubes with the pipetted reagents into the water bath for 5 minutes prior to mixing, to allow the temperatures to equilibrate. After mixing, keep the test tube with the mixture in the water bath, and note the elapsed time when it turns blue. Perform each trial twice, or until the reaction times are within 10% of each other. Record the temperature to ±0.1°C.

Part D: Determining the Effect of a Catalyst (Runs 13–16)

> ***Refer to Trials 5–8 for the amounts of reagents to be used for each run.***

- Repeat the procedure for Trials 5–8, with the following change. Add 1 drop of the catalyst, 0.0020 M $Cu(NO_3)_2$, to the test tube containing the peroxydisulfate prior to mixing. Perform each trial twice, or until the reaction times are within 10% of each other.

Calculations

> **NOTE:** ***Before coming to laboratory, calculate the initial [I^-] and [$S_2O_8^{2-}$] using the equation $M_1V_1 = M_2V_2$ for all runs, where V_1 is the volume of I^- or $S_2O_8^{2-}$ used and V_2 is the total volume. Record these on the data sheets.***

- Calculate the change in the $[S_2O_8^{2-}]$ using the equation below. The change is the same for all runs.

$$\Delta[S_2O_8^{2-}] = \frac{M(S_2O_3^{2-}) \cdot V(S_2O_3^{2-}) \cdot \dfrac{1 \text{ mol } S_2O_8^{2-}}{2 \text{ mol } S_2O_3^{2-}}}{V_{\text{total}}}$$

$V_{\text{total}} = 1.9 \times 10^{-3}$ L for all runs.

- Calculate the rate ($\Delta[S_2O_8^{2-}]/\Delta t$) for each run, using the average time for the duplicate trials.
- Calculate a for trials 1 and 2, 2 and 3, and 3 and 4. Determine the average value.
- Calculate b similarly, for runs 5–8.
- Calculate k for all runs (k should be approximately the same for all uncatalyzed runs at room temperature).

- If desired, the energy of activation may be determined in the following way, using the Arrhenius equation (which is the equation of a straight line).

$$\ln k = \frac{E_a}{R}\left(\frac{1}{T}\right) + \ln A$$

For Trials 9–12, plot ln k (y axis) versus $1/T$, K^{-1}, (x axis). The slope is equal to $-Ea/R$.
- Note the effect of adding a catalyst by comparing the average k for Trials 5–8 with that of the catalyzed Trials 13–16. If the ratio $k_{cat}/k_{uncat} = 2$, the catalyst causes the rate to double.

Name: ______________________________

Date: __________ Section: ______________________________

Pre-Laboratory Questions: Experiment 26

1. Two kinetics trials are performed at the same temperature with $[S_2O_8]^{2-}$ held constant and the $[I^-]$ in the first trial being four times that in the second. If the reaction times are 20 sec (Trial 1) and 82 sec (Trial 2), what is the order with respect to iodide ion?

2. How could one determine that the Cu^{2+} ion in the catalyst is the active species, rather than the nitrate ion?

3. Explain why the relationship

$$\Delta[S_2O_8^{2-}] = \frac{M(S_2O_3^{2-}) \cdot V(S_2O_3^{2-}) \cdot \dfrac{1 \text{ mol } S_2O_8^{2-}}{2 \text{ mol } S_2O_3^{2-}}}{V_{total}}$$

is correct.

Name: ____________________

Date: ________ Section: ____________________

Data Sheet: Experiment 26

$\Delta[S_2O_8^{2-}]$ ____________ mol L^{-1} (same for all runs)

A. Order of Reaction with Respect to Iodide

Trial	*$[I^-]$ mol/L*	*$[S_2O_8^{2-}]$ mol/L*	*Time 1, sec*	*Time 2, sec*	*Rate* *$\Delta[S_2O_8^{2-}]/\Delta t$ mol L^{-1} s^{-1}*
1	______	______	______	______	______
2	______	______	______	______	______
3	______	______	______	______	______
4	______	______	______	______	______

B. Order of Reaction with Respect to Peroxydisulfate

Trial	*$[I^-]$ mol/L*	*$[S_2O_8^{2-}]$ mol/L*	*Time 1, sec*	*Time 2, sec*	*Rate* *$\Delta[S_2O_8^{2-}]/\Delta t$ mol L^{-1} s^{-1}*
5	______	______	______	______	______
6	______	______	______	______	______
7	______	______	______	______	______
8	______	______	______	______	______

C. Determination of Activation Energy

Trial	*Temperature, °C*	*Time 1, sec*	*Time 2, sec*	*Rate* *$\Delta[S_2O_8^{2-}]/\Delta t$ mol L^{-1} s^{-1}*
9	______	______	______	______
10	______	______	______	______
11	______	______	______	______
12	______	______	______	______

Name: ______________________

Date: ________ Section: ______________________

Data Sheet: Experiment 26, p. 2

D. Run With Catalyst

Trial	*Time 1, sec*	*Time 2, sec*	*Time Uncatalyzed, sec (from Part B)*
13	________	________	________
14	________	________	________
15	________	________	________
16	________	________	________

Determine the rate law (clearly indicating the orders and rate constant) for this reaction from the data in Parts A and B.

Compare the rate constants for Trials 5–8 with that of Trials 13–16. What was the effect of adding a catalyst?

Name: ______________________________

Date: __________ Section: ______________________

Data Sheet: Experiment 26, p. 3

E. Temperature Relationships

From the data in Part C, determine the relationship between k and T. Attach a graph of ln k versus $1/T$, and from the slope, determine E_a. Show the calculation of the slope and E_a below. What are the units of the slope?

Name: ______________________________

Date: __________ Section: ______________________

Questions: Experiment 26

1. Adding a catalyst increases the rate of a reaction. If the rate increases, then either k increases, or the order of the reaction changes. What experiments might you perform to discover which is occurring?

2. In the method of initial rates, only a small amount of reagent is used up (the change in concentration is small compared to the total concentration). The elapsed time is used to calculate the rate. Why is this more accurate than calculating the rate from the amount of time required to use up all of the reagent? (Hint: What happens as the concentration decreases?)

3. If k changes due to the presence of a catalyst (at a constant temperature), what factors in the Arrhenius equation may have changed? What experiments might you perform to discover what has changed?

Chapter 8
Inorganic Chemistry

Experiment 27 Paper Chromatography: Separation of Metal Cations

OBJECTIVES

- To use paper chromatography to separate a mixture of transition metal cations.

INTRODUCTION

Whenever a chemical reaction is carried out in the laboratory on an impure substance isolated from nature, the individual compounds must be separated, purified, and identified. Over the years, chemists have developed various techniques to accomplish these tasks. The most common approaches are crystallization, filtration, and sublimation for solids, and distillation for liquids. Extraction and centrifugation are also viable options. The separation and purification of small (micro) amounts of material has become increasingly important, especially in the biological area, the synthesis of important drugs, and in environmental chemistry concerned with pollution problems, to name a few. These micro-separations have become possible due to the development of sophisticated instrumental analytical techniques that require only very tiny amounts of material to identify any particular species.

Today, the most widely used method for accomplishing chemical separation of mixtures is **chromatography.** It is often referred to as the "workhorse of the laboratory." Mikhail Semenovich Tswett, a Russian botanist, is credited with the first experimental work in this field in 1906. Tswett extracted the green color of leaves using ether, and passed the extract through a column packed with $CaCO_3$ (white chalk). The separated species appeared as colored bands on the column, thus the name chromatography (*chroma*—Greek for color, *graphein*—to write).

Color is not a necessary property to achieve separation of mixtures by this procedure. Colorless compounds can be made visible by reacting them with other reagents, or can be detected by physical means. Due to the simplicity and efficiency of the technique, chromatography has become one of the most important tools for separating and identifying compounds. There are many specialized types of chromatography, such as **liquid chromatography** (LC), which includes column and high performance (HPLC) methods, **gas chromatography** (GC), **thin layer chromatography** (TLC) including paper chromatography (this experiment), and **ion-exchange chromatography.**

Two factors are common to all types of chromatography: a *stationary phase* and a *mobile phase*. The mixture being chromatographed is separated as it is carried through the stationary phase by the flow of the mobile phase. Obviously, the components that are being separated must be soluble in the mobile phase and these components must also interact with the stationary phase based upon some type of property. Such interactions occur when the materials dissolve in the stationary phase, are absorbed by it, or chemically react with it. That is, the component must "partition" itself between the two phases.

Some generalizations can be made about the two phases: the *stationary phase* may be a liquid or a solid, and must be held in some type of a container. One simple form is when the stationary phase is a ground solid held in a narrow tube, such as a burette or, for microscale work, a Pasteur pipet. This form is referred to as **column chromatography,** and was the type originally used by Tswett. The *mobile phase* flows by gravity down through the column (see Figure E27.1).

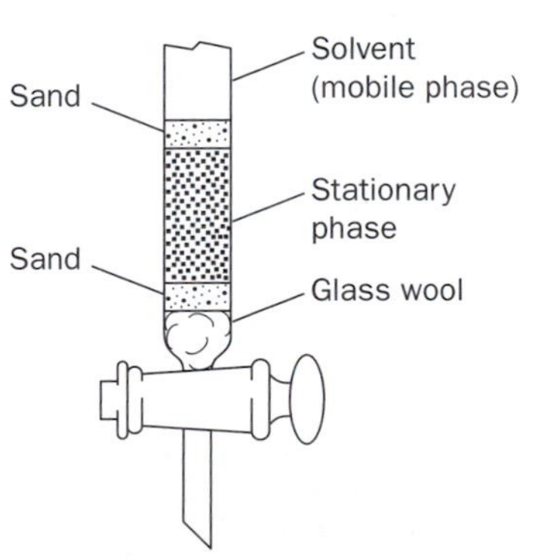

Figure E27.1 *A simple chromatography column.*

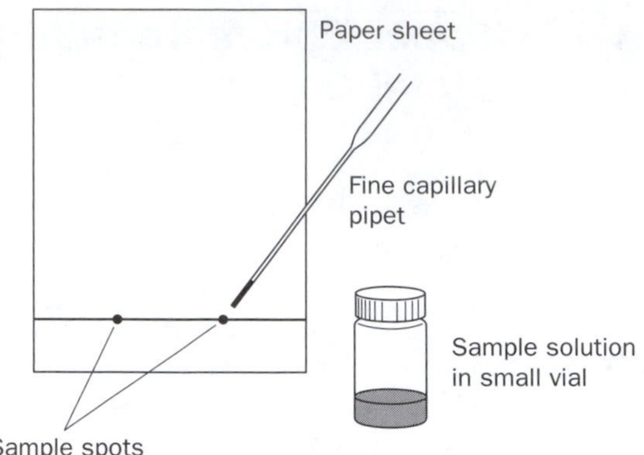

Figure E27.2 *Applying sample to the paper.*

In **high performance liquid chromatography** (HPLC), the mobile phase is forced through a column by use of a pump exerting pressure on the system. The use of a finely divided solid stationary phase, spread on a glass or plastic sheet, is referred to as planar or **thin layer chromatography** (TLC). In this technique, the mobile phase ascends the plate by capillary action. When paper is used as the stationary phase, this process is referred to as **paper chromatography.**

This experiment will explore the use of paper chromatography to separate a mixture of transition metal cations. Filter paper serves as the *stationary phase*, and the *mobile phase* consists of a mixture of acetone and 6 M hydrochloric acid in which one or more of the transition metal cations is dissolved. A single spot of the mixture to be analyzed is applied about 1.5 cm from the end of a strip of filter paper. A spot of solution containing each known cation that may be in the unknown mixture will also be individually placed on the paper strip (see Figures E27.2 and E27.3).

The treated strip is then placed in a covered jar or beaker (which acts as a developing chamber) containing a shallow layer of the solvent mixture (see Figure E27.4). Since filter paper is very permeable to the solvent, the solvent begins to rise up the strip by capillary action.

As the solvent level reaches the height where the samples were applied to the paper, various effects can occur. If a component of a given spot is very soluble in the solvent, it will be swept along with or near the solvent front. Conversely, components that are relatively insoluble in the solvent do not move any great distance up the paper. Other solutes are intermediate between

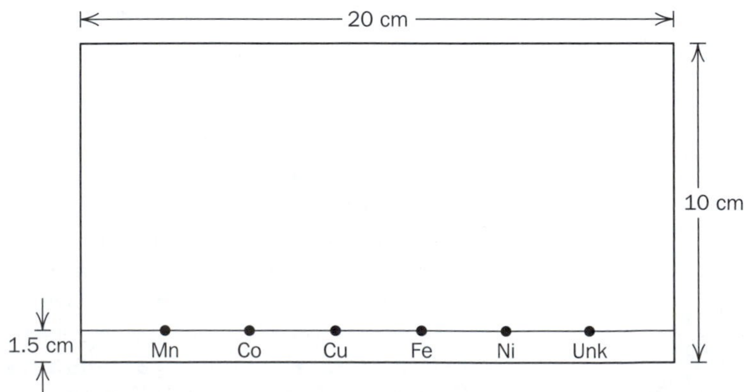

Figure E27.3 *Arrangement for paper chromatography analysis of metal ions.*

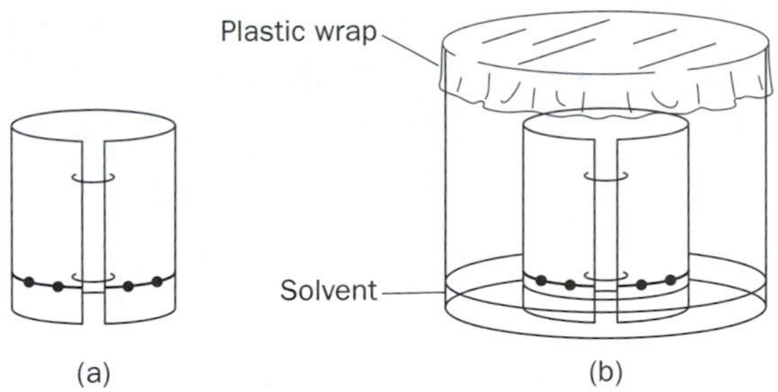

Figure E27.4 *(a) Stapled cylinder. (b) Development of the chromatogram.*

these extremes. Through this process, the original components may be separated (over the surface of the paper) into a series of spots. Each spot represents a single component of the original mixture (see Figure E27.5).

The separation of the mixture occurs because of the solubility of a given component in the mobile phase versus the interaction of that component with the paper. Conditions are worked out so that each component of the mixture will have a different "degree of partition" between the two phases. On a quantitative basis, the degree of partition is called the **retention factor,** R_f, and defined by the equation

$$R_f = \frac{\text{Distance traveled by the spot}}{\text{Distance traveled by solvent front}}$$

Figure E27.5 shows a typical chromatogram.

The retention factor depends on which solvent is used, and on the specific composition of the paper employed. Because R_f values for specific components may vary if an analysis is carried out under different conditions, a *known sample* is generally analyzed at the same time as the unknown mixture. If the unknown mixture produces spots having the same R_f value as components in the known sample, then the identification of the unknown material has been achieved.

Color is also an important observation to aid in the identification of the metal ions. Most transition metal ions are colored, and thus in addition to the R_f values obtained, the color of the spot is an added piece of evidence to establish identity. If the metal ion is not colored, the chromatogram can often be treated with a reagent that will impart color to the spots.

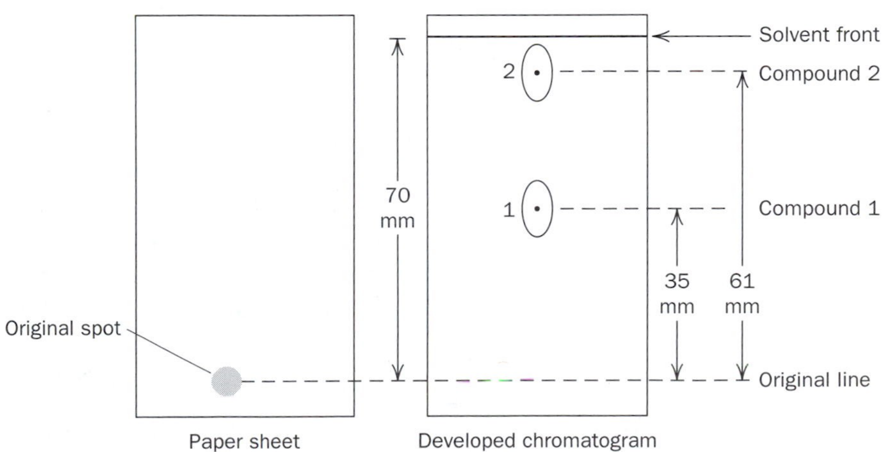

Figure E27.5 *Sample of a developed paper chromatogram R_f of spot 1 = 35 mm/70 mm; spot 2 = 61 mm/70 mm.*

In this experiment, paper chromatography will be used to analyze an unknown solution containing one or more of the following metal ions: Mn^{2+}, Co^{2+}, Cu^{2+}, Fe^{3+}, and Ni^{2+}. The mobile phase is a 90/10 (volume) mixture of acetone and 6 M HCl. The various spots on the developed chromatogram will be highlighted by treatment with several chemical reagents to enhance their color. The reagents to be used are ammonia and dimethylglyoxime. Since oils and moisture from the skin can interfere with the chromatographic separation, be careful not to touch the paper and be sure to place the paper on a clean surface when spotting the samples. Whatman #1 filter paper provides suitable results.

EXPERIMENTAL PROCEDURE

- Obtain a 10 × 20 cm piece of filter paper to serve as the solid phase, and a 600 mL beaker to serve as a developing chamber. Using a pencil and ruler, draw a light line 1.5 cm up from the bottom edge of the paper. Mark the line (even spaces) for the Mn^{2+}, Co^{2+}, Cu^{2+}, Fe^{3+}, and Ni^{2+} ion solutions and for the unknown, as shown in Figures E27.2 and E27.3. **Note:** *Do not handle the paper with your fingers, especially along the bottom edge. It is advisable to use tongs (or tweezers) to handle the paper. Do not place the paper directly on the surface of laboratory bench. Place it on a clean sheet of paper.*
- Microcapillaries may be fashioned by heating the midsection of an open-ended melting point capillary tube and drawing the glass to form a smaller capillary section. This smaller section is then broken in the middle to form two microcapillaries.
- Obtain six microcapillaries. Use a separate microcapillary for each sample solution to be applied to the paper. Selecting each of the six solutions in turn, insert the microcapillary into the solution (it will fill by capillary action) and carefully apply a single small spot of the solution at the center of the pencil mark designated for the particular metal ion. The spot should not be more than 2–3 mm in diameter, or the separation will not work well. Allow the spots to dry. Repeat the spotting–drying procedure one more time. This will increase the concentration of the metal ions on their spots (see Figures E27.2 and E27.3).
- Cut a square of plastic wrap or aluminum foil to fit over the mouth of the 600 mL beaker. Carefully add the development solvent (acetone/HCl) to the beaker to a depth of not over 6 mm, using a glass rod (see Figure 3.5, p. 33). Do not allow the solvent to wet the sides of the beaker. Make sure that the depth of solvent *will not* cover the sample spots on the filter paper when it is immersed in the beaker.
- Cover the beaker with the plastic wrap, and allow the beaker to stand for 6–9 minutes so that the air in the beaker becomes saturated with solvent vapor. *R_f values in chromatography depend strongly upon solvent saturation of the atmosphere above the liquid in the development chamber.*
- Bend the filter paper into a cylinder, *without overlapping the edges* and staple the ends as shown in Figure E27.4.a. A paper clip or tape may also be used. **Note:** *If staples or paper clips are used, only fasten the paper at the top. Otherwise the clip or staple may dissolve and thus generate a spot for iron.* Alternatively, fan fold the filter paper so that each spot has its own "track" to travel up the paper. The fan folded paper will stand on its own.
- Momentarily, remove the cover (plastic or foil) from the beaker, and gently lower the paper (spot end down) into the solvent. **Do not allow the paper to touch the sides of the beaker.** Immediately replace the cover on the beaker (see Figure E27.4b).

- **Without disturbing the beaker,** allow the system to develop until the solvent front is approximately 1 cm from the top of the paper. Once the solvent front has reached this point, *immediately* remove the paper from the beaker, remove the staples (if any), lay the paper flat on a clean surface, and mark (with your pencil) the exact location of the solvent front across the width of the paper. **This must be done before the solvent begins to evaporate.**
- Allow the chromatogram to dry completely **(HOOD),** using a hair dryer if available. Otherwise, wave the paper gently back and forth to aid the evaporation process.
- Outline (in pencil) any spots that are visible for the known samples, as well as for your unknown mixture. Record the color of these spots on the data sheet.

ANALYSIS OF THE CHROMATOGRAM

Since all the spots may not be visible at this stage, various chemical treatments will be applied to the chromatogram to make them more pronounced. NOTE: In each of the following operations, handle the chromatogram with tongs and do each operation in the HOOD.

- Hold the chromatogram, using tongs, over a small beaker of concentrated ammonia solution for 2–3 minutes. **Do not inhale the ammonia fumes! Note:** *An alternative procedure is to prepare an ammonia chamber by placing a 30 mL beaker containing 10 mL of concentrated ammonia in a 1000 mL beaker over which is placed plastic wrap as in the developing chamber prepared above. Place the chromatographic paper in the chamber, recover the beaker, and allow the paper to stand in this atmosphere for approximately 1–2 minutes.* The ammonia will neutralize any residual HCl remaining on the paper. Many metal ions form stable coordination compounds with ammonia and thus new spots may appear or the original spots may change color. Outline any new spots with pencil. Record the color of all the spots, including any that change color on your data sheet.
- Hold the chromatogram so that a light spray (fine mist is best) of 1% dimethylglyoxime solution can be applied using an aspirator bottle. *Do not completely wet the paper*. Dimethylglyoxime forms brightly colored complexes with certain transition metal ions. As before, record any changes in color of the spots and outline any new spots that may appear after this treatment. Record the results on your data sheet.
- Dry the chromatogram completely. Measure the vertical distance that the approximate center of each of the spots has traveled from the original baseline, and calculate the R_f value for each known spot and for each spot in the unknown mixture. By comparing the various colors recorded and using the respective calculated R_f values, identify which metal ions are in your unknown sample. Submit the chromatogram with your laboratory report.

Name: ____________________

Date: ________ Section: ____________________

Pre-Laboratory Questions: Experiment 27

1. In preparing the filter paper for chromatographic analysis, a pencil was used when drawing the baseline on which to position the spots of metal ion solutions. Why is ink never used?

2. If the solvent front moves 55 mm and a component in a sample being analyzed moves 35 mm from the baseline, what is the R_f value?

3. Would you expect that changing the solvent would change the R_f value obtained for a particular unknown? Why?

4. A piece of filter paper is spotted with a solution containing a mixture of two components, **X** and **Y.** The chemical affinity of X for the stationary phase is greater than that of Y, and the chemical affinity of X for the mobile phase is less than that for Y. Which substance will have the largest R_f value upon analysis of the developed chromatogram? Explain your answer.

Name: ____________________

Date: ________ Section: ____________________

Data Sheet: Experiment 27

Unknown mixture ID number: ____________

Color of unknown solution: ____________

Effect of the various chemical reagents on the spot colors:

Species	*Original Color*	*After Ammonia Treatment*	*After DMG Treatment*
Mn^{2+}	______	______	______
Co^{2+}	______	______	______
Cu^{2+}	______	______	______
Fe^{3+}	______	______	______
Ni^{2+}	______	______	______
Unknowns	______	______	______
	______	______	______

Average distance traveled by the solvent front: ____________mm

Species	*Distance Traveled by Spot, mm*	*R_f Value*
Mn^{2+}	______	______
Co^{2+}	______	______
Cu^{2+}	______	______
Fe^{3+}	______	______
Ni^{2+}	______	______
Unknown	______	______
	______	______
	______	______

Identification of the metal ion(s) in the unknown solution:

____

Name: ____________________

Date: ________ Section: ____________________

Questions: Experiment 27

1. Why is it important to keep the spots applied to the filter paper as small as possible?

2. Suppose two metal cations have the same R_f value. What might you do to resolve the identity of the two cations using paper chromatography?

3. Why is the beaker used in developing the chromatogram kept covered during the experiment?

4. When you placed the chromatographic sheet into the developing chamber, why was it important that the developing solvent did not come above the pencil line on the paper sheet?

Experiment 28 Synthesis of Iron(III) Acetylacetonate

OBJECTIVES

- To synthesize a coordination metal complex.

PRIOR READING

Section 3.4: Liquid Volumes
Section 3.7: Filtration

INTRODUCTION

Coordination compounds (also called **complexes**) consist of a central atom surrounded by various other atoms, ions or small molecules called **ligands.** There is only a small distinction between coordination complexes and "normal" molecular compounds. The most common dividing line is that complexes have more ligands than the central atom oxidation number. SiF_4 would not be a coordination compound, since there are four ligands on the Si^{4+}. SiF_6^{2-} would be considered a coordination compound, as there are six ligands on the Si^{4+}. In this experiment, the coordination compound *tris*-(2,4-pentanedionato)iron(III) will be synthesized. This is more commonly known as iron(III) acetylacetonate or iron(III) acac.

In the presence of base, 2,4-pentanedione (acetylacetone, acacH) easily loses a proton to form the acetylacetonate anion ($acac^-$), $CH_3COCHCOCH_3^-$, as shown below. In this experiment, sodium acetate trihydrate is used to make the solution basic.

$$\underset{\text{acacH}}{CH_3COCH_2COCH_3} \xrightarrow{-H^+} \underset{\text{acac}-}{CH_3COCH^-COCH_3 \longleftrightarrow \cdots \longleftrightarrow \cdots}$$

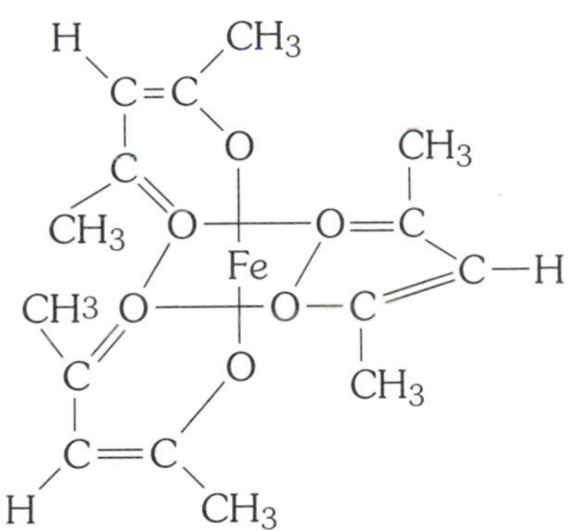

Figure E28.1 *Structure of Fe(acac)₃.*

$Acac^-$ is an example of a **bidentate** (*bi*—two, *dent*—teeth) ligand, since it bonds to metals at **both** of its oxygens. Ligands of this type are also called **chelating** (*chelos*—claw) ligands. Since most metals like to have six ligands around them, three $acac^-$ ligands are needed (remember: $acac^-$ bonds in two places!). Since $acac^-$ has a charge of -1, three $acac^-$'s will have a total charge of -3, and form compounds nicely with metals having a charge of $+3$, of formula $M(acac)_3$. An example of a $+3$ metal is Fe^{3+}, and the coordination compound we will prepare in this experiment is $Fe(acac)_3$, whose structure is shown in Figure E28.1. Since the outer part of the complex consists of organic groups, most metal acetylacetonates are hydrophobic, and insoluble in water—thus the product will precipitate from solution.

EXPERIMENTAL PROCEDURE

Preparation of Iron(III) Acetylacetonate

- Record observations on the data sheet (colors of reagents, products, formation of solid, etc.)
- Obtain a 10 mL Erlenmeyer flask, a micro-magnetic stir bar, a magnetic stirring hot plate, a 10 mL beaker, a 10 mL graduated cylinder, a Pasteur pipet, a small watch glass, a Hirsch funnel, a 25 mL suction flask, filter paper, and a clay tile.

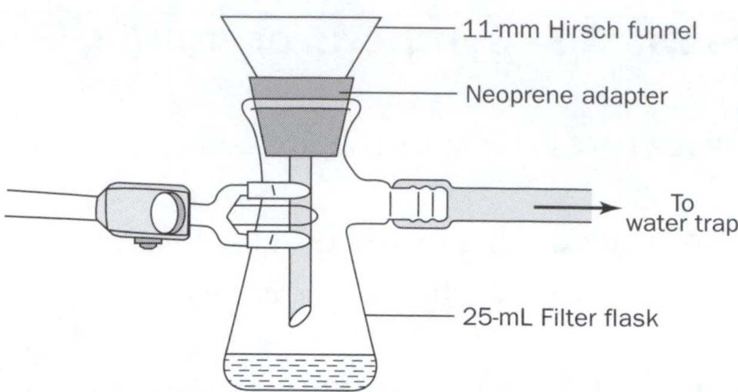

Figure E28.2 *Vacuum filtration apparatus.*

- Place the 10 mL Erlenmeyer flask (containing a magnetic spin bar) on the balance, and tare out its weight. Add about 270 mg of $FeCl_3{\cdot}6\ H_2O$ to the flask. Add 25 mL of water using a 10 mL graduated cylinder. Place the flask on the magnetic stirrer, and stir the contents of the flask to dissolve the iron chloride.
- Dissolve 500 mg of sodium acetate trihydrate in 2.5 mL of water. With stirring, add this to the iron solution using a Pasteur pipet. With an automatic delivery pipet, add 500 μL of acetylacetone **(in the HOOD).** Cover the flask with a watch glass and allow this mixture to stir at room temperature for 10 minutes.
- Collect the dark solid product by suction filtration (Figure E28.2; also see Figure 3.19) using a Hirsch funnel. **Be sure to clamp the flask!** Wash the filter cake with 0.5 mL of water and dry the material on a clay tile or on filter paper. Weigh the dried product.
- Calculate the percent yield of this initial product [based on the amount of the iron(III) chloride]. Obtain a melting point of the product.
- If time permits, recrystallize a portion of the material using a 10 mL Erlenmeyer flask. This purifies the product. Take approximately 80 mg of the product and dissolve it in 2.0 mL of hot 2-propanol. Allow the solution to cool to room temperature, and then cool the solution in an ice bath. Crystals of product should form. Collect the crystals by suction filtration as before. Obtain a melting point of the recrystallized product and compare it to that of the crude material.

Name: ______________________________

Date: __________ Section: ______________________________

Pre-Laboratory Questions: Experiment 28

1. Iron has several common oxidation states other than Fe^{3+}. What are they? What color are solutions of these species? Suggest an easy way of determining whether an oxidation or reduction of an iron-containing solution has taken place.

2. Rusting is the process of a metal reacting with oxygen, forming the metal oxide, and thereby degrading the metal surface. Iron metal will rust, whereas aluminum metal does not rust. Why?

3. Many coordination compounds are very water soluble, such as $K_3[Fe(C_2O_4)_3] \cdot 3\ H_2O$, $[Co(NH_3)_5Cl]Cl_2$, and $[Cr(H_2O)_6]Cl_3$. Suggest why $[Fe(acac)_3]$ is not very soluble in water, and the others are.

Name: ______________________________

Date: __________ Section: ______________________

Data Sheet: Experiment 28

Iron(III) Acetylacetonate

1. Mass of iron(III) chloride hexahydrate used ______________ g

2. Moles of iron(III) chloride hexahydrate used ______________ mol

3. Moles of acetylacetone used (ρ = 0.97 g mL^{-1}) ______________ mol

4. Mass of iron(III) acetylacetonate obtained ______________ g

5. Moles of iron(III) acetylacetonate obtained ______________ mol

6. Mol percentage yield ______________ %

7. Melting point of crude product ______________ °C

8. Melting point of recrystallized product ______________ °C

Record observations (colors, formation of solids, evolution of gases, etc.)

Name: ______________________________

Date: __________ Section: ______________________________

Questions: Experiment 28

1. The π electrons in $acac^-$ are delocalized over the carbon–oxygen framework. Simpler examples of electron delocalization occur in the NO_2^-, NO_3^-, and CO_3^{2-} ions. Draw the Lewis dot structures for these three ions, and determine the bond order in each.

2. The $Fe(acac)_3$ complex synthesized in this experiment resembles hemoglobin in several ways, in its appearance, properties and in its structure. Give several examples of this.

Experiment 29 Chemical Transformations of Copper

OBJECTIVES

- To prepare a series of copper compounds.
- To recover the original copper after a series of transformations.

PRIOR READING **Section 3.7:** Filtration

INTRODUCTION

Copper was one of the first metals to be isolated, due to the ease of separating it from its ores. It is believed that the process **(metallurgy)** was known as early as 4500 B.C.E. It is a ductile, malleable metal being easily pounded and/or drawn into different shapes for use as wire, ornaments, and implements of various types. Alloys of copper (bronze, brass) were discovered quite early in history, and were among the first commodities for international trade. Pure copper is a good electrical conductor, but not as good as silver. However, it is used extensively as a conductor due to its much lower price. It has good thermal conductivity and is corrosion resistant. Copper as a metal is second in commercial importance only to iron.

In this experiment, a weighed amount of copper metal is transformed, through a series of reactions, into other copper-containing compounds, and eventually returned to the metal state. The series involves the use of reactions classified as metathesis, decomposition, displacement and oxidation-reduction reactions. The following equations (unbalanced) outline the series of transformations to be performed.

A. $Cu_{(s)} + HNO_{3(aq)} \rightarrow Cu(NO_3)_{2(aq)} + NO_{2(g)} + H_2O_{(l)}$

B. $Cu(NO_3)_{2(aq)} + NaOH_{(aq)} \rightarrow Cu(OH)_{2(s)} + NaNO_{3(aq)}$

C. $Cu(OH)_{2(s)} \rightarrow CuO_{(s)} + H_2O_{(l)}$

D. $CuO_{(s)} + H_2SO_{4(aq)} \rightarrow CuSO_{4(aq)} + H_2O_{(l)}$

E. $CuSO_{4(aq)} + Zn_{(s)} \rightarrow Cu_{(s)} + ZnSO_{4(aq)}$

The series of reactions begins and ends with copper metal. Since no copper is added or removed between reactions A and E, and since each reaction nearly goes to completion, you should be able to quantitatively recover all the copper metal you started with. The sequence of reactions also shows the wide variety of colors often observed for inorganic compounds.

EXPERIMENTAL PROCEDURE

Part A: Preparation of Copper(II) Nitrate Solution

NOTE: ***This reaction should be carried out in a well ventilated area (HOOD). DO NOT USE CONCENTRATED NITRIC ACID. Use of the dilute acid generates a brown haze over the surface of the solution, the bulk of the gas being dissolved in the aqueous phase.***

- Record all observations (colors, formation/dissolution of solids, evolution of gases, etc.) on the Data Sheet.
- Place about 100 mg of copper wire (~1 inch of 18 gauge wire), weighed to the nearest mg, in a 10 mL Erlenmeyer flask. In the **HOOD,** add 2 mL of 6 M nitric acid to the flask, cover the Erlenmeyer flask with an inverted 10 mL beaker, and warm the contents **gently** on a hot plate. Continue heating until the copper metal has completely dissolved, and the evolution of brown fumes of nitrogen dioxide is no longer observed.

- Allow the resulting blue solution of copper(II) nitrate to cool to room temperature. This cooling process may be hastened by running cold tap water over the outside of the flask. When cool, add 2 mL of distilled water to the blue solution.

Part B: Preparation of Copper(II) Hydroxide

- Add a magnetic stirring bar to the Erlenmeyer flask, and place it in an ice/water bath on a magnetic-stirring hot plate. With stirring, carefully add 6 M NaOH in 0.5 mL increments, until the solution is basic to red litmus paper (red litmus turns blue). Add an additional 0.5 mL of 6 M NaOH. A light blue precipitate of copper(II) hydroxide is formed as this reaction is carried out.

Part C: Preparation of Copper(II) Oxide

- With stirring, heat the Erlenmeyer flask containing the copper(II) hydroxide on the magnetic-stirring hot plate using a sand bath temperature of 110–115°C. **NOTE: Stirring is essential to prevent bumping of the mixture and loss of CuO.** During this time, the copper(II) hydroxide is transformed into copper(II) oxide, a black precipitate.
- Allow the mixture to cool to room temperature, and remove the magnetic stirring bar using forceps. If necessary, rinse the bar with a small amount of water, collecting the rinse in the Erlenmeyer flask. Isolate the black solid by suction filtration (see Section 3.7, p. 42) using a Hirsch funnel. Rinse the Erlenmeyer flask with 1–2 mL of distilled water to complete the transfer. **NOTE: If the solid proves hard to filter, a centrifuge may be used to isolate the solid.** Discard the filtrate. Wash the collected filter cake with an additional 1–2 mL of distilled water.

Part D: Preparation of Copper(II) Sulfate Pentahydrate Solution

- Warm 6 mL of 3 M sulfuric acid in a 30 mL beaker. Pour the warm sulfuric acid over the black copper(II) oxide and filter paper, collecting the light blue filtrate of copper(II) sulfate pentahydrate in a clean 30 mL beaker. Continue to recycle the acid solution until the black solid has completely dissolved. Gently rinse the filter paper with 1–2 mL of water. Once the transfer is complete, rinse the funnel with 1–2 mL of water. This rinse is also added to the beaker.

Part E: Regeneration of Copper Metal

- Place the beaker containing the blue copper(II) sulfate pentahydrate solution in the **HOOD** and add, in small portions, about 800 mg of zinc powder. Stir the mixture with a glass rod, until the blue color of the original solution disappears. A metallic precipitate of copper metal forms during this period. Vigorous evolution of hydrogen gas is also observed during the addition.
- Verify that no copper(II) ions are left in the solution by adding a drop of the solution to 1 mL of concentrated ammonia in a small test tube. If a deep midnight blue color of $Cu(NH_3)^{2+}$ forms, the reaction is not yet complete.
- After the reaction is complete, add 5 mL of 3 M hydrochloric solution, and stir the mixture with a glass rod. This process removes any unreacted zinc metal. The copper metal does not react under these conditions.
- Allow the copper precipitate to settle, decant the aqueous solution, and wash the solid three times with 2 mL portions of distilled water. Decant the rinse solution between washings. Repeat this process using acetone as the wash solvent.

- Spread the copper solid on a piece of filter paper or clay tile, and allow it to air-dry. Alternatively, the Cu may be dried by placing the material on a watch glass, which is then heated over a hot water bath. Weigh the copper and calculate the percentage recovery.
- Place a small portion of the dry copper metal on a clay tile, and slide the flat part of a spatula over the copper while pushing down hard on the spatula. The copper will transform from the amorphous state to the metallic state, with its characteristic bright metallic luster.

Name: ____________________

Date: ________ Section: ____________________

Pre-Laboratory Questions: Experiment 29

1. Balance each of the following equations.

A. __$Cu_{(s)}$ + __$HNO_{3(aq)}$ → __$Cu(NO_3)_{2(aq)}$ + __$NO_{2(g)}$ + __H_2O

B. __$Cu(NO_3)_{2(aq)}$ + __$NaOH_{(aq)}$ → __$Cu(OH)_{2(s)}$ + __$NaNO_{3(aq)}$

C. __$Cu(OH)_{2(s)}$ → __$CuO_{(s)}$ + __H_2O

D. __$CuO_{(s)}$ + __$H_2SO_{4(aq)}$ → __$CuSO_{4(aq)}$ + __H_2O

E. __$CuSO_{4(aq)}$ + __$Zn_{(s)}$ → __$Cu_{(s)}$ + __$ZnSO_{4(aq)}$

2. Based on the balanced equations of Question 1, if you used 100.0 mg of copper in Equation A, calculate the exact amount of zinc required to complete the reaction in Equation E.

Name: ______________________________

Date: __________ Section: ______________________

Data Sheet: Experiment 29

1. Initial mass of copper wire __________ g

2. Mass of copper recovered __________ g

3. Percentage recovery (show calculation) __________ %

Record observations:

Name: ______________________________

Date: __________ Section: ______________________________

Questions: Experiment 29

1. How many milliliters of 6.0 M NaOH are required to react with 100.0 mg of copper(II) ion to form copper(II) hydroxide? Why was more than this used in the experimental procedure?

2. Would the percentage recovery of copper metal in the experiment be high, low, or the same if insufficient NaOH was added to react with both the unreacted nitric acid and the Cu^{2+} generated in the first step of the sequence? Explain.

3. Two of the reactions carried out in the transformation scheme (A to E) involve an oxidation-reduction sequence.

a. Which of the reactions are of the redox type?

b. In each case, identify what species is oxidized and which is reduced.

c. For each reaction, identify the oxidizing and reducing reagent.

4. In Part E of the experiment, the zinc metal is added to undergo a redox reaction with the copper(II):

$$Zn + Cu^{2+} = Cu + Zn^{2+}$$

Hydrogen gas is also generated in this step. How is the hydrogen gas formed?

5. Why is a large excess of zinc added in the reaction of Part E?

Chapter 9
Inorganic Qualitative Analysis

Chapter Nine Introduction to Inorganic Qualitative Analysis

There are many instances in chemistry when the composition of a substance must be determined, or the identity of a material dissolved in a solution must be discovered. The answers to the questions

- What elements are present?
- What cations and anions are present?
- What functional groups are present?

lie in the domain of **qualitative analysis.** Determining the absolute or relative amounts of these components is the area of **quantitative analysis.** Together with **instrumental analysis,** these form the basic elements of **analytical chemistry.**

The following experiments in inorganic qualitative analysis are concerned with discovering which cations or anions are present. Several experiments will include the problems of separating ions from a mixture before identifying them. We will also determine the identity of solid inorganic salts. A different approach is used in identifying organic compounds qualitatively, and may be found in Experiment 36.

PART A: REACTIONS INVOLVED IN INORGANIC QUALITATIVE ANALYSIS

Most reactions taking place in aqueous solution occur between ions, including individual charged atoms (e.g., Na^+) and polyatomic ions carrying a net charge (e.g., NO_3^-). Some inorganic compounds, however, exist as neutral molecules in solution. The general principles are as follows:

1. Dissolved metal compounds and the strong acids HCl, HBr, HI, H_2SO_4, and HNO_3 will be completely dissociated in solution as the corresponding ions.
2. Water, ammonia, dissolved gases, insoluble compounds, and weak acids such as acetic acid, CH_3COOH, will be present as neutral molecules.

It is important to realize that many chemical reactions do not go to completion. The extent to which a reaction occurs depends on the magnitude of the **equilibrium constant** (K_{eq}) for the reaction, and the relative amounts of reagents present. Several types of reactions encountered in qualitative analysis are listed below, as well as examples of the equilibrium reactions involved in each. The equilibria can be shifted by adding or removing other reagents and changing physical conditions, in accordance with **Le Chatelier's principle.** By applying this principle, we can force precipitation to occur, cause some sparingly soluble compounds to dissolve, or complex particular ions so that they will not interfere with tests for other ions of interest.

Reaction A: Hydrolysis

Many ions or molecules react with water, resulting in the formation of either H_3O^+ or OH^- ions, thus affecting the pH of the solution. This type of reaction is known as **hydrolysis.** A common example is the hydrolysis of the ammonium ion to form ammonia:

$$NH^+_{4\,(aq)} + H_2O_{(l)} \rightleftharpoons NH_{3(aq)} + H_3O^+_{(aq)}$$

When acid is added to an ammonia solution, it reacts with ammonia, forming ammonium ion. This causes the equilibrium to shift to the left, increasing the amount of NH_4^+ in solution. The equilibrium can be driven to the right by heating to evolve NH_3 (g). Also, the equilibrium can be driven to the right by adding base.

The concentration of the species undergoing hydrolysis also has an effect on the pH. A solution of 0.1 M NH_3 in deionized water has a pH of 8.9, whereas concentrated ammonia (14.5 M) has a pH of 10.0, an increase of hydroxide concentration by a factor of approximately 10. The pH and concentration of ammonia can determine whether an ion precipitates as a hydroxide salt, forms a soluble ammonia or hydroxide complex, or doesn't react at all.

Reaction B: Brønsted Acid–Base Reactions

In a Brønsted acid–base reaction, there is a change in the concentration of H_3O^+ and OH^- in solution. Many hydrolysis reactions fall into this category as well. The following reactions are examples of this diverse class.

$$2\,H_3O^+_{(aq)} + CO_3^{2-}{}_{(aq)} \rightleftharpoons 3\,H_2O_{(aq)} + CO_{2(aq)} \quad (1)$$

$$Al(OH)_{3(s)} + OH^-_{(aq)} \rightleftharpoons Al(OH)^-_{4\,(aq)} \quad (2)$$

$$Al(OH)_{3(s)} + 3\,H_3O^+_{(aq)} \rightleftharpoons Al^{3+}_{(aq)} + 6\,H_2O_{(l)} \quad (3)$$

In Reaction (1), H_3O^+ reacts with carbonate ion, CO_3^{2-}, to produce water and carbon dioxide, thereby raising the pH. In Reaction (2), aluminum hydroxide, $Al(OH)_3$, dissolves in basic solution to form the tetrahydroxoaluminate ion, $Al(OH)_4^-$. This lowers the pH, as hydroxide ion is consumed. In Reaction (3), $Al(OH)_3$ dissolves in acidic solution to form the Al^{3+} ion. This consumes protons, thereby raising the pH. Compounds such as $Al(OH)_3$, which can react with either acid or base, are said to be **amphoteric.**

Buffers are a special case of acid–base reactions (see Experiment 20). A solution is said to be buffered when it has (approximately) equal amounts of a weak acid or base and its conjugate salt. Common examples are CH_3COOH/CH_3COONa and NH_3/NH_4Cl. When buffered, a solution resists changes in pH when small amounts of a strong acid or base are added. The weak acid (or base) or the conjugate salt will react to absorb the added OH^- or H_3O^+. Buffers are used as a means of controlling the pH, usually in order to determine the nature of the species in solution. Consider the equilibrium between chromate (CrO_4^{2-}) and dichromate ($Cr_2O_7^{2-}$).

$$H_2O_{(l)} + 2\,CrO_4^{2-}{}_{(aq)} \rightleftharpoons Cr_2O_7^{2-}{}_{(aq)} + H_3O^+_{(aq)}$$

Careful pH control using a buffer allows one to control the concentration of chromate ion in solution. We can therefore selectively precipitate specific metal ions as their insoluble chromates, leaving more soluble ones in solution.

Reaction C: Precipitation Reactions

When the ion product of ions in solution exceeds the solubility product constant of a particular salt, **precipitation** will occur (see Experiment 22). Determining the presence of a precipitate is not always trivial. Some solutions become cloudy, which indicates a precipitate has formed. In some cases, you may be able to see grains of solid falling from solution. In other cases, the solution may become milky in appearance. The solution and the precipitate may or may not change color. The presence of a precipitate is difficult to detect when a solution is dark. In such cases, it helps to dilute a small sample of the solution, or to examine a thin layer of it against a light. A note on terminology: clear means not cloudy; colorless means uncolored (like water). The two terms are not synonymous (see Experiment 2, Introduction).

Table 9.1 Solubility Rules for Common Ions in Aqueous Solution

1. All nitrates **(NO_3^-)** are soluble.
2. Salts of the group 1 cations **(Na^+, K^+, Rb^+, and Cs^+)** or of ammonium **(NH_4^+)** are soluble, with obscure exceptions.
3. Halides **(Cl^-, Br^-, and I^-)** and thiocyanate **(SCN^-)** are soluble, except for Ag^+, Cu^+, Tl^+, Pb^{2+}, and Hg_2^{2+}. Bromide and iodide are oxidized by some cations.
4. Sulfates **(SO_4^{2-})** are all soluble except for Pb^{2+}, Ba^{2+}, and Sr^{2+}. (Ca^{2+}, Hg_2^{2+}, and Ag^+ are somewhat soluble).
5. Nitrites **(NO_2^-)** and permanganates **(MnO_4^-)** are all soluble, except for $AgNO_2$. These anions are powerful oxidizing agents, so they are unstable with ions that are easily oxidized.
6. Thiosulfates **($S_2O_3^{2-}$)** are soluble, except for Pb^{2+}, Ba^{2+}, and Ag^+. $Ag_2S_2O_3$ decomposes in excess thiosulfate with reduction of Ag^+ to Ag^0.
7. Sulfites **(SO_3^{2-})**, carbonates **(CO_3^{2-})**, phosphates **(PO_4^{3-})**, oxalates **($C_2O_4^{2-}$)**, and chromates **(CrO_4^{2-})** are only soluble in acidic solution. Exceptions: calcium oxalate is insoluble even in acid and the salts with Mg^{2+} and ions listed in Rule 2 are soluble in acidic, neutral, and basic solution. Sulfite and oxalate can form soluble complexes.
8. Fluorides **(F^-)** are insoluble, except for Ag^+, Fe^{3+}, and the ions listed in Rule 2. Some transition metal fluorides are soluble, especially in excess fluoride, due to complex ion formation.
9. Ferrocyanides **[$Fe(CN)_6^{4-}$]** are insoluble, except for ions listed in Rule 2.
10. Hydroxides **(OH^-)** are insoluble, except for Sr^{2+}, Ba^{2+}, and the ions listed in Rule 2. Many metals are soluble in excess hydroxide, due to complex ion formation.

The solubility of any sparingly soluble compound is determined by its solubility product constant (K_{sp}). Tables of K_{sp} values are available (see Appendix F), but even more valuable to qualitative analysis are the **solubility rules,** summarized in Table 9.1 and Appendix G.

The solubility of a particular species may be affected by adding reagents that cause a competing reaction. For instance, silver chloride, AgCl, is insoluble in deionized water, but can be made to dissolve by adding aqueous ammonia. This is due to the extremely favorable formation of the soluble silver amine complex, $Ag(NH_3)_2^+$:

$$AgCl_{(s)} \rightleftharpoons Ag^+_{(aq)} + Cl^-_{(aq)}$$

$$Ag^+_{(aq)} + 2\ NH_{3(aq)} \rightleftharpoons Ag(NH_3)^+_{2(aq)}$$

This ties up the silver ion in solution by complex formation, causing the first reaction to proceed further to the right, allowing more AgCl solid to dissolve.

Reaction D: Decomposition

A **decomposition** reaction occurs when one chemical species decomposes into one or more different products. An example is the fizzing that results when acid is added to a carbonate:

$$CO_3^{2-}{}_{(aq)} + 2\ H_3O^+_{(aq)} \rightleftharpoons H_2CO_{3(aq)} + 2\ H_2O_{(l)}$$

$$H_2CO_{3(aq)} \rightleftharpoons H_2O_{(l)} + CO_{2(g)}$$

The carbonate ion undergoes an acid–base reaction, producing carbonic acid. Carbonic acid is very unstable, and spontaneously decomposes into water and carbon dioxide gas. The evolution of carbon dioxide from solution causes the fizzing, and shifts both equilibria further to the right.

Reaction E: Oxidation-Reduction Reactions

Oxidation-reduction, or redox, reactions (see Experiment 21) are used frequently in qualitative analysis as an indication of the presence or absence of an ion.

These reactions are often accompanied by color changes. Redox reactions are also used as a means of dissolving very insoluble compounds and for converting an ion to a different oxidation state, in which case it may be more easily separated or identified. In the following example, pale green chromium(III) hydroxide, $Cr(OH)_3$, can be separated from a mix of hydroxide solids where its pale color is often masked, by oxidizing it with hydrogen peroxide, H_2O_2, to form the soluble chromate ion, CrO_4^{2-}, which is bright yellow in solution.

$$2\ Cr(OH)_{3(s)} + 3\ H_2O_{2(aq)} + 4\ OH^-_{(aq)} \rightleftharpoons 2\ CrO_4^{2-}{}_{(aq)} + 8\ H_2O_{(l)}$$

The most common oxidizing agents are nitric acid, HNO_3, and basic hydrogen peroxide solution, H_2O_2. Reducing agents include the ferrous [iron(II), Fe^{2+}], stannous [tin(II), Sn^{2+}], thiosulfate ($S_2O_3^{2-}$), oxalate ($C_2O_4^{2-}$), and iodide (I^-) ions. Hydrogen peroxide in acidic solution is also used as a reducing agent.

Disproportionation (self-redox) reactions are a special case of redox reactions, wherein part of the reagent is oxidized, and an equivalent part according to the stoichiometry is reduced. An example is the spontaneous decomposition of hydrogen peroxide, where oxygen in peroxide (O in the $^-1$ state) disproportionates into oxygen gas (O_2, O in the 0 state) and water (H_2O, O in the $^-2$ state):

$$2\ H_2O_{2(aq)} \rightleftharpoons 2\ H_2O_{(l)} + O_{2(aq)}$$

The opposite, where two different oxidation states of the same element undergo redox and end up at the same intermediate oxidation state, is called **conproportionation.**

Reaction F: Formation of Complex Ions

Many common anions and neutral molecules can donate one or more lone pairs of electrons (thus acting as **Lewis bases**) to a **Lewis acid** to form a bond. Metal cations are usually Lewis acids, and as such accept electron pairs from Lewis bases. This type of reaction can result in the formation of a complex ion formed from a central metal cation bonded to 2–6 lone pairs of electrons on surrounding Lewis base species (see Experiment 28). In a complex ion, the Lewis bases are known as **ligands.** Water and ammonia are examples of neutral ligands. Anions that readily act as ligands include Cl^-, Br^-, I^-, SCN^-, $C_2O_4^{2-}$, OH^-, F^-, and CN^-.

Metal cations in aqueous solution tend to have a fixed number of water molecules acting as ligands, although the water molecules are often not included when writing reactions. Square brackets, [□], are often used to indicate the complex ion.

For example, the reaction

$$CoSO_{4(s)} \rightleftharpoons Co^{2+}_{(aq)} + SO_4^{2-}{}_{(aq)}$$

should more properly be written as

$$CoSO_{4(s)} + 6\ H_2O_{(l)} \rightleftharpoons [Co(H_2O)_6]^{2+}_{(aq)} + SO_4^{2-}{}_{(aq)}$$

The presence of water (or other such ligands) causes many of the transition metals to be colored in solution. When a ligand that has a higher affinity for the metal ion is added to a solution, ligand replacement often occurs. A new complex ion forms, often accompanied by a change in color of the solution. The color of the solution after adding a particular ligand is often a diagnostic test for the presence of a given metal cation. This is a common way of confirming the presence or absence of a cation in solution. In some cases, a metal cation will be tied up as a complex, to prevent reaction with a reagent being used to test for a different cation.

Both Fe^{3+} and Co^{2+}, for example, react with thiocyanate (SCN^-) to form colored complexes. The $[FeSCN]^{2+}$ complex is dark red-brown, while the $[Co(SCN)_4]^{2-}$ complex is a less intense aqua blue. If both metal ions are present, only the Fe^{3+} will be identified as its dark color will mask the lighter color from the cobalt complex. This can be prevented by addition of fluoride ions, F^-, complexing the Fe^{3+} as the colorless $[FeF_6]^{3-}$ complex. The fluoride does not interfere with the formation of the $Co(SCN)_4^{2-}$ complex, so that when thiocyanate is added, the blue color of the cobalt complex can be seen.

G. Extraction

Extraction is more of a physical process than an actual reaction. It allows the segregation of chemical species in order to purify them or to observe a diagnostic color. Extraction involves the use of two immiscible solvents, so that a two-phase liquid system results. In some cases, the test reaction occuring in one layer may be obscured by the color of the reagents present. If the reaction product can be isolated in a different phase, and has a diagnostic color in that phase, the positive result can be seen. The identification of the halides (see Experiment 34) is a good example.

The chemistry involves the relative polarities of the product and the two liquid phases. The rule of polarity is that "like dissolves like," i.e., polar compounds dissolve better in polar solvents than in nonpolar solvents. Organic solvents tend to be nonpolar, whereas water and acids are polar.

PART B: SEPARATION OF CATION GROUPS A, B1, B2, C, D, AND E

Introduction

Qualitative chemical analysis remains important in the introductory inorganic chemistry laboratory. It introduces the chemical behavior of ions in solution and how that behavior can be utilized. We will discuss 23 of the more common metal cations. They are listed in Table 9.2 (the ammonium ion, NH_4^+ is not a metal cation, but is included as an important cation).

There is one central problem in inorganic qualitative analysis: Given an aqueous solution, how does one go about identifying, without doubt, what cations and anions are present? Unique tests for all cations in the possible presence of many other cations are virtually impossible to devise. Many tests give similar results with several cations. A mixture is not easily analyzed directly for each cation. However, if a solution is treated such that cations are subdivided into smaller groups, identification is simplified.

The most common way to subdivide into smaller groups is by selective precipitation, in which a small group of cations is precipitated chemically. They can then be physically separated from those remaining in solution by centrifuging. The **precipitate** (solid) settles out and the solution **(supernatant liquid)** is transferred into another container. In this fashion, the initial large group is separated into smaller and smaller groups, until finally, a definitive test can be performed to verify the presence or absence of each specific cation.

It is important that one major distinction be recognized: the groups of the Periodic Table (alkali metals, halogens, etc.) and the groups referred to in a qualitative analysis scheme are not the same. Periodic Table groups are based on similarities in electronic configuration. While many properties of elements do fall in groups that coincide with the periodic table groups, the groups used in qualitative analysis do not necessarily match those in the periodic table groups. To aid in distinguishing the groups in the qualitative analysis scheme, the designations A, B1, B2, C, D, and E are used.

Table 9.2 K_{sp} Values of Metal Compounds[a]

Cation	Precipitating Agents					
	Cl^-	OH^-	$C_2O_4^{2-}$	CO_3^{2-}	CrO_4^{2-}	SO_4^{2-}
Group A	**(i. chloride)**					
Ag^+	1.7×10^{-10}	1.5×10^{-8}	1.3×10^{-11}	6.2×10^{-12}	1.1×10^{-12}	7.0×10^{-5}
Hg_2^{2+}	1.1×10^{-18}	Decomp.		8.9×10^{-17}	2.0×10^{-9}	—
Pb^{2+}	1.6×10^{-5}	4.0×10^{-15}	2.7×10^{-11}	4.0×10^{-14}	1.8×10^{-11}	1.0×10^{-8}
Subgroup B1	**(i. alkali)**					
Mn^{2+}	Soluble	4.5×10^{-14}	Insoluble	9.8×10^{-11}	Soluble	
Fe^{3+}	Soluble	1.1×10^{-36}	Soluble	—	Soluble	
Bi^{3+}	Soluble	Insoluble	Soluble			
ZrO^{2+}	Soluble	i (ZrO_2)				
Subgroup B2	**(sol. alkali)**					
Al^{3+}	Soluble	3.7×10^{-15}	Sol. alkali	—	Soluble	
Cr^{3+}	Soluble	6.7×10^{-31}	—	Soluble	—	Soluble
Sn^{4+}	Soluble	1.0×10^{-26}	Sol. alkali			
Group C	**(i. NH_3)**					
Ba^{2+}	Soluble	5.0×10^{-3}	1.6×10^{-7}	1.9×10^{-8}	2.3×10^{-10}	1×10^{-10}
Ca^{2+}	Soluble	7.9×10^{-6}	2.6×10^{-9}	9.3×10^{-9}	7.1×10^{-4}	2.3×10^{-4}
Co^{2+}	Soluble	2.5×10^{-16}	insol	insol	1.0×10^{-12}	Soluble
Sr^{2+}	Soluble	3.2×10^{-4}	1.6×10^{-8}	1.6×10^{-9}	3.6×10^{-5}	2.9×10^{-7}
Group D	**(sol. NH_3)**					
Cu^{2+}	Soluble	6×10^{-20}	Sol. NH_3	1×10^{-10}	3.6×10^{-6}	Soluble
Mg^{2+}	Soluble	1×10^{-11}				
Ni^{2+}	Soluble	1.6×10^{-16}	Sol. NH_3			
Cd^{2+}	Soluble	1.2×10^{-14}	Sol. NH_3			
Zn^{2+}	Soluble	1.0×10^{-14}	Sol. NH_3			
Group E						
Li^+, Na^+, K^+	Soluble	Soluble	Soluble			
NH_4^+	Soluble	Soluble	Soluble			

[a]Abbreviations: i = insoluble, s NH_3 = soluble in NH_3, alk = soluble in alkali, i neut = insol. in deionized water.

The separation scheme historically used in qualitative inorganic analysis is based on the solubilities of the metal chlorides and sulfides. In that scheme, the metal cations are divided into five groups, numbered Groups I, II, III, IV, and V. Group I is separated out as insoluble chlorides, while groups II and III are precipitated as insoluble sulfides. The precipitating agent originally was H_2S, a toxic and foul-smelling gas. Later modifications of this classical method generated H_2S more safely by using thioacetamide. This was somewhat of an improvement, but thioacetamide is classified as a carcinogen, and still has a foul smell. In this laboratory textbook, neither material is used, thus avoiding some potential safety and environmental problems.

Equilibrium and the Separation of Cations

The problem now becomes what reagents to use, in what order, and how much? To answer these questions, the solubility product constants of various precipitates of these cations must be examined. It is useful to consider the formation constants of various complex ions formed by the metal cations.

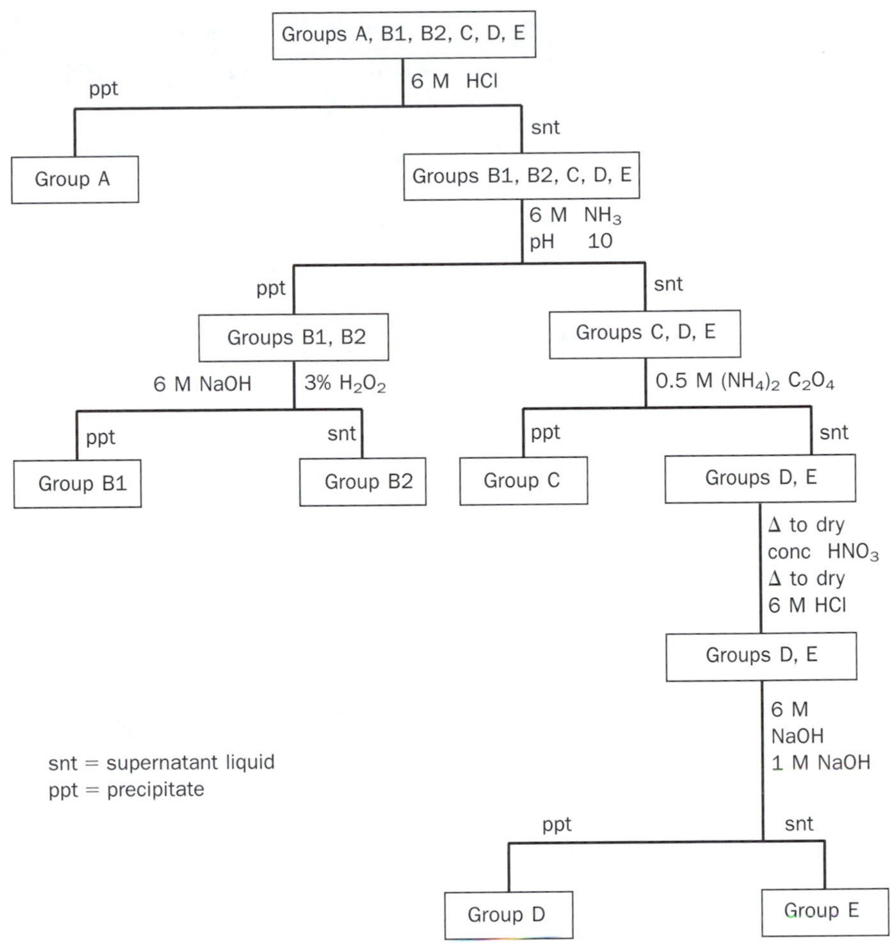

Figure 9.1 *Flow chart for separation of groups.*

Table 9.2 lists the K_{sp} values of precipitates under consideration in this qualitative analysis scheme. The cations are organized according to the subgroups used in this procedure. The object is to select reagents and conditions that result in the precipitation of a small group of 3–6 cations. A flowsheet of the separation procedure is shown in Figure 9.1.

Group A: The Insoluble Chlorides

From Table 9.2, it is readily seen that a small group, Ag^+, Hg_2^{2+}, and Pb^{2+}, are the only subset of cations that form **insoluble chlorides.** To separate this chloride group, one need only add a source of chloride ions to the solution. Any chloride could be used, but the best would be aqueous hydrochloric acid, as it does not add any other metal cations. It has the further advantage of making the concentration of hydroxide ion very low, preventing the possible co-precipitation of any hydroxide or oxide. Further separation of these cations from each other is based on the chemistry of this smaller group of cations and is discussed in Experiment 30.

Group B: Hydroxides Insoluble in Ammonium Hydroxide

Of the 20 remaining cations, it can be seen that many of them form insoluble hydroxides. On closer inspection, note that they can be subdivided based on hydroxide solubility in NH_3 and oxidizing alkaline (e.g., $NaOH/H_2O_2$) solution. The Group B cations are a rather large group, (Mn^{2+}, Fe^{3+}, Bi^{3+}, ZrO^{2+}, Al^{3+}, Cr^{3+}, and Sn^{4+}), all having hydroxides (or oxides) that are insoluble in NH_3

Table 9.3 Soluble Complexes of Metal Cations

Cation	NH_3	OH^-	O^{2-}	Other
Ag^+	$Ag(NH_3)_2^+$			$Ag(S_2O_3)_2^{3-}$
Pb^{2+}		$Pb(OH)_4^{2-}$		
Cu^{2+}	$Cu(NH_3)_4^{2+}$	$Cu(OH)_4^{2-}$		
Co^{2+}	$Co(NH_3)_6^{2+}$			$Co(SCN)_4^{2-}$
Co^{3+}	$Co(NH_3)_6^{3+}$			$Co(NO_2)_6^{3-}$
Cr^{3+}	$Cr(NH_3)_5Cl^{2+}$	$Cr(OH)_4^-$		
Cr^{6+}			CrO_4^{2-}, $Cr_2O_7^{2-}$	
Ni^{2+}	$Ni(NH_3)_6^{2+}$			$Ni(DMG)_2$ insoluble
Zn^{2+}	$Zn(NH_3)_4^{2+}$	$Zn(OH)_4^{2-}$		
Al^{3+}		$Al(OH)_4^-$		
Cd^{2+}	$Cd(NH_3)_4^{2+}$			
Sn^{4+}		$Sn(OH)_6^{2-}$		$SnCl_6^{2-}$
Fe^{3+}				$FeCl_4^-$, $FeSCN^{2+}$, FeF_6^{3-}, $Fe(CN)_6^{3-}$
Mn^{7+}			MnO_4^-	

solution. After they have been precipitated and separated by centrifuging, they can be further divided into two subgroups of 4 and 3 cations, respectively.

This subdivision is based on the fact that the hydroxides of four of these cations (Subgroup B1: Mn^{2+}, Fe^{3+}, Bi^{3+}, ZrO^{2+}) remain insoluble in oxidizing alkaline solution. The other three (Subgroup B2: Al^{3+}, Cr^{3+}, Sn^{4+}) are soluble in oxidizing alkaline solution. These cations are soluble because they form soluble hydroxide or oxide complexes (see Table 9.3). In the case of Cr^{3+}, an oxidizing agent, such as H_2O_2, must be present to convert the chromic hydroxide, $Cr(OH)_3$, to the soluble chromate ion, CrO_4^{2-}.

The insoluble hydroxides of Groups B are therefore handled in the following way:

1. By maintaining the pH at about 9–10 (using a NH_3/NH_4^+ buffer), the hydroxides or oxides of all Group B cations will precipitate. The Group C and Group D cations remain in solution, to be dealt with later (Experiments 32 and 33).
2. The precipitate containing the Group B cations is first treated with 6 M NaOH and H_2O_2. This dissolves the Subgroup B2 cations.
3. The precipitate containing the Group B1 cations is then treated with hot water to destroy any remaining H_2O_2, and is then analyzed (Experiment 31, Part A).
4. The liquid solution from step 2 is also heated to destroy any remaining H_2O_2, and is then analyzed for Group B2 (Experiment 31, Part B).

Group C: Oxalates Insoluble in Ammonium Hydroxide

Of the 13 cations remaining in solution after removal of groups A and B, we can see that four (Ba^{2+}, Ca^{2+}, Co^{2+}, and Sr^{2+}) form oxalates that are insoluble in NH_3 solution. Other cations form insoluble oxalates in water, but they are soluble in NH_3 solution.

Thus, the Group C cations are precipitated from solution by adding NH_3 to adjust the pH to about 9–10, and then adding ammonium oxalate solution. The solution is centrifuged and the precipitate separated (Experiment 32).

Group D: Hyroxides and Oxides Soluble in Ammonia but Insoluble in NaOH

After Groups A, B, and C are removed 9 cations remain in solution. Five of these (the Group D cations mentioned earlier, Cu^{2+}, Mg^{2+}, Ni^{2+}, Zn^{2+}, and Cd^{2+}) form hydroxides or oxides that are insoluble in basic (NaOH) solution. Since most of these are soluble in NH_3 solution, it is necessary to first remove any NH_3 that might be present in solution. If ammonia or oxalate are present from the previous precipitation of the Group C cations, they can be simultaneously decomposed by evaporating the solution to dryness, in a crucible, over a low flame. Any oxalate present is converted to CO_2, and any ammonium ion to NH_3. These are eliminated as gases. A small quantity of concentrated HNO_3 is then added to complete the decomposition to the cooled crucible, which is heated to dryness once again.

The Group D cations remain behind as solid oxides. Any remaining cations, including Group E, would also be present at this stage as oxides or hydroxides. This hydroxide-oxide precipitate is dissolved in HCl to yield the dissolved cations (Groups D and E). The Group D cations are then precipitated by the addition of NaOH, forming insoluble hydroxides (Experiment 33). The Group E cations remain in solution.

Group E: The Soluble Cations

After Groups A–D are removed, four cations remain in solution: Li^+, Na^+, K^+, and NH_4^+. These are the Group E cations. They are characterized by their great stabilities (i.e., they are nonreactive). The first three are in the alkali metals group. The ammonium ion is not a metal cation, but since it is commonly encountered, it is included in our analysis.

Subsequent Experiments

The following experiments examine each of the groups (A–D) discussed previously in more detail. The reactions that are used to achieve separations are presented within the discussion for each group. A blank flow chart accompanies each experiment. The reagents and conditions for each step are on the flow charts, but the fate of the cations at each step should be filled in as a pre-laboratory exercise.

After reading through the discussion and procedure for each experiment, fill in the flow chart with the correct form of each cation at each stage, indicating ionic charges, phases, and colors. The best way to accomplish this is to follow the steps in the procedure, and simultaneously follow the reactions in the discussion to ascertain what each cation is doing at each step. Divided page problem solving techniques (see Exp. 30, page 399) are especially helpful here.

PART C: QUALITATIVE ANALYSIS TECHNIQUES

Cleanliness

Make sure that all test tubes and stirring rods are clean. Rinse the test tubes with deionized water and shake out as much of the liquid as possible before use. Rinse stirring rods before using them with a second solution. For the anion tests, rinse the spot plate thoroughly with tap water, then give it a final rinse with deionized water. Shake it to remove as much deionized water as possible before use.

Adding Reagents

Use clean medicine droppers. Generally, all liquid reagents will be dispensed from bottles equipped with dropper caps. Be sure to replace the cap on the correct bottle. Screw it on firmly if it is a screw cap. Never place the tip of a dispensing medicine dropper into the test solution in the test tube. Insert the tip ~0.5 cm below the top of the test tube, and release the indicated number of drops.

Mixing

If a small amount of liquid is present in a test tube, it may be mixed by flicking the base of the test tube with a finger while holding the test tube lightly by the top. Never shake a test tube that is capped with a finger or cork. Getting chemicals on the fingers is an excellent means of introducing them into your body. Even if gloves are used, it is an easy means of contaminating other solutions. If a test tube is capped with a finger or a cork, pressure may build up due to the evolution of heat or a gas in the test tube. When the pressure is released, chemicals can spray out of the test tube.

If the flicking technique is unsuccessful, or if the test tube is more than one-third full, a thin glass stirring rod should be used to mix the contents. Unless otherwise directed, always mix thoroughly after adding each reagent before making observations, checking pH or proceeding to the next step.

How to Describe Mixtures

Always describe the color and clarity of mixtures and reagents before mixing and what the mixture looks like after mixing, heating, centrifuging, etc. "Starting solution—clear and colorless, 6 M HCl—clear and colorless. Add 10 drops and stir mixture—cloudy white, precipitate formed. After centrifuging, white precipitate, clear and colorless supernatant."

Centrifuging

Be sure that the test tubes in use are the appropriate size for the centrifuge. Each tube must be balanced by a tube of approximately the same mass in the opposite slot of the centrifuge. This is easily accomplished by using a test tube of the same size, which is filled to approximately the same height. If you are simultaneously testing a known and an unknown, they can usually be centrifuged against each other. Other test tubes in the centrifuge may be of different masses, but each opposite pair should be matched. If the centrifuge is too unbalanced, it may "walk" around the countertop while it is spinning.

Be sure that test tubes being centrifuged are neither cracked nor chipped. The stress applied by the centrifuge can cause damaged test tubes to shatter, resulting in chemicals and pieces of glass being flung around the room. A centrifuge without a top is dangerous—if there is a top for the centrifuge, USE IT. If not, a rigid plastic bucket is an acceptable alternative.

Do not slow centrifuges down with your hands. They are spinning at a high rate of speed, and if there is any imperfection on the spinning surface, it can catch the flesh and do a great deal of damage in only an instant. Likewise, long hair must be tied back.

Decanting

After centrifuging, the supernatant is usually decanted into a clean test tube. Carefully tip the test tube, and pour off the supernatant without disturbing the pellet of solid. It may be poured directly, or a stirring rod may be placed across the mouth of the test tube to direct the supernatant into a clean test tube. If the supernatant has some solid, it may be clarified in the following manner: Twist a small piece of clean cotton batting and insert it into the end of a clean eyedropper from the outside (not pushed down the barrel), leaving a tuft on the outside. Draw the supernatant up through the cotton, remove the cotton, and release the clear supernatant into a clean test tube. The small amount of precipitate on the cotton is discarded.

Washing a Precipitate

After separation from the supernatant, a precipitate is often washed to free it from reagents that might interfere at a later stage. Usually, the rinse is deionized water, but other solutions may be used. Add the indicated amount of the wash liquid and stir the contents of the test tube thoroughly. The pellet of solid must be broken up and mixed well with the wash liquid.

Heating

Due to the small quantity of material being heated, these test tubes should NEVER be heated directly in a flame. Within a few seconds, the material will reach the boiling point and will be ejected violently from the test tube. All heating should be done in a water bath (a 100 mL beaker for 10×75 mm test tubes) on a hot plate. Be careful that the tops of the test tubes are well above the water. The water may be boiling at times and could spatter into the test tubes, contaminating the contents.

Testing pH

When directed to check the pH of a solution, stir the solution thoroughly with a clean glass stirring rod or Pasteur pipet, and then touch the tip of the rod to a piece of litmus paper (wide range pH paper is the most useful). Several such tests may be performed on each strip of paper. Never insert the test paper into the test tube, as the chemicals on the paper could contaminate the contents.

General Safety Tips

Add all reagents gradually. Heat may be evolved, and the solution could get hot enough to boil. This is most likely to occur when neutralizing strong acids and bases. If a gas is evolved, as when dissolving a carbonate solid in acid, the solution could bubble out of the test tube. Never situate a test tube so it is pointing at anyone. Never smell the contents of a test tube directly (Figure 1.6). If directed to check an odor, hold the test tube about 15 cm from the face, and gently waft any fumes from the top of the test tube toward the nose.

Experiment 30 Separation and Identification of Group A Cations (Pb^{2+}, Hg_2^{2+} and Ag^+)

OBJECTIVE

- To use a qualitative analysis scheme to separate and identify the Group A cations.

PRIOR READING

Chapter 9: Introduction
Page 399: Divided Page Problem Solving

INTRODUCTION

Of the 23 commonly encountered inorganic cations discussed in the Introduction to Quantitative Analysis, only three form insoluble chlorides: lead (II), Pb^{2+}; mercury(I) (mercurous), Hg_2^{2+}; and silver(I), Ag^+. They may be separated from a mixture of other cations by adding aqueous hydrochloric acid, HCl, to the solution, causing them to precipitate. They are then physically separated by centrifuging and decanting the supernatant liquid into another test tube for further work. If centrifuges are not available, separation may often be accomplished by filtering through filter paper in a funnel (see Fig. 3.20, p. 43).

We must now separate the three cations from each other in order to determine which ones are present and which are absent.

Separation and Identification of Group A Cations

The three insoluble chlorides formed upon addition of HCl are lead(II) chloride ($PbCl_2$), silver(I) chloride (AgCl), and mercury(I) chloride (Hg_2Cl_2). All are white solids. Of the three, only lead chloride is soluble in hot deionized water. Silver chloride is soluble in aqueous ammonia and mercury(I) chloride reacts with aqueous ammonia.

The following reactions apply:

$$PbCl_{2(s)} + \text{hot } H_2O_{(l)} \rightleftharpoons Pb^{2+}_{(aq)} + 2\ Cl^-_{(aq)}$$

$$AgCl_{(s)} + 2\ NH_{3(aq)} \rightleftharpoons Ag(NH_3)^+_{2(aq)} + Cl^-_{(aq)}$$

$$Hg_2Cl_{2(s)} + 2\ NH_{3(aq)} \rightleftharpoons \underset{\text{white}}{HgNH_2Cl_{(s)}} + \underset{\text{black}}{Hg^\circ} + NH^+_{4(aq)} + Cl^-_{(aq)}$$

From this information a separation scheme can be derived:

1. The precipitate is treated with hot deionized water, which will dissolve any lead(II) chloride present. The presence of lead(II) ion can be detected in the supernatant liquid by adding KI solution. The golden lead iodide will precipitate if lead is present. This colored precipitate is the **confirming test for lead ion.**
2. The remaining precipitate is treated with aqueous ammonia. Any silver chloride present will dissolve, due to the formation of the soluble complex $[Ag(NH_3)_2]^+$. Simultaneously, if any mercurous chloride is present, it will react with the ammonia to yield a new precipitate, a mixture of mercuric amidochloride, $HgNH_2Cl$ and elemental mercury. The former is a white compound, and elemental mercury, when finely divided, will appear gray to black. This reaction is an example of a disproportionation reaction.

The supernatant liquid, containing any dissolved $[Ag(NH_3)_2]^+$, is saved to test for silver. If the precipitate is gray to black, **this is the confirming test for mercury(I) ion.**

3. The supernatant liquid from step 2 is tested for silver by adding nitric acid. The acid decomposes the $[Ag(NH_3)_2]^+$ complex. If silver is present, it will react with the chloride ion (present from dissolving AgCl) and precipitate. **This is the confirming test for silver ion.**

EXPERIMENTAL PROCEDURE

Although the directions are written for testing a known mixture all the way through, followed by testing an unknown mixture, it is usually convenient to do Parts A (the known mixture) and B (your unknown) simultaneously. NOTE: All of the Group A cations are toxic. Wear gloves. Avoid contact and wash up immediately if any is spilled or splashed on you. Wear eye protection at all times.

Part A: Precipitation of the Group A Cations in a Known Mixture

- Obtain 10 drops of a solution that is 0.1 M in each of the Group A cations (Pb^{2+}, Hg_2^{2+}, and Ag^+) in a centrifuge tube. Add 3 drops of 6 M HCl and stir. When precipitation is complete, centrifuge the mixture for about two minutes. (Alternatively, filter the entire mixture through filter paper.)
- If you are working only with Group A cations, discard the supernatant liquid (filtrate). If cations from other groups may be present, decant (collect) the supernatant liquid (filtrate) into a clean 10 × 75 mm test tube, labeled Groups B–E, and set this aside. Record all observations.

Separation and Identification of Pb^{2+}

- Prepare a boiling water bath. Wash the precipitate saved from the previous step twice with 10 drops of deionized water. Centrifuge and discard the washes. Add about 20 drops of water to the precipitate, and heat it in the boiling water bath for at least 3 minutes, stirring constantly. Centrifuge quickly, and decant the hot supernatant liquid into a clean 10 × 75 mm test tube. Label the centrifuge tube containing the precipitate "Hg_2Cl_2, and AgCl," and set it aside. Record all observations on the data sheet.
- Confirm the presence of lead by adding three drops of 1 M KI (potassium iodide) solution to the test tube containing the supernatant liquid (filtrate). Record all observations. Dispose of the lead waste (if any) as directed by your laboratory instructor.

Separation and Identification of Hg_2^{2+}

NOTE: ***6 M ammonia is an irritant.***

- Add 10 drops of deionized water to the centrifuge tube containing the precipitate "Hg_2Cl_2, and AgCl" and heat for 2 minutes in the boiling water bath, stirring continuously. Centrifuge quickly, and discard the supernatant liquid. Add an additional 10 drops of deionized water to the precipitate, and repeat this step.
- Add 10 drops of 6 M NH_3, stir, and centrifuge. Record all observations. Decant the supernatant liquid into a clean 10 × 75 mm test tube labeled

"$[Ag(NH_3)_2]^+$" and set it aside. Record all observations. Dispose of the mercury waste (if any) as directed by your laboratory instructor.

Separation and Identification of Ag^+

- Add 6 M HNO_3 (nitric acid) dropwise, with stirring, to the supernatant liquid saved from the previous section (labeled "$[Ag(NH_3)_2]^+$") until the solution just tests acidic to litmus. Record all observations. Dispose of silver waste (if any) as directed by your laboratory instructor.

Part B: Analysis of an Unknown Containing the Group A Cations

- Obtain 10 drops of an unknown solution containing one or more of the Group A cations in a clean centrifuge tube. Repeat the procedures outlined in Part A, using the unknown solution. Report which cations are present in the unknown, along with observations to verify the identification.

Divided Page Problem Solving

The use of divided page problem solving is especially helpful when running a series of tests on an unknown, such as in this experiment. The left-hand column can list the test being performed. Use the center columns to keep track of results (observations) for the known and the unknown. Use the right-hand column for your conclusions. This is illustrated below for the 3 cations and the tests of this experiment. (ppt. = Precipitate.)

Reaction	*Result-known*	*Result-Unknown*	*Conclusion*
1. Add 3 drops 6 M HCl. This precipitates Group A cations.	Solution becomes cloudy white.	Solution becomes cloudy white.	At least one Group A cation is present in the unknown.
2. Centrifuge and decant the supernatant liquid.	White ppt. is seen. Supernatant is clear, colorless.	White ppt. is seen. Supernatant is clear, colorless.	At least one Group A cation is present in the unknown.
3. Add water, boil. Centrifuge. Decant supernatant liquid, which may contain Pb^{2+}.	White ppt. is seen. Supernatant is clear, colorless.	White ppt. is seen. Supernatant is clear, colorless.	Can't tell if any ppt dissolved.
4. Test supernatant liquid for Pb^{2+}: Add 3 drops 1 M KI. Yellow ppt forms if Pb^{2+} present.	Yellow ppt. is seen	Yellow ppt. is seen.	Both known and unknown have Pb^{2+}. **Lead is present.**
5. Wash the ppt. with water. Add 10 drops. 6 M NH_3. Ag^+ will dissolve, Hg_2^{2+} will form gray-black ppt.	Ppt. turns gray—supernatant is colorless.	Ppt. all dissolves. Supernatant is colorless.	Known has Hg_2^{2+}. **No mercury in the unknown.**
6. Add 6 M HNO_3 to supernatant liquid until soln tests acidic. White ppt if Ag^+ present.	Took 12 drops to become acidic. White ppt. is seen.	Took 13 drops to become acidic. White ppt. is seen.	Both known and Unknown have Ag^+. **Ag is present.**

Name: ______________________________

Date: __________ Section: ______________________________

Pre-Laboratory Questions: Experiment 30

1. Why is the precipitate washed with hot water after separating out Pb^{2+}? What is the purpose of the washes?

2. An unknown is treated according to the procedure of this experiment. At each stage below, state what cations may be present, which are definitely present, and which are definitely absent.

a. The white chloride precipitate is washed and treated with hot water. A white precipitate and colorless supernatant results.

b. The supernatant from (a) is treated with KI. A clear, colorless solution results.

c. The precipitate from (a) is treated with $NH_{3(aq)}$. A black precipitate and a colorless solution result.

d. The solution from (c) is treated with HNO_3. A white precipitate results.

Name: ______________________________

Date: __________ Section: ______________________

Pre-Laboratory Flow Sheet: Experiment 30

3. Fill in the following flowsheet.

Flow chart for Group A separation.

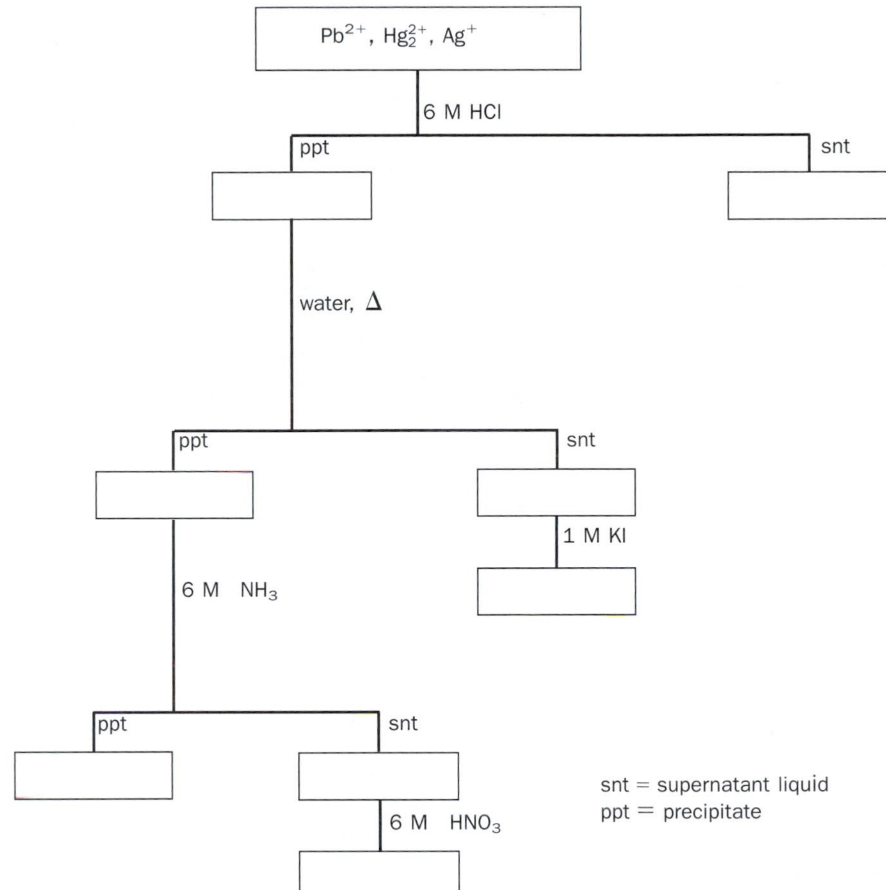

Name: ______________________________

Date: __________ Section: ______________________________

Data Sheet: Experiment 30

1. Group A Cations: Analysis of a Known Cation Mixture

A. Observations: Separation of Group A:

B. Observations: Separation and Identification of Pb^{2+}:

C. Observations: Separation and Identification of Hg_2^{2+}:

D. Observations: Separation and Identification of Ag^{+}:

Name: ____________________

Date: ________ Section: ____________________

Data Sheet: Experiment 30, p. 2

Group A Cations: Unknown Analysis

Unknown Number ________

Cations Detected: ________ ________ ________

Observations to Support the Above Conclusions:

Draw a Flowsheet Showing the Steps and Products in Your Unknown Analysis

Name: ______________________________

Date: __________ Section: ______________________________

Questions: Experiment 30

1. Could aqueous ammonia be used instead of hot water to separate lead ion from the mixed chloride precipitate? Explain.

2. The test for Ag^{+} is to add HNO_3 and see if a precipitate of AgCl forms. Where does the Cl^{-} come from? Why does adding HNO_3 make AgCl precipitate? Write an appropriate balanced reaction.

3. In the step where Ag^{+} and Hg_2^{2+} are being separated by addition of 6 M NH_3, a student obtains a black precipitate and a clear, colorless supernatant for both the known and the unknown. The student concludes that both cations are present because the results are the same. Assess the student's conclusions for their accuracy, being sure to explain why the student is or is not correct for each cation.

Experiment 31 Separation and Identification of the Group B Cations (Bi^{3+}, Fe^{3+}, ZrO^{2+}, Mn^{2+}, Al^{3+}, Cr^{3+}, and Sn^{4+})

OBJECTIVE

- To use a qualitative analysis scheme to separate and identify the Group B cations.

PRIOR READING

Chapter 9: Introduction

Page 399: Divided Page Problem Solving

INTRODUCTION

The Group B cations are characterized by having hydroxides and oxides that are insoluble in ammonia solution. These cations are

Subgroup B1: bismuth(III), Bi^{3+}; iron(III), Fe^{3+}; zirconyl, ZrO^{2+}; manganese(II), Mn^{2+};

Subgroup B2: aluminum, Al^{3+}; chromium(III), Cr^{3+} and tin(IV), Sn^{4+}.

Separation of the Group B Cations

Refer to the Introduction to Inorganic Qualitative Analysis for a discussion of the separation scheme for the six groups of cations, A, B1, B2, C, D, and E. Initially, Groups B1 and B2 are simultaneously precipitated as hydroxides or oxides that are insoluble in ammonia solution (pH 9–10—this prevents precipitation of the more soluble Group C and D cations as oxides and hydroxides). Group B1 is then separated from Group B2 by adding a strongly alkaline solution containing hydrogen peroxide (H_2O_2), which will dissolve the Group B2 oxides and hydroxides. After centrifuging and decanting, the precipitate containing the Group B1 cations is washed with deionized water and heated to destroy excess peroxide.

The chemistry discussed below assumes that the unknown is being analyzed for the presence of all six groups.

PART A: PRECIPITATION OF THE GROUP B1 CATIONS

In order to determine whether a B1 Group cation is present in an unknown, the presence of Group A cations must first be determined. As discussed in Experiment 30, this is done by the addition of hydrochloric acid. The Group B1 cations do not precipitate when HCl is added.

Upon the addition of ammonia, the Group B1 cations precipitate as white bismuth(III) hydroxide, $Bi(OH)_3$; red-brown iron(III) hydroxide, $Fe(OH)_3$; white zirconyl hydroxide, $ZrO(OH)_2$; and pale pink manganese(II)hydroxide, $Mn(OH)_2$, as shown in Equations 1 through 4 (not balanced):

$$Bi^{3+}_{(aq)} + NH_{3(aq)} + H_2O \rightleftharpoons Bi(OH)_{3(s)} + NH^+_{4\,(aq)} \quad (1)$$

$$Fe^{3+}_{(aq)} + NH_{3(aq)} + H_2O \rightleftharpoons Fe(OH)_{3(s)} + NH^+_{4\,(aq)} \quad (2)$$

$$ZrO^{2+}_{(aq)} + NH_{3(aq)} + H_2O \rightleftharpoons ZrO(OH)_{2(s)} + NH^+_{4\,(aq)} \quad (3)$$

$$Mn^{2+}_{(aq)} + NH_{3(aq)} + H_2O \rightleftharpoons Mn(OH)_{2(s)} + NH^+_{4\,(aq)} \quad (4)$$

$Bi(OH)_3$, $Fe(OH)_3$, and $ZrO(OH)_2$ do not react further with either NaOH or H_2O_2; hence the bismuth, zirconyl, and iron are present as $Bi(OH)_3$, $ZrO(OH)_2$, and $Fe(OH)_3$ at the start of the Group B1 separation.

Upon addition of NaOH and H_2O_2, the Mn(II) is oxidized to form the brown-black manganese dioxide, MnO_2, according to Equation 5 (not balanced):

$$Mn(OH)_{2(s)} + H_2O_{2(aq)} \rightleftharpoons MnO_{2(s)} + H_2O_{(l)} \quad (5)$$

Thus, after separating the Group B1 cations from all other groups, the cations are present as a precipitate containing $Bi(OH)_3$, $Fe(OH)_3$, $ZrO(OH)_2$, and MnO_2. At this point, the Group B1 cations are physically separated from the other groups by centrifuging. The precipitate is then washed and heated to destroy excess peroxide.

The next step is to dissolve the precipitate in hot HCl solution. The oxides and hydroxides dissolve, regenerating the cations Bi^{3+}, Fe^{3+}, ZrO^{2+}, and Mn^{4+} (some may be reduced back to Mn^{2+}). The tests for each of these four cations can be carried out without further separation, although some chemical manipulation is necessary to mask similar reactions from one or more of the other cations in the group. An aliquot of this acidic solution will be used for each test.

Separation and Identification of the Group B1 Cations

Zirconyl Ion

The test for zirconyl ion involves the formation of a red-violet complex with Alizarin Red S (HARS), an organic molecule with one acidic hydrogen. HARS undergoes an acid–base reaction with $Zr(OH)_4$, as shown in Equations 6a and 6b (not balanced).

$$ZrO^{2+} + 3\ H_2O_{(l)} \rightleftharpoons Zr(OH)_{4(s)} + 2\ H^+_{(aq)} \quad (6a)$$

$$Zr(OH)_{4(s)} + HARS_{(aq)} \rightleftharpoons Zr(OH)_3ARS_{(aq)} + H_2O \quad (6b)$$

The acidic supernatant liquid (containing the dissolved Group B1 cations) is treated with hydrogen peroxide to reduce any Mn^{4+} to Mn^{2+}, as shown in Equation 7 below, to prevent oxidation reactions and to avoid color confusion. After heating to destroy excess peroxide, a drop of this solution is added to a drop of HARS solution and heated briefly in boiling water. **The appearance of a red-violet color signals the presence of ZrO^{2+} ion.** The product is sufficiently stable to form in moderately acidic solution. The pH must be acidic, as HARS is violet in basic solution, and is indistinguishable from the color of the Zr-HARS product.

Manganese(II) Ion

A portion of the remaining solution is tested for manganese. The intensely purple-colored permanganate ion, MnO_4^-, is the best way to identify the presence of manganese. Permanganate is produced by oxidizing Mn(II) using bismuthate, BiO_3^-, as the oxidizing agent. At this stage, the manganese must be present as Mn(II), as Mn(IV) does not react with bismuthate. This is accomplished by using hydrogen peroxide in acidic solution, followed by adding bismuthate, as shown in Equations 7 and 8 (not balanced).

$$Mn^{4+}_{(aq)} + H_2O_{2(aq)} \rightleftharpoons Mn^{2+}_{(aq)} + O_{2(aq)} + H^+_{(aq)} \quad (7)$$

$$Mn^{2+}_{(aq)} + H^+_{(aq)} + BiO^-_{3\,(aq)} \rightleftharpoons Bi^{3+}_{(aq)} + MnO^-_{4\,(aq)} \quad (8)$$

The bismuthate should be added in several small portions, as the effervescence may be vigorous, until no further reaction occurs. If no Mn^{2+} is present, the bismuthate may react with any undestroyed peroxide or other reducing agents. **It is the color of the supernatant liquid (after adding the bismuthate) that determines the presence or absence of Mn^{2+}. A purple supernatant confirms the presence of Mn^{2+}.**

Bismuth(III) Ion

A portion of the remaining solution is tested for bismuth. One test for the presence of bismuth involves reducing Bi^{3+} [present as $Bi(OH)_3$] to elemental bismuth, Bi°, in basic solution, using tin(II), Sn^{2+}, as the reducing agent. In basic solution, Sn^{2+} exists as the trihydroxostannate(II) ion, $Sn(OH)_3^-$. The following reactions occur (not balanced):

$$Bi^{3+}_{(aq)} + OH^-_{(aq)} \rightleftharpoons Bi(OH)_{3(s)} \tag{9}$$

$$Sn^{2+}_{(aq)} + OH^-_{(aq)} \rightleftharpoons Sn(OH)^-_{3\,(aq)} \tag{10}$$

$$Bi(OH)_{3(s)} + Sn(OH)^-_{3(aq)} + OH^-_{(aq)} \rightleftharpoons Bi^\circ + Sn(OH)_6^{2-}{}_{(aq)} \tag{11}$$

The immediate appearance of a black precipitate of elemental bismuth confirms the presence of Bi(III).

Iron(III) Ion (Ferric Ion)

The last portion of the test solution is divided in two, and two different tests are performed. Potassium thiocyanate solution (KSCN) is added to the first portion. If ferric ion is present, it is identified by the formation of the dark red-brown thiocyanatoiron(III), $FeSCN^{2+}$. The second portion is made less acidic with NH_3 and tested with potassium ferrocyanide solution, $K_4Fe(CN)_6$. This step is necessary because if ferrocyanide was added to the strongly acidic solution, toxic HCN (hydrogen cyanide) gas could be released. If ferric ion is present, it is identified by the formation of the dark blue dye ferric ferrocyanide, $Fe_4[Fe(CN)_6]_3$, also known as Prussian blue. The following reactions occur in acid solution (not balanced):

$$Fe^{3+}_{(aq)} + SCN^-_{(aq)} \rightleftharpoons FeSCN^{2+}_{(aq)} \tag{12}$$

$$Fe^{3+}_{(aq)} + Fe(CN)_6^{4-}{}_{(aq)} \rightleftharpoons Fe_4[Fe(CN)_6]_{3(s)} \tag{13}$$

In the absence of iron, a white, yellow or greenish precipitate may be observed with ferrocyanide, as the other Group B1 cations form insoluble ferrocyanides.

PART B: SEPARATION OF THE GROUP B2 CATIONS

Initially, in Part A, Group B1 and Group B2 were both precipitated as hydroxides and oxides insoluble in ammonia solution (pH 9–10). The two groups were then separated by adding a strongly alkaline oxidizing solution. The Group B2 cations redissolve in these solutions as oxide, peroxide, and hydroxide complexes, leaving the Group B1 cations as a precipitate. The chemistry discussed below for Group B2 assumes that the solution is being analyzed for all 6 analytical groups.

The solution containing the Group B2 cations was first treated to remove any Group A cations by adding aqueous HCl. The Group B1 and B2 cations were then precipitated with the addition of aqueous ammonia. The following unbalanced reactions occurred for the Group B2 cations (simultaneously with reactions 1–4 above):

$$Al^{3+}_{(aq)} + NH_{3(aq)} \rightleftharpoons Al(OH)_{3(s)} + NH^+_{4(aq)} \tag{14}$$

$$Cr^{3+}_{(aq)} + NH_{3(aq)} \rightleftharpoons Cr(OH)_{3(s)} + NH^+_{4(aq)} \tag{15}$$

$$Sn^{4+}_{(aq)} + NH_{3(aq)} \rightleftharpoons SnO_{2(s)} + NH^+_{4(aq)} \tag{16}$$

The Group B2 cations were then redissolved in basic solution containing H_2O_2 to form oxide, peroxide or hydroxide complexes. The following reactions (unbalanced) occur:

$$Al(OH)_{3(s)} + OH^-_{(aq)} \rightleftharpoons Al(OH)^-_{4\,(aq)} \tag{17}$$

$$Cr(OH)_{3(s)} + 4\ OH^-_{(aq)} + H_2O_{2(aq)} \rightleftharpoons CrO_4^{2-}{}_{(aq)} \tag{18}$$

$$SnO_{2(s)} + OH^-_{(aq)} \rightleftharpoons Sn(OH)_6^{2-}{}_{(aq)} \tag{19}$$

The Group B1 cations remain as precipitates as discussed earlier.

Separation of the Group B2 Cations

After centrifugation and separation from the B1 cations, the supernatant containing the Group B2 cations is heated to destroy excess peroxide. The Group B2 cations are now present as tetrahydroxoaluminate(III), $Al(OH)_4^-$; hexahydroxostannate(IV), $Sn(OH)_6^{2-}$; and chromate, CrO_4^{2-}. The only colored species at this stage is the chromate, which is yellow.

Tin (IV) Ion (Stannic Ion)

The supernatant liquid is made acidic with HCl, taking care to watch for the appearance of a dark blue to purple color that rapidly fades. This may happen if chromate is present and the hydrogen peroxide has not all been destroyed (see Equations 25 and 26). The appearance and fading of the dark blue color is also the diagnostic test for chromium—make a note of it if it occurs here. The other cations remain soluble, but do undergo reactions as shown in Equations 20–22 (unbalanced).

$$Al(OH)^-_{4\,(aq)} + H^+_{(aq)} \rightleftharpoons Al^{3+}_{(aq)} \tag{20}$$

$$Sn(OH)_6^{2-}{}_{(aq)} + H^+_{(aq)} + Cl^-_{(aq)} \rightleftharpoons SnCl_6^{2-}{}_{(aq)} \tag{21}$$

$$CrO_4^{2-}{}_{(aq)} + H^+_{(aq)} \rightleftharpoons Cr_2O_7^{2-}{}_{(aq)} \tag{22}$$

The cations in acid solution are present as aluminum(III), hexachlorostannate(IV), $SnCl_6^{2-}$, and dichromate, $Cr_2O_7^{2-}$. The solution is now treated with sodium bicarbonate and brought to neutrality. The tin ions precipitate as stannic oxide (see Equation 16). The aluminum and dichromate ions are unaffected (dichromate may be present as a mixture of yellow chromate and orange dichromate, depending on the pH). After centrifuging, the precipitate is tested for tin. The supernatant liquid is saved and tested for Al^{3+} and $Cr_2O_4^{2-}$.

The precipitate containing any tin is dissolved in HCl, generating $SnCl_6^{2-}$. Solid aluminum, Al^0 is added, which reduces Sn(IV) to Sn(II), which remains in solution (Equation 23, unbalanced).

$$SnCl_6^{2-}{}_{(aq)} + Al^0_{(s)} \rightleftharpoons SnCl_4^{2-}{}_{(aq)} + Al^{3+}_{(aq)} + Cl^-_{(aq)} \tag{23}$$

Mercury(II) is now added to the supernatant. If Sn^{2+} is present, it will be oxidized back to Sn^{4+}, and the Hg^{2+} will be reduced to Hg_2^{2+}, which precipitates as the insoluble white Hg_2Cl_2 (Equation 24, unbalanced). **The appearance of the white mercury(I) chloride is the confirming test for the presence of tin ion:**

$$SnCl_4^{2-}{}_{(aq)} + Hg^{2+} + Cl^- \rightleftharpoons SnCl_6^{2-}{}_{(aq)} + Hg_2Cl_{2(s)} \tag{24}$$

Aluminum

The supernatant liquid remaining from the separation of tin may be tested directly for the presence of aluminum and chromium without further separation. The supernatant liquid is divided into two portions.

The test for aluminum involves the adsorption of the red dye, aluminon by the aluminum hydroxide precipitate, changing the color of the precipitate from white to red. Portion #1 is first made acidic to destroy any carbonate ion present. The aluminon dye is added, which colors the solution red. Ammonia is then added to make the solution basic, allowing the hydroxide precipitate to form. In forming, the dye is adsorbed onto the precipitate. Since the solution is red, any precipitate that forms must be centrifuged to determine its color. **A red precipitate after centrifuging is the confirming test for the presence of the aluminum ion.**

Chromium

The test for chromium involves the reduction of dichromate by hydrogen peroxide in acidic solution. Hydrogen peroxide is added to portion #2, and HCl is added. The solution must be watched carefully, as the identifying species, peroxychromate, (CrO_5), decomposes quickly. There is a sudden appearance of dark blue-purple peroxychromate, which fades rapidly as the chromium(VI) is reduced to chromium(III) (see Equations 25 and 26, unbalanced). The peroxychromate has chromium in the VI oxidation state, as it contains one oxide (O^{2-}) ligand and two peroxide (O_2^{2-}) ligands. **The appearance of a dark blue-purple color followed by rapid fading is the confirming test for the presence of the chromium ion.**

$$Cr_2O_7^{2-}{}_{(aq)} + H_2O_{2(aq)} + H^+_{(aq)} \rightleftharpoons CrO_5 \quad (25)$$

$$CrO_5 + H^+_{(aq)} \rightleftharpoons Cr^{3+}_{(aq)} + O_{2(aq)} \quad (26)$$

EXPERIMENTAL PROCEDURE

It is usually convenient to do the known mixture and your unknown simultaneously. Unless otherwise indicated, discarded solutions (including washes) should be collected for disposal. Some solutions with special hazards should be collected separately, according to your instructor's directions.

Part A: Precipitation of the Group B1 Cations in a Known Mixture and Unknown

Be sure to record each reagent added and all observations on the data sheet.

- **[If you are continuing the analysis of a mixture from Experiment 30, skip to the next step and proceed, using the supernatant liquid saved after precipitating the Group A cations.]** Obtain 10 drops of a known solution, which is 0.1 M in each of the following seven cations: Bi^{3+}, Fe^{3+}, ZrO^{2+}, Mn^{2+}, Al^{3+}, Sn^{4+}, and Cr^{3+}. Add 4 drops of 6 M HCl **(Caution: HCl is corrosive).** If any precipitate forms, centrifuge and use the supernatant liquid to continue. The precipitate may be analyzed for the presence of Group A cations.
- Add 6 M NH_3 **(Caution: Irritant)** dropwise, with stirring, to bring the pH of the solution to 9–10. Stir thoroughly before testing the pH. Centrifuge for 2 minutes. If the solution may contain any Group C, D, or E cations, decant the supernatant liquid into a clean, **labeled** test tube and set it aside for later analysis. If the original solution contains only Group B1 and B2

cations, discard this supernatant liquid. Wash the precipitate twice with 10 drops of deionized water, discarding the washes.

- Prepare a boiling water bath. Add 10 drops of 6 M NaOH **(Caution: corrosive and toxic)** and 2 drops of 3% H_2O_2 **(Caution: oxidizer)** to the precipitate, stir thoroughly and allow the solution to sit for 2 minutes. Centrifuge for 2 minutes. Decant the supernatant liquid into a clean, **labeled** test tube and set it aside for analysis in Part B for the Group B2 cations.
- Add 2 mL of deionized water to the remaining precipitate, stir, and heat for 10 minutes in a boiling water bath to destroy any excess peroxide. Centrifuge for 2 minutes, and discard the supernatant liquid. Wash the precipitate with 2 mL of deionized water and centrifuge, discarding the wash.
- Add 10 drops of 6 M HCl to the precipitate, stir, and heat for 2 minutes in the boiling water bath. Centrifuge for 2 minutes, and decant the supernatant liquid into a clean, **labeled** test tube. This supernatant liquid will be tested for all of the Group B1 cations, and is referred to as the **TEST SOLUTION.** Record all observations.

Identification of Zirconyl Ion

- Prepare a boiling water bath. Add 2 drops of 3% H_2O_2 to the test solution. Let the solution sit for 30 sec, then heat the test tube and contents for 2 minutes in the boiling water bath. Cool to room temperature.
- Place 1 drop of Alizarin Red S solution in a clean test tube, and add 1 drop of the cooled test solution. Check that the pH is moderately acidic (pH 4–6). **NOTE: Do not dip the pH paper into the solution. Use a glass rod to transfer one drop of solution to the pH paper.** Heat the solution for 2 minutes in the boiling water bath. Record all observations.

Identification of Manganese(II) Ion

- Place 4 drops of the test solution in a clean test tube. Add several small portions of solid $NaBiO_3$ until no further reaction occurs. If the solution is cloudy, centrifuge to see the color of the supernatant liquid. Record all observations.

Identification of Bismuth(III) Ion

- Place 2 drops of the test solution in a clean test tube. Add 3 drops of 6 M NaOH, and then add a small quantity of solid $SnCl_2$. Record all observations.

Identification of Iron(III) (Ferric) Ion

- Divide the remaining test solution into two parts. To one, add 3 drops of 0.1 M KSCN solution. Record all observations.
- To the second, add 6 M NH_3 until the strongly acidic solution tests only weakly acidic. If it becomes basic, add 6 M acetic acid until it is weakly acidic. Then add 3 drops of 0.1 M $K_4Fe(CN)_6$. Record all observations.

Part B: Precipitation of the Group B2 Cations in a Known Mixture and Unknown

- Retrieve the test tube saved from Part A, containing the solution of group B2 cations. Heat the tube for 10 minutes in a boiling water bath. Cool the test tube and contents for 2 minutes in a cold water bath.

Identification of Tin(IV) Ion

- Add 6 M HCl until the solution tests just acidic, and note any color changes which may occur here.
- Add 1 M sodium bicarbonate ($NaHCO_3$) dropwise (fizzing will occur) until the solution tests neutral (pH paper).

- Centrifuge the test tube and contents for 2 minutes. Transfer the supernatant to a clean test tube for use in the next two sections.
- Wash the precipitate with 10 drops of deionized water, centrifuge, and discard the wash. Record all observations.
- Dissolve the precipitate in 4 drops of 6 M HCl, and add a small piece of solid aluminum. If necessary, add another drop of 6 M HCl to dissolve any excess aluminum. Add 2 drops of 0.1 M $HgCl_2$ **(Caution: $HgCl_2$ is highly toxic)** to the supernatant liquid. Record all observations.

Identification of Aluminum(III) Ion

- Place 2 drops of the supernatant saved from the previous section into a clean test tube. Add two drops of 6 M HCl (fizzing will occur). Add 2 drops of aluminon, and then add 6 M NH_3 until the solution tests basic to pH paper. Centrifuge for one minute. Record all observations.

Identification of Chromium(III) Ion

- Add 1 drop of 3% H_2O_2 to the remaining supernatant from the Identification of Tin(IV) Ion section. Add 6 M HCl (fizzing will occur) until the solution tests acidic to pH paper, observing closely for a flash of color. Record all observations.

PART C: ANALYSIS OF UNKNOWN SOLUTION

- Obtain an unknown solution, and repeat the above procedure. Report which cations are present in your unknown, along with your observations to confirm your analysis.

Name: ____________________

Date: __________ Section: ____________________

Pre-Laboratory Questions: Experiment 31

Part A: Group B1

1. A Group B1 unknown is treated according to the procedure of Part A. At each stage below, state what cations may be present and which are absent.

a. The initial Group B1 oxide/hydroxide precipitate is treated with HCl and a clear yellow solution results.

b. This solution is treated with H_2O_2 and heated. One drop of this solution is added to 1 drop of HARS solution, and a clear yellow solution results.

c. The solution from (a) is treated with $NaBiO_3$. A clear purple solution results.

d. Two drops of the solution from (a) is treated with 6 M NaOH and solid $SnCl_2$. A black precipitate is observed.

e. The remaining solution from (a) is divided in two portions. KSCN is added to one, and a dark red solution results. $K_4Fe(CN)_6$ is added to the other, and a dark blue precipitate results.

Name: ____________________

Date: ________ Section: ____________________

Pre-Laboratory Questions: Experiment 31, p. 2

Part B: Group B2

2. A solution containing Group B2 cations is analyzed using the procedure of Part B, and gives the following results. At each stage, indicate which cations may be present and which are absent.

a. After adding $NaOH/H_2O_2$, centrifuging and heating, a yellow supernatant liquid results.

b. The supernatant from (a) is treated with HCl and $NaHCO_3$, yielding a white precipitate and an orange-yellow supernatant.

c. The precipitate from (b) is treated with HCl, solid aluminum, and yields a white precipitate.

d. The supernatant from (b) is divided in two. One part is treated with HCl, aluminon, and NH_3, yielding a clear, red solution. The second part is treated with 3% H_2O_2, yielding a dark blue solution which quickly fades to colorless.

3. Fill in the following flowsheets, indicating the chemical form of the Group B cations at each stage.

Name: ______________________________

Date: __________ Section: ______________________

Pre-Laboratory Flow Sheet: Experiment 31

PART A Flow Chart for Group B1 separation.

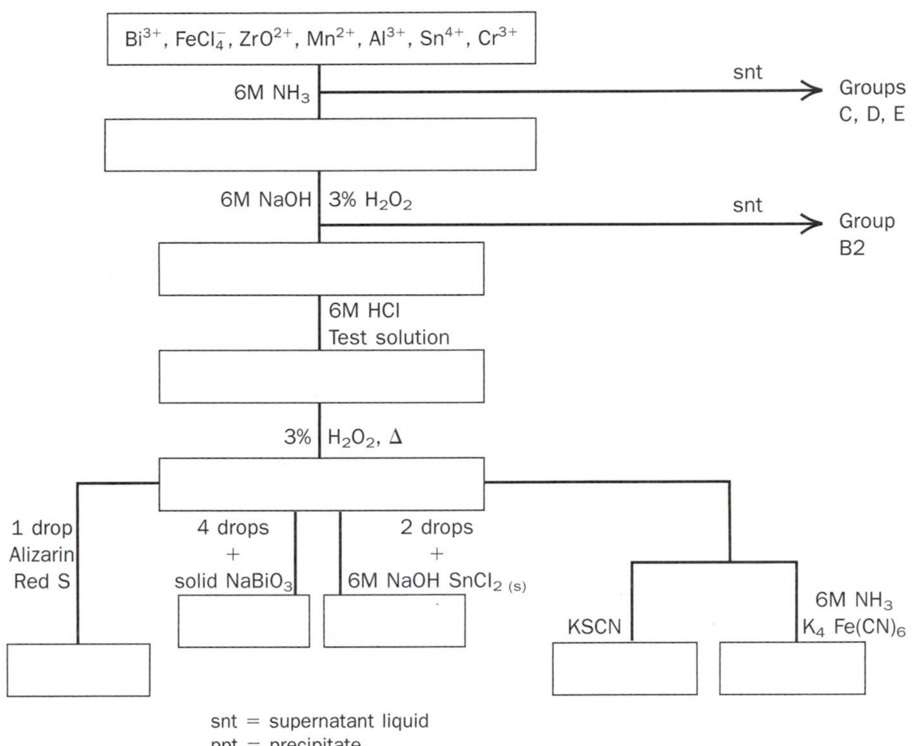

PART B Flow Chart for Group B2 separation.

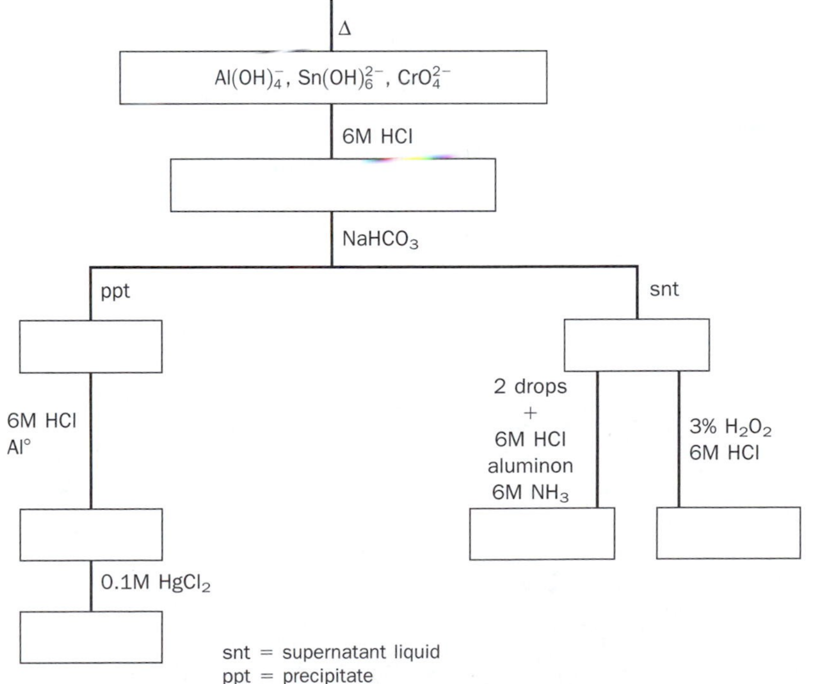

Name: ______________________________

Date: __________ Section: ______________________________

Data Sheet: Experiment 31

Group B1 Cations: Analysis of Known and Unknown Cation Mixtures

Unknown Number ____________

Actions, Reagents	*Result, Known*	*Result, Unknown*	*Conclusion*

Cations detected in unknown: __________ __________ __________

Name: ______________________________

Date: __________ Section: ______________________

Data Sheet: Experiment 31, p. 2

Group B2 Cations: Analysis of Known and Unknown Cation Mixtures

Actions, Reagents	*Result, Known*	*Result, Unknown*	*Conclusion*

Cations detected in unknown: __________ __________ __________

Name: ______________________________

Date: __________ Section: ______________________

Questions: Experiment 31

1. Balance redox Reactions 7, 8, 11, and 18.

2. If a solution is known to contain only Group B1 cations, could the H_2O_2 be left out of the procedure in Part A, which is designed to dissolve out any B2 cations present? Explain.

3. Part of the test for the presence of tin involves adding aluminum. In the known, at least, aluminum is already present. Why does it need to be added to test for tin?

4. Suppose the chromium in a mixture is already present as chromate, CrO_4^{2-}. Would it precipitate out with the Group B cations? Explain. How could the original unknown be checked to see if Cr in this form is present?

Experiment 32 Separation and Identification of the Group C Cations (Ba^{2+}, Ca^{2+}, Sr^{2+}, and, Co^{2+})

OBJECTIVE

- To use a qualitative analysis scheme to separate and identify the Group C cations.

PRIOR READING

Chapter 9: Introduction

Page 399: Divided Page Problem Solving

INTRODUCTION

The Group C cations are characterized by having hydroxides and oxides that are **soluble** in ammonia solutions and oxalates which are **insoluble** in ammonia solution. These cations are barium, Ba^{2+}; calcium, Ca^{2+}; strontium, Sr^{2+}; and cobalt(II), Co^{2+}.

Separation of the Group C Cations

> **NOTE:** ***The redox reactions given below are not balanced. This is an exercise in the post lab questions.***

The solution containing the Group C cations is first treated to remove any Group A cations as insoluble chlorides. The Group B cations are then precipitated as oxides and hydroxides, with the addition of aqueous ammonia. These steps have been previously discussed in Experiments 30 and 31. The Group C cations are then precipitated by adding ammonium oxalate solution.

The cobalt(II) ion is the only Group C cation that reacts prior to the addition of the ammonium oxalate. Cobalt(II) reacts with NH_3 as shown in Reaction (1). The formation of the hexammine-cobalt(II) ion, $Co(NH_3)_6^{2+}$, prevents the precipitation of cobalt(II) hydroxide in ammonia solution. When ammonium oxalate solution is added, cobalt(II) precipitates as cobalt(II) oxalate as shown in Reaction (2).

$$Co^{2+}_{(aq)} + 6\ NH_{3(aq)} \rightarrow Co(NH_3)_6^{2+}{}_{(aq)} \quad (1)$$

$$Co(NH_3)_6^{2+}{}_{(aq)} + C_2O_4^{2-}{}_{(aq)} \rightarrow CoC_2O_{4(s)} + 6\ NH_{3(aq)} \quad (2)$$

The other Group C cations precipitate directly as oxalates:

$$Ba^{2+}_{(aq)} + C_2O_4^{2-}{}_{(aq)} \rightarrow BaC_2O_{4(s)} \quad (3)$$

$$Ca^{2+}_{(aq)} + C_2O_4^{2-}{}_{(aq)} \rightarrow CaC_2O_{4(s)} \quad (4)$$

$$Sr^{2+}_{(aq)} + C_2O_4^{2-}{}_{(aq)} \rightarrow SrC_2O_{4(s)} \quad (5)$$

Cobalt oxalate is light pink in color, and the other oxalates are white.

In order to prevent the oxalate in the precipitate from interfering at a later stage, the oxalate must be destroyed. This is accomplished by transferring the precipitate to a crucible and heating to dryness. Aqueous nitric acid, HNO_3, is then added, and the solution is taken to dryness again. Oxalate is thereby converted to CO_2 and the ammonium ions are oxidized to nitrous oxide (N_2O). The following reactions occur:

$$NH_{4\,(aq)}^{+} + Cl_{(aq)}^{-} + \Delta \rightarrow NH_4Cl_{(s)} \rightarrow NH_{3(g)} + HCl_{(g)} \tag{6}$$

$$NH_{4\,(aq)}^{+} + NO_{3\,(aq)}^{-} + \Delta \rightarrow NH_4NO_{3(s)} \rightarrow N_2O_{(g)} + H_2O_{(g)} \tag{7}$$

$$NO_{3\,(aq)}^{-} + C_2O_4^{2-} + H_{(aq)}^{+} \rightarrow NO_{2(g)} + 2\ CO_{2(g)} \tag{8}$$

Aqueous HCl is added to the residue (oxides of the Group C cations). The residue dissolves, yielding a solution containing the Group C cations. The solution may be blue due to the presence of tetrachlorocobaltate(II) ions, $CoCl_4^{2-}$.

Separation and Confirmation of Group C Cations

Cobalt(II)

Cobalt(II) ion is separated and identified in one step. The addition of nitrite ion (NO_2^-) to an acidified solution containing cobalt(II) ions results in the oxidation of cobalt(II) to cobalt(III), which precipitates as the insoluble yellow compound potassium hexanitritocobaltate(III), $K_3Co(NO_2)_6$ (see Equations 9 and 10). This compound is used as a means of quantitative analysis for cobalt (or potassium), due to its extreme insolubility.

The supernatant liquid is retained to test for the other Group C ions. Reaction (11) also occurs as nitrogen(II) oxide, NO, evolves from solution and reacts with atmospheric oxygen. Red-brown fumes of nitrogen dioxide, NO_2, will be observed, regardless of whether cobalt or any other Group C cation is present. Nitric oxide will be produced when nitrite is added to the acid solution, as nitrite will undergo disproportionation in the presence of acid as shown in Reaction 12:

$$NO_{2\,(aq)}^{-} + Co_{(aq)}^{2+} + H_{(aq)}^{+} \rightarrow NO_{(g)} + Co_{(aq)}^{3+} \tag{9}$$

$$6NO_{2\,(aq)}^{-} + Co_{(aq)}^{3+} + 3K^{+} \rightarrow K_3Co(NO_2)_{6(s)} \tag{10}$$

$$NO_{(g)} + O_{2(g)} \rightarrow NO_{2(g)} \tag{11}$$

$$NO_{2\,(aq)}^{-} + H_{(aq)}^{+} \rightarrow NO_{(g)} + NO_{3\,(aq)}^{-} \tag{12}$$

Barium Ion

Ammonia is added dropwise to the supernatant liquid until the solution is basic, and then ammonium carbonate solution is added to precipitate the cations as the carbonates, $BaCO_3$, $CaCO_3$, and $SrCO_3$. The precipitate is washed and dissolved in acetic acid buffered at pH 5, and diluted. Chromate ion is then added. The buffered pH provides a steady, low concentration of chromate, according to Reaction 13. Although all three remaining Group C cations form fairly insoluble chromate compounds, barium chromate is several orders of magnitude less soluble. Under these conditions, only the yellow barium chromate, $BaCrO_4$, will precipitate.

$$2CrO_{4\,(aq)}^{2-} + 2H_{(aq)}^{+} \rightarrow Cr_2O_{7\,(aq)}^{2-} + H_2O_{(l)} \tag{13}$$

Calcium and Strontium

To remove excess chromate ions from the remaining two cations, calcium and strontium are precipitated once again as carbonates. The washed precipitate is dissolved in HCl and diluted, and ammonium sulfate solution is then added. Upon standing, white strontium sulfate ($SrSO_4$) will precipitate ($CaSO_4$ is much more soluble). The supernatant liquid remaining from this precipitate is tested for the presence of calcium ions by the addition of ammonium oxalate solution. If calcium is present, it will precipitate as calcium oxalate, CaC_2O_4.

If the procedure for separating barium, calcium, and strontium is followed carefully, the presence or absence of a precipitate indicates the presence or absence of the corresponding cation. However, if experimental conditions (pH, degree of dilution, etc.) are not quite what they should be, the three cations will not behave as indicated and false results may be obtained. Since all three chromates are pale yellow and the carbonates, sulfates, and oxalates of all three are white, the color of the precipitate is no indication that the correct cation has precipitated.

Further verification of the identity of the cation in each precipitate can be had by performing a **flame test.** This relies on the excitation of valence electrons in the metal cations when subjected to a hot flame. The electrons are excited to higher energy levels. In collapsing back to lower energy levels, the electrons of each cation emit energy of a characteristic wavelength. For these three cations, some of the wavelengths of light emitted are in the visible region of the spectrum. Barium ions emit a yellow-green color, strontium a crimson color and calcium a brick red color.

EXPERIMENTAL PROCEDURE

It is usually convenient to do Parts A (the known mixture) and B (your unknown) simultaneously. Unless otherwise indicated, discarded solutions (including washes) should be collected for disposal. Some solutions with special hazards should be collected separately, according to your instructor's directions. Be sure to record all reagents added and all observations (including before and after centrifuging) on the data sheets.

PART A: ANALYSIS OF A KNOWN MIXTURE

Removal of Groups A and B

- **(NOTE: If you are analyzing a solution as part of a continuing experiment, use the supernatant left from separating the Group B cations and start at the "Precipitation of the Group C Cations" section of this procedure.)** In a clean 10 × 75 mm test tube, obtain 10 drops of a solution that is approximately 0.1 M in each of the four Group C cations. Add 4 drops of 6 M HCl (**Caution:** HCl is corrosive). Any precipitate here contains Group A cations and should be centrifuged and removed.
- Add 6 M ammonia **(Caution: irritant)** dropwise to the supernatant, with stirring and counting the drops, until the solution tests neutral to pH paper. Add an equal number of drops of 6 M ammonia in excess. Any precipitate here contains Group B cations and should be centrifuged and removed. Continue with the supernatant in the next section.

Precipitation of the Group C Cations

- Add 5 drops of 0.5 M ammonium oxalate solution to the supernatant from above. Stir thoroughly, and centrifuge for 2 minutes. If the supernatant liquid contains any Group D or E cations, decant the supernatant liquid into a clean, **labeled,** 10 × 75 mm test tube, otherwise discard the supernatant liquid. Wash the precipitate with 1 mL of deionized water, centrifuge, and discard the wash. Record all observations.

- Transfer the precipitate into a clean crucible using several drops of concentrated nitric acid **(Caution: corrosive and oxidizer),** in a **HOOD.** Place the lid on the crucible, slightly ajar, and heat the precipitate to dryness over a low flame.
- Cool the crucible for 5 minutes. Add 6 drops of concentrated nitric acid, washing the inside of the crucible, replace the lid and heat to dryness again.
- Cool the crucible for 5 minutes. Dissolve the precipitate in 4 drops of 6 M HCl.
- Using a clean eyedropper, transfer the solution to a clean 10 × 75 mm test tube. Rinse the crucible with 4 drops of deionized water and add this rinse to the same test tube.

Separation and Identification of the Cobalt(II) Ion

- Add 10 drops of 6 M KNO_2 **(Caution: oxidizer)** to the solution from above. When reaction ceases, centrifuge for 2 minutes. Decant the supernatant liquid into a clean 10 × 75 mm test tube. Record all observations.

Separation and Identification of Barium Ion

- Dropwise, add 6 M NH_3 to the supernatant liquid from the previous section until the solution is basic. Add 10 drops of 3 M $(NH_4)_2CO_3$ solution (or a small amount of the solid), centrifuge thoroughly, and discard the supernatant liquid.
- Wash the precipitate with 10 drops of deionized water, stirring thoroughly and then centrifuging. Discard the wash. Record all observations.
- Add 6 M acetic acid (CH_3COOH) dropwise (counting drops) to the precipitate until it just dissolves, and then add an equal number of drops in excess. Dilute to 2 mL with deionized water (a total of 40 drops of acetic acid and deionized water should be in the test tube).
- Add 5 drops of 0.1 M K_2CrO_4 potassium chromate **(Caution: carcinogen and oxidizer)** mix, centrifuge for 2 minutes, and decant the supernatant liquid into a clean test tube labeled **"A"** for use in the Strontium section below. Record all observations.

Flame Test for Barium

- Dissolve the precipitate in 2 drops of concentrated HCl. Clean a piece of nichrome wire with a loop on the end by dipping the loop in concentrated HCl, and then inserting the loop into a microburner flame. Once clean, the wire should impart no color to the flame.
- Dip the clean wire loop into the solution containing the dissolved precipitate, insert it in the flame and note any color it imparts to the flame. Best results will be obtained if a film of solution is caught on the loop. Clean the loop again after this flame test.

Separation and Identification of Strontium Ion

- To the supernatant liquid in test tube **A,** add 6 M NH_3 until the solution is basic (litmus). Add 10 drops of 3 M $(NH_4)_2CO_3$ solution (or a small amount of the solid), centrifuge for 2 minutes, and discard the supernatant liquid. Wash the precipitate with 10 drops of deionized water, stirring thoroughly, and centrifuge. Discard the wash. Record all observations.
- Dissolve the precipitate in 4 drops of 6 M HCl. Bring the volume to 2 mL by adding 36 drops of deionized water. Add 10 drops of 1 M $(NH_4)_2SO_4$ solution (ammonium sulfate), mix, and let the solution stand for 10 minutes.
- Centrifuge for 2 minutes, and decant the supernatant liquid into a clean 10 × 75 mm test tube labeled **"B"** for use in the next section. Record all observations.

- Dissolve the precipitate in 2 drops of concentrated nitric acid. Perform a flame test in the same manner as done for barium ion. Clean the loop again after this flame test.

Separation and Identification of Calcium Ion

- Add 2 drops of 6 M NH_3 , then 5 drops of 0.5 M $(NH_4)_2C_2O_4$ solution to the supernatant liquid in test tube **B.** Centrifuge and decant the supernatant liquid into a clean test tube. Record all observations.
- Dissolve the precipitate in 2 drops of 6 M HCl. Perform a flame test as done previously. Clean the loop again after this flame test.

PART B: ANALYSIS OF AN UNKNOWN CONTAINING THE GROUP C CATIONS

- Obtain 10 drops of an unknown solution containing one or more of the Group C cations in a clean centrifuge tube. Repeat the procedures outlined in Part A, using the unknown solution. Report which cations are present in the unknown, along with observations to verify the identification.

Name: ______________________________

Date: __________ Section: ______________________

Pre-Laboratory Questions: Experiment 32

1. Why are the flame tests recommended as an additional identification step?

2. A solution containing the Group C cations is analyzed according to the procedure of this experiment with the following results. At each stage, indicate what cations may be present and which are absent.

a. The residue, after destroying oxalate and ammonium ions with HNO_3, is dissolved in HCl. A clear colorless solution results.

b. After adding 6 M KNO_2, red brown fumes evolve and a clear yellow solution results.

c. Adding 3 M $(NH_4)_2CO_3$ to the supernatant from (b) yields a white precipitate.

d. Addition of 0.1 M K_2CrO_4 to the dissolved precipitate from (c) yields a clear, yellow supernatant and a yellow precipitate. The flame test for the precipitate is green.

e. Addition of 3 M $(NH_4)_2CO_3$ to the solution from (d) yields a white precipitate and a yellow supernatant.

f. Addition of 1 M $(NH_4)_2SO_4$ to the dissolved precipitate from (e) yields a white precipitate and a colorless supernatant. The flame test for the precipitate is crimson.

g. Addition of 0.5 M $(NH_4)_2C_2O_4$ to the supernatant from (f) gives a colorless supernatant.

Name: ______________________________

Date: __________ Section: ______________________

Pre-Laboratory Flow Sheet: Experiment 32

3. Fill in the following flow sheet.

Flow chart for Group C separation.

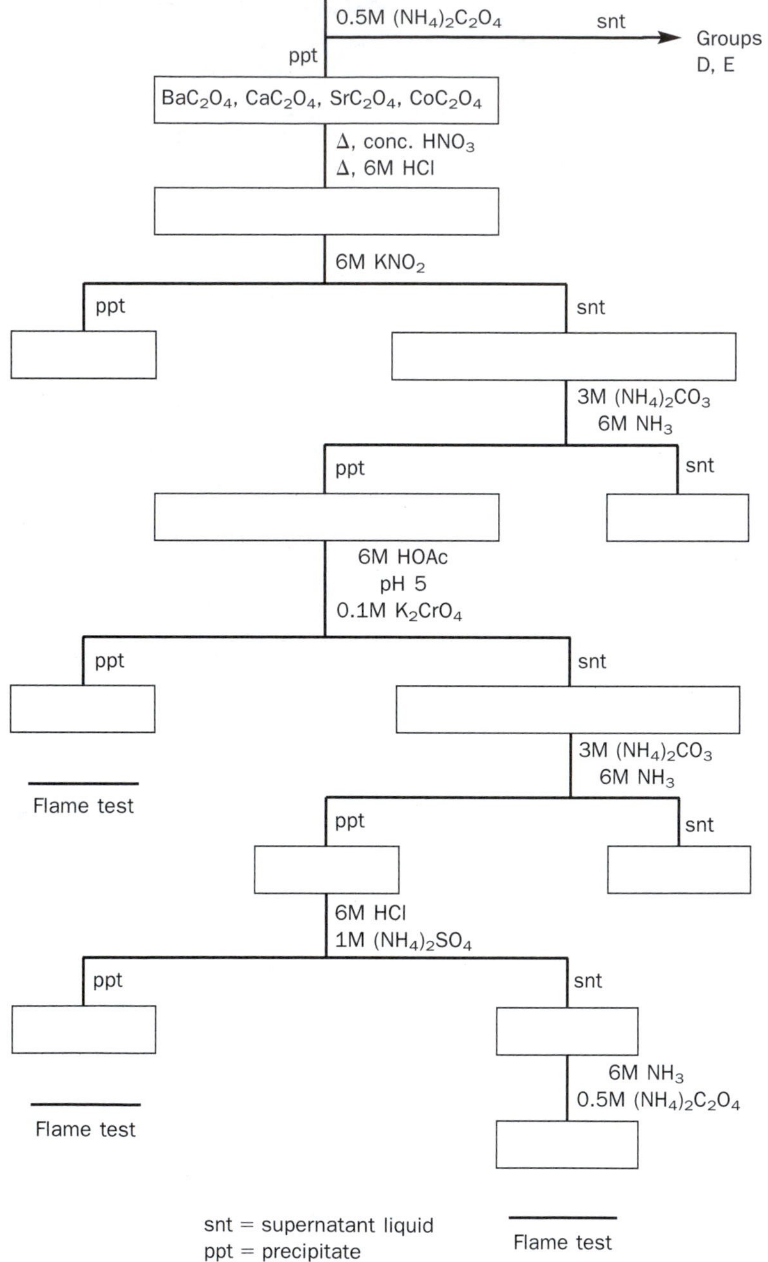

Name: ______________________________

Date: __________ Section: ______________________

Data Sheet: Experiment 32

Group C Cations: Analysis of Known and Unknown Cation Mixtures

Unknown Number ____________

Actions, Reagents	*Result, Known*	*Result, Unknown*	*Conclusion*

Cations detected in unknown: ________ ________ ________

Name: ______________________________

Date: __________ Section: ______________________________

Questions: Experiment 32

1. Balance each of the redox reactions (7, 8, 9, 11, 12) in the Introduction section.

2. A student gets a precipitate after the final addition of $(NH_4)_2C_2O_4$ (last section of the procedure). The flame test for that precipitate is crimson. What ion is present? Explain.

3. After separation of the Group B cations with NH_3, Co^{2+} is present as the ammine complex, $Co(NH_3)_6^{2+}$ and several of the Group D cations (Cu^{2+}, Ni^{2+}, Cd^{2+}) are also present as ammine complexes. Although the oxalates of Group D are relatively insoluble, when ammonium oxalate is added, only Co^{2+} (of the ammine complexes) precipitates as the oxalate. What can be said about the relative stability of the cobalt ammine complex relative to the Group D ammine complexes?

4. Solution pH is used to control chromate ion concentration for what purpose? What would happen in the separation of barium if the solution were too basic?

Experiment 33 Separation and Identification of the Group D Cations (Mg^{2+}, Ni^{2+}, Cu^{2+}, Zn^{2+}, and Cd^{2+})

OBJECTIVES

- To separate and identify the Group D cations.

PRIOR READING

Chapter 9: Introduction

Page 399: Divided Page Problem Solving

INTRODUCTION

The Group D cations are characterized by having hydroxides, oxides, and oxalates that are soluble in ammonia solution. Their oxides are insoluble in strongly basic solutions that do not contain ammonia. These cations are magnesium, Mg^{2+}; nickel(II), Ni^{2+}; copper(II), Cu^{2+}; zinc(II), Zn^{2+}; and cadmium(II), Cd^{2+}.

Separation of the Group D Cations

> **NOTE: *The redox reactions below are not balanced. This is left as an exercise (see Pre-Laboratory Question 1).***

Metal cations dissolved in aqueous solution are usually present as aquo complexes, that is, some fixed number of H_2O molecules are formally bonded to the metal [e.g., $Ni(H_2O)_6^{2+}$]. Many transition metal aquo complexes are colored, while those of the s-block metals are colorless. Of the Group D cations, only Ni^{2+} and Cu^{2+} have colored aquo complexes (green and blue, respectively). In the following reactions, the starting form of the cations is written without the H_2O.

The solution containing the Group D cations is first treated to remove any Group A cations as insoluble chlorides. The Group B cations are then precipitated as oxides and hydroxides with the addition of aqueous ammonia. The Group C cations are then precipitated by adding ammonium oxalate solution.

Upon addition of ammonia, four of the Group D cations form ammonia complexes. Of these, the copper complex is royal blue, the nickel complex is medium blue, and the cadmium and zinc complexes are colorless. These complexes are very stable, thus the oxalates do not precipitate in ammonia solution, even though they are relatively insoluble in deionized water. The magnesium ion does not form a complex with ammonia, but its hydroxide and oxalate are more soluble and do not precipitate. Thus, the Group D cations remain in the supernatant liquid.

$$Cu^{2+}_{(aq)} + 4\ NH_{3(aq)} \rightarrow Cu(NH_3)_4^{2+}{}_{(aq)} \quad (1)$$

$$Cd^{2+}_{(aq)} + 4\ NH_{3(aq)} \rightarrow Cd(NH_3)_4^{2+}{}_{(aq)} \quad (2)$$

$$Ni^{2+}_{(aq)} + 6\ NH_{3(aq)} \rightarrow Ni(NH_3)_6^{2+}{}_{(aq)} \quad (3)$$

$$Zn^{2+}_{(aq)} + 4\ NH_{3(aq)} \rightarrow Zn(NH_3)_4^{2+}{}_{(aq)} \quad (4)$$

The supernatant liquid containing Group D and possibly Group E cations must be treated to remove the ammonium and oxalate ions that would interfere with characterization tests for these ions. This is accomplished by transferring the solution to a crucible and heating to dryness. Aqueous nitric acid, HNO_3, is then added, and the solution is taken to dryness once again. Oxalate is converted to CO_2 and the ammonium ions are oxidized to nitrous oxide. The following reactions occur:

$$NH_{4(aq)}^{+} + Cl_{(aq)}^{-} + \Delta \rightarrow NH_4Cl_{(s)} \rightarrow NH_{3(g)} + HCl_{(g)} \quad (5)$$

$$NH_{4(aq)}^{+} + NO_{3(aq)}^{-} + \Delta \rightarrow NH_4NO_{3(s)} \rightarrow N_2O_{(g)} + H_2O_{(g)} \quad (6)$$

$$NO_{3(aq)}^{-} + C_2O_4^{2-}{}_{(aq)} + H_{(aq)}^{+} \rightarrow NO_{2(g)} + CO_{2(g)} + H_2O_{(l)} \quad (7)$$

The residue is then dissolved in aqueous HCl. The solution may be blue to green, due to the presence of copper and nickel chloro- or aquo- complexes. The Group D cations are then precipitated as hydroxides in strong base (NaOH) solution, leaving any Group E cations in solution. Nickel(II) hydroxide is green, copper(II) hydroxide is blue, while cadmium(II) hydroxide, zinc(II) hydroxide, and magnesium(II) hydroxide are all white. Care must be taken not to use too much base as some of the cations can form soluble hydroxide complexes [tetrahydroxocuprate(II), $Cu(OH)_4^{2-}$, and tetrahydroxozincate(II), $Zn(OH)_4^{2-}$]. After separating the precipitate from the supernatant, the precipitate is redissolved in aqueous HCl.

Confirmation of Group D Cations

Nickel Ion

Nickel is confirmed directly from the dissolved hydroxide residue. A drop of this solution is mixed with a drop of dimethylglyoxime solution. The mixture is made basic with aqueous ammonia. A cherry red nickel dimethylglyoxime complex (abbreviated DMGH, the H signifying that one acidic hydrogen is easily removed) precipitate forms if nickel is present. Nickel forms the only highly colored precipitate with dimethylglyoximate ions. Note that the solution must be basic in order for the precipitate to form after the DMGH deprotonates to DMG^-.

$$Ni(H_2O)_6^{2+}{}_{(aq)} + 6\ NH_{3(aq)} \rightarrow Ni(NH_3)_6^{2+}{}_{(aq)} + 6\ H_2O_{(l)} \quad (8)$$

$$Ni(NH_3)_6^{2+}{}_{(aq)} + 2\ DMGH \rightarrow Ni(DMG)_{2(s)} + 4\ NH_{3(aq)} + 2\ NH_{4(aq)}^{+} \quad (9)$$

Copper (II) Ion

Copper ion is also confirmed directly from the dissolved hydroxide residue. The confirmation test is the formation of a maroon precipitate of copper hexacyanoferrate, $Cu_2Fe(CN)_6$, from slightly acidic solution. In the absence of Cu^{2+}, a white, yellow, or pale green precipitate may be observed, as ferrocyanide forms precipitates with most of the Group D cations.

$$2\ Cu(NH_3)_4^{2+}{}_{(aq)} + Fe(CN)_6^{4-}{}_{(aq)} \rightarrow Cu_2Fe(CN)_{6(s)} + 8\ NH_{4(aq)}^{+} \quad (10)$$

The strongly acidic solution is treated with aqueous ammonia until it is only slightly acidic. The purpose of this is to avoid the addition of hexacyanoferrate to a strongly acidic solution, which could result in the formation of hydrogen cyanide (HCN), a very toxic gas. By reducing the acidity of the solution, this possibility is avoided.

Cadmium

Cadmium is particularly difficult to identify, as it must be isolated from virtually all other cations. This is due to the fact that most cadmium compounds are white to pale yellow, and their presence is easily masked by the presence of even traces of copper and nickel cations that form highly colored compounds. In addition, zinc compounds are also mostly white or pale yellow.

Cadmium can be separated from nickel, zinc, and copper based on the solubilities of their sulfides. Cadmium sulfide is considerably more soluble than either copper or nickel sulfide. Zinc sulfide is much more soluble and under the conditions used, it does not precipitate. Magnesium does not form a sulfide. Copper and nickel sulfides are black, whereas cadmium sulfide is yellow. (For an unknown, if the sulfide precipitate is yellow, no further test for cadmium is necessary.)

In hot, weakly acidic solution, sodium thiosulfate disproportionates to form sulfate and sulfide ions. This is the source of sulfide ions to precipitate the Cu^{2+}, Ni^{2+}, and Cd^{2+} ions, leaving Mg^{2+} and Zn^{2+} in solution. The Ni^{2+} ions may not completely precipitate as the sulfide in weakly acidic solution, so some Ni^{2+} may remain dissolved, along with the Zn^{2+} and Mg^{2+}. This supernatant is used later to test for the presence of Zn^{2+} and Mg^{2+}.

After centrifuging, the precipitate is treated briefly with HCl, then centrifuged. The HCl dissolves the CdS, leaving NiS and CuS as solids. This treatment must be done quickly and at room temperature—if the exposure to HCl is too long or at an elevated temperature, CuS and NiS may begin to dissolve, negating the separation. The remaining NiS and CuS solids may be discarded and the supernatant liquid is tested for the presence of Cd^{2+}. The diagnostic test for Cd^{2+} is the precipitation of white cadmium hexacyanoferrate in weakly acidic solution.

$$S_2O_3^{2-}{}_{(aq)} + H_2O_{(l)} \rightarrow S^{2-}_{(aq)} + SO_4^{2-}{}_{(aq)} + H^+_{(aq)} \tag{11}$$

$$2\ Cd^{2+}_{(aq)} + Fe(CN)_6^{4-}{}_{(aq)} \rightarrow Cd_2Fe(CN)_{6(s)} \tag{12}$$

The supernatant containing the Cd^{2+} is initially very acidic from the HCl. As in the identification of Cu^{2+}, the solution is made weakly acidic by adding ammonia, and then is treated with hexacyanoferrate.

Magnesium

The identification of magnesium is accomplished using the supernatant remaining after the precipitation of Cu^{2+}, Ni^{2+}, and Cd^{2+} as sulfides. This supernatant contains Zn^{2+}, Mg^{2+}, and probably some Ni^{2+} (the presence of Ni^{2+} will give a blue color to the solution). The solution is made basic with aqueous ammonia, and disodium hydrogen phosphate, Na_2HPO_4, is added. A white precipitate of zinc phosphate, $Zn_3(PO_4)_2$, and magnesium ammonium phosphate, $MgNH_4PO_4$, forms. Any other Group D cations present at this point remain dissolved as their phosphates are soluble in ammonia.

$$H^+_{(aq)} + NH_{3(aq)} \rightarrow NH_4^+{}_{(aq)} \tag{13}$$

$$Mg^{2+}_{(aq)} + NH_4^+{}_{(aq)} + HPO_4^{2-}{}_{(aq)} \rightarrow MgNH_4PO_{4(s)} + H^+_{(aq)} \tag{14}$$

$$3\ Zn^{2+}_{(aq)} + 2\ HPO_4^{2-}{}_{(aq)} + 2\ NH_3 \rightarrow Zn_3(PO_4)_{2(s)} + 2\ NH_4^+{}_{(aq)} \tag{15}$$

If any Ni^{2+} is present, it will undergo reaction according to Reaction (8), yielding a blue nickel hexammine complex.

After being centrifuged, the supernatant is discarded and the precipitate is treated with NaOH. Any zinc phosphate will dissolve to form the colorless tetrahydroxozincate(II) species, $Zn(OH)_4^{2-}$. Any remaining white precipitate is magnesium ammonium phosphate that is insoluble in base solution, proving the presence of magnesium.

$$Zn_3(PO_4)_{2(s)} + 12\ OH^-_{(aq)} \rightarrow 3\ Zn(OH)_4^{2-}{}_{(aq)} + 2\ PO_4^{3-}{}_{(aq)} \tag{16}$$

Identification of Zinc

The basic supernatant is tested for the presence of zinc by making the solution slightly acidic with aqueous acetic acid (CH_3COOH). A white precipitate of $Zn_2Fe(CN)_6$ forming on the addition of $K_4Fe(CN)_6$ proves the presence of Zn^{2+}.

EXPERIMENTAL PROCEDURE

It is usually convenient to do Parts A (the known mixture) and B (your unknown) simultaneously. Unless otherwise indicated, discarded solutions (including washes) should be collected for disposal. Some solutions with special hazards should be collected separately, according to your instructor's directions. Be sure to record all reagents added and all observations (including before and after centrifuging) on the data sheets.

PART A: PRECIPITATION OF THE GROUP D CATIONS IN A KNOWN MIXTURE

- **(NOTE: If you are analyzing a solution as part of a continuing experiment, use the supernatant left from separating the Group C cations and begin at the Precipitation of Group D cations section below.)** In a clean 10 × 75 mm test tube, obtain 10 drops of a solution that is approximately 0.1 M in each of the four Group D cations. Add 4 drops of 6 M HCl **(Caution: corrosive).** Any precipitate here contains Group A cations and should be centrifuged and removed.
- Add 6 M ammonia **(Caution: irritant)** dropwise to the supernatant, with stirring and counting the drops, until the solution tests neutral to pH paper. Add an equal number of drops of 6 M ammonia in excess and stir. Any precipitate here contains Group B cations and should be centrifuged and removed.
- Add 5 drops of 0.5 M ammonium oxalate solution, stir thoroughly, and centrifuge for 2 minutes. Any precipitate present will contain cations from Group C. The supernatant contains the Group D and E cations. Record all observations.

Precipitation of the Group D Cations

- Decant the supernatant liquid containing the Group D and E cations into a clean crucible. Place the lid on the crucible, slightly ajar, and heat the precipitate to dryness over a low flame.
- Cool the crucible for 5 minutes, and then add 6 drops of concentrated nitric acid **(Caution: corrosive and oxidizer),** washing the inside of the crucible. Replace the lid and heat to dryness once again **(HOOD).**
- Cool the crucible for 5 minutes, repeat the addition of concentrated nitric acid and heat to dryness **(HOOD).**
- Cool the crucible for 5 minutes, and dissolve the residue in 5 drops of 6 M HCl. Using a clean Pasteur pipet, transfer the solution to a clean 10 × 75 mm test tube labeled **"A."** Rinse the crucible with 5 drops of deionized water and add this rinse to the same test tube. Record all observations.

Identification of Nickel(II) Ion

- Transfer one drop of solution **A** to a clean test tube. Add one drop of 1% dimethylglyoxime. Add 6 M NH_3 with stirring, until the solution tests basic to litmus paper. Record all observations.

Identification of Copper(II) Ion

- Transfer a second drop of solution **A** to a clean test tube. Add 6 M NH_3 until the solution tests only weakly acidic. If the solution becomes basic, use 6 M acetic acid to make it slightly acidic. Add 3–4 drops of 0.1 M $K_4Fe(CN)_6$ solution. Record all observations.

Separation and Identification of Cadmium(II) Ion

- Add 6 M NH_3 to the remainder of solution **A** until it tests only weakly acidic. If the solution becomes basic, use 6 M acetic acid to make it slightly acidic.
- Add about 200–300 mg of solid sodium thiosulfate, $Na_2S_2O_3$, and heat for 5 minutes in a boiling water bath. Cool for 1 minute by swirling the test tube in cold tap water.
- Centrifuge and decant the supernatant into a clean test tube labeled **"B"** for magnesium analysis.
- Wash the precipitate with 10 drops of deionized water. Discard the wash. To the precipitate, add 5 drops of 6 M HCl and stir for 10–15 sec. Immediately centrifuge for 1 minute, and quickly decant the supernatant into a clean test tube.
- Add 6 M NH_3 to the liquid, until the solution tests only weakly acidic. If the solution becomes basic, use 6 M acetic acid to make it slightly acidic. Add 5 drops of 0.1 M $K_4Fe(CN)_6$ solution. Record all observations.

Identification of Magnesium Ion

- To solution **B,** add 6 M NH_3 until the solution tests basic to litmus paper (pH 9–10). Add 6–8 drops of 0.1 M Na_2HPO_4 solution. Let the solution sit several minutes.
- Centrifuge and discard the supernatant. Wash the precipitate with 10 drops of deionized water. Add 6 drops of 6 M NaOH **(Caution: corrosive and toxic)** and stir thoroughly. Centrifuge and decant the supernatant into a clean test tube labeled **"C."** Record all observations.

Identification of Zinc Ion

- To the supernatant in test tube **"C,"** add 6 M acetic acid with stirring until the solution tests weakly acidic to litmus paper (pH 4–6). Add 4 drops of 0.1 M $K_4Fe(CN)_6$. Record all observations.

PART B: ANALYSIS OF AN UNKNOWN CONTAINING THE GROUP D CATIONS

- Obtain 10 drops of an unknown solution containing one or more of the Group D cations in a clean centrifuge tube. Repeat the procedures outlined in Part A, using the unknown solution. Report which cations are present in the unknown, along with observations to verify the identification.

Name: ______________________________

Date: __________ Section: ______________________________

Pre-Laboratory Questions: Experiment 33

1. On a separate sheet, balance each of the redox Reactions (6, 7, and 11) in the experimental introduction.

2. A solution containing Group D cations is subjected to the procedure of this experiment. At each stage, indicate what cations may be present and which are absent.

a. The blue Group D hydroxide precipitate is dissolved in 6 M HCl, and a clear green solution results.

b. One drop of the solution from (a) is tested with DMG, and made basic with ammonia. A red precipitate results.

c. A second drop of the solution from (a) is made weakly acidic. When 0.1 M $K_4Fe(CN)_6$ is added, a light green precipitate results.

d. The remainder of the solution from (a) is made weakly acidic. When $Na_2S_2O_3 \cdot 5\ H_2O$ is added and warmed, a black precipitate and a colorless supernatant result.

e. The precipitate from (d) is treated briefly with 6 M HCl. A black precipitate and a colorless supernatant results.

f. The supernatant from (e) is made weakly acidic. When 0.1 M $K_4Fe(CN)_6$ is added, a colorless solution results.

g. The supernatant from (d) is made basic with ammonia. Adding Na_2HPO_4 yields a white precipitate.

h. The precipitate from (g) is treated with NaOH. A clear, colorless solution and a white precipitate result.

i. The solution from (h) is made weakly acidic. When 0.1 M $K_4Fe(CN)_6$ is added, a colorless solution results.

Name: ______________________________

Date: __________ Section: ______________________

Pre-Laboratory Flow Sheet: Experiment 33

3. Fill in the attached flow sheet.

Flow Chart for group D Separation.

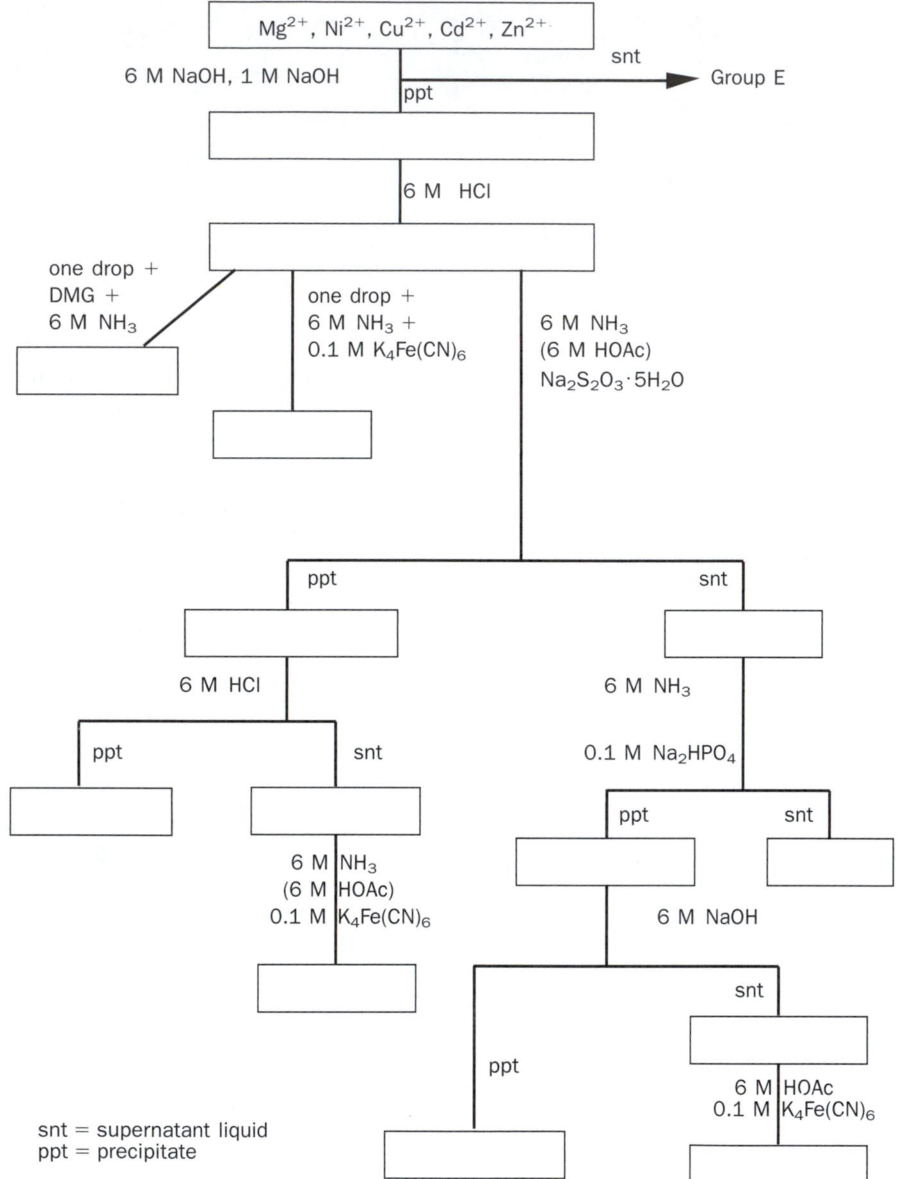

Name: ____________________

Date: ________ Section: ____________________

Data Sheet: Experiment 33

Unknown Number ____________

Actions, Reagents	*Result, Known*	*Result, Unknown*	*Conclusion*

Cations Detected: ______ ______ ______ ______ ______

Name: ______________________________

Date: __________ Section: ______________________________

Questions: Experiment 33

1. Although the five Group D cations have hydroxides which are insoluble in basic solution, they do not precipitate out with the Group B cations when the solution is made basic with ammonia. Explain.

2. Two of the Group D cations are identified by the formation of white ferrocyanide precipitates. Which are they? When a white ferrocyanide precipitate is obtained, how can you tell which cation is present?

3. What is a disproportionation reaction? Where is one used in this experiment?

4. Most ferrocyanides will not precipitate except in acidic solution. If a solution is already strongly acidic, why is it made weakly acidic before adding $K_4Fe(CN)_6$?

5. Search the internet for information on the uses and hazards of cadmium. An MSDS is a good place to start.

Experiment 34 Identification of Anions (Br^-, Cl^-, I^-, CO_3^{2}, OH^-, NO_3^-, $C_2O_4^{2-}$, PO_4^{3-}, SO_4^{2-}, $S_2O_3^{2-}$)

OBJECTIVES

- To identify a series of anions using qualitative analysis.

PRIOR READING

Chapter 9: Introduction

INTRODUCTION

In this experiment, tests will be performed on several known anions in order to learn how they react with each of three different reagents. These results will allow you to determine which results are unique for each anion. The tests will be done on spot plates.

The anions to be investigated are:

Bromide	Br^-	Nitrate	NO_3^-
Carbonate	CO_3^{2-}	Oxalate	$C_2O_4^{2-}$
Chloride	Cl^-	Phosphate	PO_4^{3-}
Hydroxide	OH^-	Sulfate	SO_4^{2-}
Iodide	I^-	Thiosulfate	$S_2O_3^{2-}$

The reagents to be used are Fe^{3+} (dissolved $FeCl_3$), Fe^{2+} (dissolved $FeSO_4$), and Ag^+ (dissolved $AgNO_3$). The Fe^{2+} reagent should be followed with H_2SO_4 if no reaction is observed.

Iron(III)—the Ferric Ion

Iron(III) forms a large number of complex ions, many of them highly colored. It is a moderately strong oxidizing agent, and therefore can react with anions that are reducing agents. The hydroxide is rust colored.

Iron(II)—The Ferrous Ion

Iron(II) forms fewer complexes with inorganic anions, but forms highly colored complexes with several organic ligands. Most iron(II) salts are soluble, the major exceptions being oxalate, hydroxide, ferrocyanide, and chromate. Iron(II) is easily oxidized to iron(III),—hence it can react with anions that are oxidizing agents.

Silver Ion

Most silver salts are insoluble in distilled water, the major exceptions being the nitrate and fluoride. The permanganate, sulfate, and acetate are somewhat soluble. The thiosulfate and sulfite are soluble in the presence of an excess of the anion, due to the formation of complex ions. Most silver salts are soluble in aqueous ammonia, due to the formation of the complex ion, $Ag(NH_3)_2^+$. Silver nitrate and silver salts are toxic and cause brown to black stains on skin and clothing due to **photoreduction** (reduction caused by light) to elemental silver. Avoid contact and wash up immediately if contact is suspected.

DISCUSSION OF ANIONS

Bromide, Chloride, Iodide

The three common halides are very similar in their chemical reactions. Most of their salts are soluble, with the common exceptions of silver, lead, thallium(I), and mercury(I). Bromide and iodide are moderately strong reducing agents, being oxidized to bromine (orange-brown in aqueous solution), and iodine (red-brown in aqueous solution). Chloride is a poor reducing agent. All three can act as ligands in forming complex ions, chloride being the best of the three.

Carbonate, Sulfate, Phosphate, Oxalate

Carbonates, phosphates, and oxalates are generally soluble only with the alkali metals plus Mg^{2+}, while most sulfates are soluble. Virtually all are soluble if the mixture is made to be acidic. Barium sulfate and copper(II) oxalate are the two major exceptions. The solubility in acid is due to extensive acid–base reactions involving the anion. The carbonate ion reacts to give off carbon dioxide, which is colorless and odorless. The phosphate ion reacts to form weak acids (HPO_4^{2-}). The sulfate and oxalate ions do not react appreciably with acid. In neutral solution, carbonate and phosphate undergo hydrolysis to form their conjugate weak acids, resulting in a basic solution. Thus, a precipitate with one of these anions present may actually be a hydroxide, instead of the carbonate or phosphate. Oxalate is a good ligand in forming complex ions, while carbonate and phosphate are moderately good ligands.

Nitrates

All common nitrates are soluble. Nitrate is an excellent oxidizing agent in acid solution. It is reduced to NO, a dissolved gas, which reacts rapidly with atmospheric oxygen to form the reddish-brown gas, NO_2. A reducing agent must be present for this to occur. A common test for nitrate is to add a source of iron(II), Fe^{2+} (reducing agent), make the solution strongly acidic and note the presence of a brown solution of $Fe(NO)^{2+}$, followed by the evolution of the red-brown, toxic gas NO_2.

Thiosulfates

The thiosulfate ion, $S_2O_3^{2-}$, is soluble except with Pb^{2+}, Ba^{2+}, and Ag^+. $Ag_2S_2O_3$ is soluble in excess thiosulfate, due to the formation of a complex ion. Thiosulfate is a good reducing agent.

Hydroxides

Most hydroxides are insoluble except for with the alkali metals, Sr^{2+} and Ba^{2+}. Some others are soluble in excess hydroxide, due to the formation of complexes. Zinc hydroxide, $Zn(OH)_2$, will dissolve in very basic solutions, for example, to form the soluble tetrahydroxozincate(II) ion, $Zn(OH)_4^{2-}$.

TECHNIQUES USED IN THE ANALYSIS OF INDIVIDUAL ANIONS

1. The tests will be performed on spot plates (or well plates). A 30-hole (6×5) plate is a convenient size. A small quantity of the solid salt (the size of a grain of rice) should be placed in a well and dissolved in 3–4 drops of distilled water. The reagent (2–3 drops) is then added, and if necessary, the solution is stirred.
2. Watch carefully for the evolution of a gas with those anions that are known to evolve gases. You should learn the conditions under which they evolve gas. Be on the lookout when these conditions exist. It might help to look across the spot plate at a sheet of white paper while adding the reagent.
3. White precipitates are difficult to observe against the white color of the spot plate. If no reaction appears to have occurred, insert a clean glass stirring rod into the solution. If the stirring rod "disappears," a white precipitate has formed. (Think of putting your finger into milk—your finger "disappears" because milk is a suspension of solids in water.) A better method of observing a precipitate is to take up the solution in a clean Pasteur pipet.

4. The three reagents used are 0.1 M silver nitrate ($AgNO_3$), 0.1 M iron(III) nitrate [$Fe(NO_3)_3$], and 1.0 M iron(II) sulfate ($FeSO_4$) solution. If no reaction is observed between a particular anion and any of the three reagents, 2 drops of concentrated sulfuric acid should be added to the test mixture containing iron(II) sulfate and any further observations recorded. For iodide and bromide and any unknown that you suspect may be bromide or iodide, add 6–7 drops of hexane to the test with $Fe(NO_3)_3$. Use a clean eyedropper to mix the solution by drawing it up and down in the dropper. Then observe the color of the upper (hexane) layer.
5. The silver nitrate tests should be performed separately from the other tests, as silver salts are extremely toxic. The silver waste should not be mixed with that from the other tests, so that proper disposal can be effected. Silver wastes should be brought to pH 2–5, and filtered to remove the solids. The liquid should be disposed of separately and the silver solids (now small in volume) can be sent to a chemical landfill as relatively harmless waste, or, if desired, recovered.
6. In observing color changes of solutions, remember that the iron(III) solution is yellow in color to begin with. If it gets slightly paler yellow on mixing it with an anion in solution, this is due to dilution, not a reaction. If it changes color, becomes colorless or gets deeper yellow in color, a reaction has occurred. It might help to put some of the reactant in a well by itself for comparison and perhaps add a few drops of distilled water to account for color changes due to dilution. The iron(II) solution is pale green and faint color changes are more difficult to detect.
7. If you are unsure about the identity of an unknown, it might help to repeat the tests with a known compound containing that anion.

EXPERIMENTAL PROCEDURE

Part A: Identification of Anions in a Known Mixture

Obtain a 6 × 5 spot plate. Clean the spot plate with distilled water before use.

- Using a clean spatula, place a small amount of the correct salt in the first well. Using a clean spatula each time, place small amounts of the other salts in other wells, being very careful to keep track of where each salt is. Dissolve each salt in 3–4 drops of distilled water. If necessary, stir the mixture with a clean glass stirring rod to dissolve, being sure to rinse the stirring rod before using it in a different well.
- Add 2–3 drops of 0.1 M $AgNO_3$ solution to each well, noting the results on the chart for each as you add the test reagent. Note any color changes that occur in the solutions and the formation of any precipitates. If necessary, draw the solution up into a clean Pasteur pipet to confirm the color. Release the solution back into the same well afterward.
- When finished, rinse the spot plate into the waste container marked "SILVER WASTE." If necessary, use a test tube brush to loosen precipitates. Rinse thoroughly before using the spot plate for the next test.
- Repeat the above procedure, using fresh samples of the salts, adding 0.1 M $Fe(NO_3)_3 \cdot 9\,H_2O$ solution **instead of** the $AgNO_3$ solution. Note all results on the chart. Dispose of these solutions in a "GENERAL WASTE" container.
- Repeat the procedure, using fresh samples of the salts, adding 1.0 M $FeSO_4 \cdot 7\,H_2O$ solution **instead of** the $AgNO_3$ solution. If no apparent reaction occurs for a particular anion with the 1.0 M $FeSO_4 \cdot 7H_2O$ reagent, check the results for that anion with the other two reagents. If no reaction

was observed for that anion with either of the other two reagents, add 2–3 drops of concentrated sulfuric acid (CAUTION: Corrosive and Oxidizer) to the well containing that anion and the Fe^{2+} reagent. Do this for at least sulfate and nitrate. Note all results on the Chart. Dispose of these solutions in a "GENERAL WASTE" container.

Part B: Analysis of an Unknown Anion

- Obtain solid unknowns from the laboratory instructor and repeat the procedures outlined in Part A for each unknown. Report the identity of each unknown, along with observations to verify the identification.

Name: ____________________

Date: __________ Section: ____________________

Pre-Laboratory Questions: Experiment 34

1. From the information given in the experimental discussion, fill in the "Characteristics" portion of the chart below for each anion (the reagent cations have been done). Use the abbreviation "C" if it forms complex ions readily, "R" if it is a reducing agent, "O" if it is an oxidizing agent, and "col" if it is a colored ion by itself. For example, for bromide, you would enter only "R."

For each block corresponding to a combination of a reagent cation and an anion, use the following abbreviations: R-O (redox reaction likely), gas (gas evolves), P (precipitate likely to form), C (complex ion is likely to form). For example, the entries for Br^- would be: [Fe^{2+} column: nothing], [Fe^{3+} column, R-O], and [Ag^+ column, P]. Follow these with a "?" if the outcome is not certain (e.g., R-O?). For the Fe^{2+} reagent, a separate section is provided for the second step where acid (H^+) is added.

		Fe^{2+}	**Fe^{2+}, then H^+**	**Fe^{3+}**	**Ag^+**
	Characteristics	**R, col**	**R, col**	**O, C, col**	**—**
Br^-	R	—	—	R-O	P
CO_3^{2-}					
Cl^-					
OH^-					
I^-					
NO_3^-					
$C_2O_4^{2-}$					
PO_4^{3-}					
SO_4^{2-}					
$S_2O_3^{2-}$					

Name: ______________________________

Date: __________ Section: ______________________

Data Sheet: Experiment 34

Record your observations on the chart below. Identify the unknowns in the spaces provided.

Unknown Number ________ ________ ________

	Fe^{2+}	Fe^{2+}, then H^+	Fe^{3+}	Ag^+
Br^-				
CO_3^{2-}				
Cl^-				
OH^-				
I^-				
NO_3^-				
$C_2O_4^{2-}$				
PO_4^{3-}				
SO_4^{2-}				
$S_2O_3^{2-}$				
Unknown 1				
Unknown 2				
Unknown 3				

My unknowns contains the following anions:

Unknown 1 ____________

Unknown 2 ____________

Unknown 3 ____________

Name: ______________________________

Date: __________ Section: ______________________

Questions: Experiment 34

1. List three ways in which wells in the spot plate could become contaminated, leading to inaccurate results, and discuss how to prevent each.

2. List four types of observations that would lead you to believe that a chemical reaction has occurred.

3. For unknowns containing the following mixtures of anions, indicate which anions could be definitely identified and which test(s) would identify each. Indicate if an anion could not be distinguished by the tests used in this experiment.

a. Cl^-, I^-, OH^-

b. SO_4^{2-}, $C_2O_4^{2-}$, SCN^-

Chapter 10
Organic Chemistry

Experiment 35: Organic Compounds: Structures Using Models (Dry Lab)

Experiment 36: Introduction to Organic Qualitative Analysis

Experiment 37: Synthesis and Analysis of Aspirin

Experiment 35 Organic Compounds: Structures Using Models (Dry Lab)

OBJECTIVES

- To gain familiarity with organic structures and bonding through the use of models.

INTRODUCTION

The classification of chemical compounds into the general areas of organic and inorganic derives from the use of the "mineral, vegetable, and animal" designation by the early workers in chemistry. Those compounds derived from living systems were termed **organic,** while those derived from mineral sources were termed **inorganic.** In modern terms, organic compounds are classified as compounds of carbon containing either carbon–carbon or carbon–hydrogen bonds, or both.

Originally, organic compounds were thought to be imbued with a "vital essence," attainable only from God. Thus, it was believed that organic compounds could only be prepared from sources that had once lived, as this would be the only way that this vital essence could be obtained by man. In 1828, Freidrich Wöhler prepared the organic compound urea (found in human urine) from entirely nonliving sources, thereby destroying the theory of organic vitalism. Since Wöhler's time, approximately five million organic compounds have been synthesized and characterized, many of which are not found in nature.

Why are there so many organic compounds? The main reason is that carbon atoms have the ability to link to other carbon atoms **(concatenate)** to produce chains or rings of almost infinite size. Other elements do not concatenate nearly as well, due to such factors as poor orbital overlap and lone pair–lone pair electronic repulsions. Some elements can also combine with carbon to from hetero-species. Hydrogen, oxygen, nitrogen, sulfur, and the halogens are the most common examples.

The distinction between the organic and inorganic disciplines is not very sharp. The bonding of metals to carbon has resulted in the large, important, and fast-growing area of **organometallic** chemistry. Organometallic compounds containing metals such as lithium, magnesium, copper, iron, boron, and silicon (the latter two being metalloids) play major roles as synthetic reagents.

The purpose of this experiment is to prepare models of the more common organic compound types, so that a visual picture of these materials will demonstrate their three dimensional nature. Bond angles between atom groupings will also become apparent. You should consult your textbook for a more complete treatment of the nature and formation of organic systems.

Saturated Hydrocarbons: The Alkanes

Alkanes are hydrocarbons containing carbon atoms linked with single bonds. All the carbon atoms are sp^3 hybridized, and are tetrahedrally bonded to four other carbon or hydrogen atoms. Members of this class have the general formula C_nH_{2n+2}, where n is an integer (1, 2, 3, . . .). Examples of this class of organic compounds include ethane (C_2H_6), propane (C_3H_8), butane (C_4H_{10}), and pentane (C_5H_{12}). Structures of these as straight chain compounds are shown below. Longer chain hydrocarbons are named using Latin prefixes for the number of carbons, followed by the ending *-ane*. Thus, the eight-carbon alkane is known as octane.

Methane 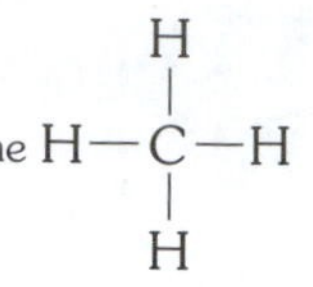 CH_4

Ethane 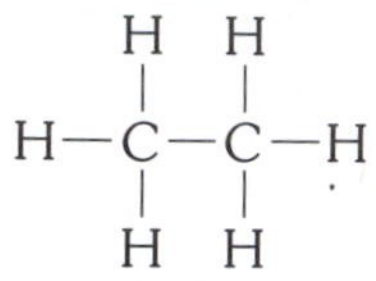 CH_3CH_3

Propane 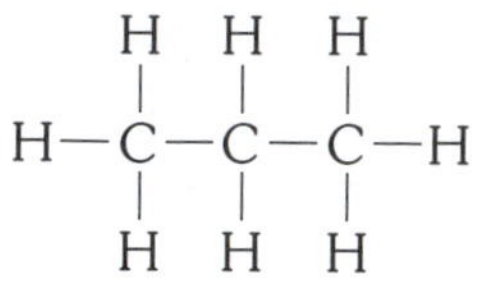 $CH_3CH_2CH_3$

Butane 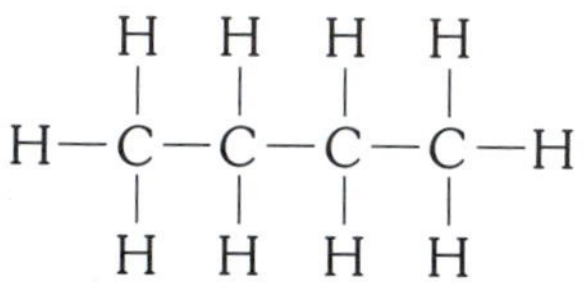 $CH_3CH_2CH_2CH_3$

Pentane 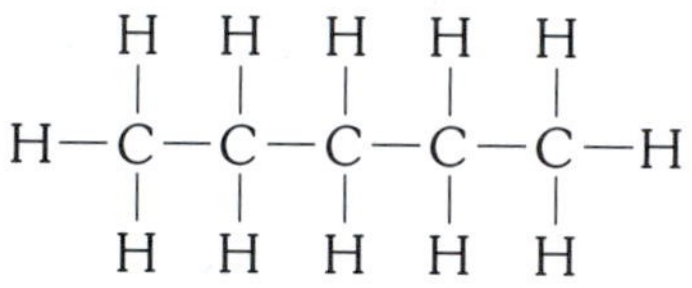 $CH_3CH_2CH_2CH_2CH_3$

These representations do not show the actual geometrical structure of these compounds. Due to the sp^3 hybridization of the carbon atom, all the bond angles in the molecules are close to 109.5° (tetrahedral), and thus the carbon chain is non-linear, as shown below for butane.

Ethane

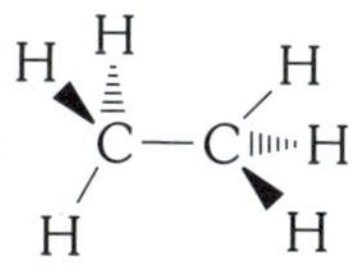

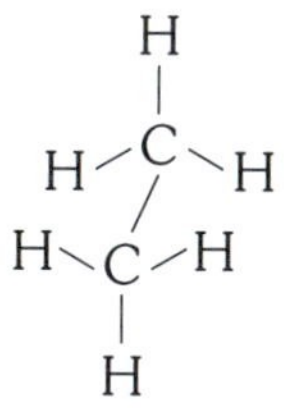

Butane

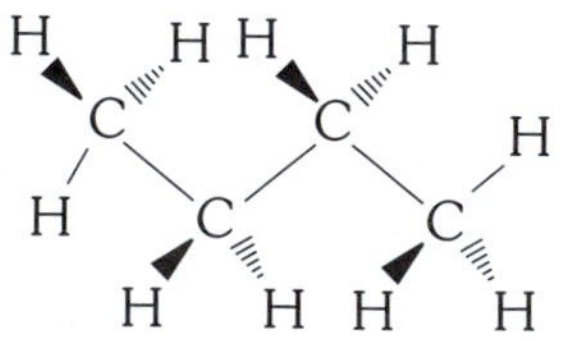

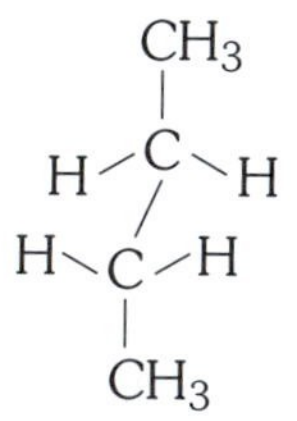

Wedge-dash Sawhorse

As you prepare models of the alkanes, note that each carbon atom can rotate about its respective carbon–carbon bond. Some conformations (arrangements of the bonds and groups relative to each other) are more stable than others, since in these arrangements there is less interference (steric repulsion) between the hydrogen atoms attached to nearby carbons.

An interesting aspect of the alkanes is that **structural isomers** can exist in compounds having four or more carbon atoms. Structural isomers are species that have the same molecular formula, but possess different physical properties due to different arrangements of the carbon backbone. The two compounds below are an example of an isomeric pair of hydrocarbons. Both are butanes, and have formula C_4H_{10}. The structures, however, are different, and are shown below.

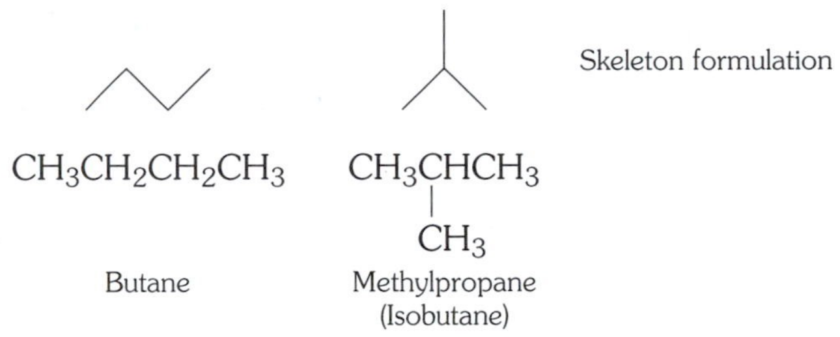

Skeleton formulation

$CH_3CH_2CH_2CH_3$ $\quad$ $CH_3CH(CH_3)CH_3$

Butane $\quad$ Methylpropane (Isobutane)

This figure relates several new points. First is the use of the skeleton formula. These structures represent the carbon backbone without the hydrogen atoms being shown, and are often used as a shorthand method of representing the structures. The second point is the use of substituents in naming compounds. We can see that in methylpropane, one of the CH_3 groups is a "twig" off of the main "branch" of the compound. Such twigs are called alkyl groups. In the figure, the methyl group (CH_3—) name is derived from the hydrocarbon methane (CH_4) having lost a hydrogen. The number of possible isomers increases rapidly as the number of carbon atoms increases in a compound. For example, the pentane system (five carbons) has 3 isomers, the heptane system (seven carbons) has 9 isomers, the octane system (eight carbons) has 18 isomers, and the decane system (ten carbons) has no fewer than 75 isomers.

Alkanes can also exist as **cyclic** hydrocarbons, wherein the carbon atoms are arranged in rings. The general molecular formula is C_nH_{2n}, where n is an integer. These carbons are named identically to the alkanes, except for the additional prefix *cyclo-*. The structures for several cyclic hydrocarbons are given below.

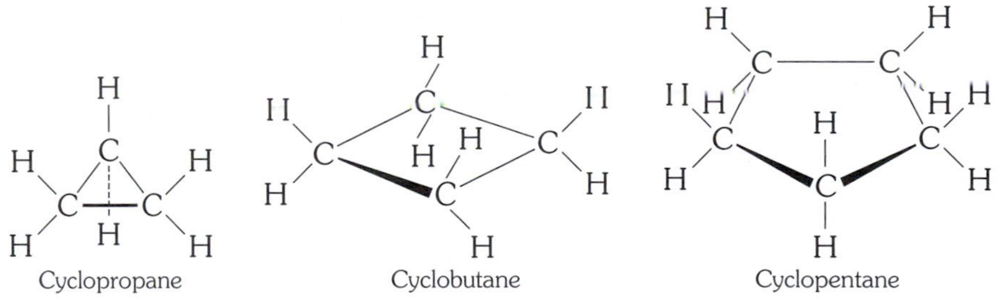

Cyclopropane $\quad$ Cyclobutane $\quad$ Cyclopentane

In the smaller rings such as cyclopropane and cyclobutane, smaller bond angles are evident (60° and 90°, respectively). Such angles are seen as being strained from the normal tetrahedral angle, and these compounds have less stability than their larger counterparts. The most stable cyclic compounds contain rings of five and six carbons.

Unsaturated Hydrocarbons: The Alkenes

Alkenes are hydrocarbons in which there are one or more carbon–carbon double bonds, C=C. The carbon atoms attached to the double bond are sp^2 hybridized. One of the bonds in the double bond is a **sigma** bond (oriented along the internuclear axis), while the other a **pi** bond (oriented perpendicular to the internuclear axis). Members of this class that have one such double bond, have the general formula of C_nH_{2n}, where n is an integer (this is the same general formula as for the cycloalkanes). Examples of alkenes include ethene (C_2H_4, also known as ethylene), propene (C_3H_6, also known as propylene), and the butenes (C_4H_8). Structures of several of these species are shown below. Alkenes are named in the same way as alkanes, except that the ending *-ene* replaces the ending *-ane*. In cases where more than one structural isomer exists (such as butene, below), it is necessary to indicate the location of the double bond by numbering the carbon atoms in the longest chain containing the double bond, and then giving it the lowest possible number.

```
              H       H
               \     /
Ethene          C=C                    CH2=CH2
(Ethylene)     /     \
              H       H

                H  H   H
                |  |  /
Propene     H—C—C=C                 CH3CH=CH2
(Propylene)     |      \
                H       H

             H  H  H   H
             |  |  |  /
1-Butene H—C—C—C=C                CH3CH2CH=CH2
             |  |      \
             H  H       H

             H  H  H  H
             |  |  |  |
2-Butene H—C—C=C—C—H            CH3CH=CHCH3
             |        |
             H        H
```

Many kinds of isomerism exist in organic compounds. Since carbon atoms are free to rotate about C—C single bonds, but <u>not</u> around a C=C double bond, (this disrupts the overlap of the pi part of the double bond), it is possible to have two separate geometrical isomers of 2-butene. These are known as the *cis-* and *trans-* isomers, and are shown below.

```
 H       H         H       CH3
  \     /           \     /
   C=C               C=C
  /     \           /     \
CH3      CH3      CH3      H
   cis-isomer        trans-isomer
```

Cyclic structures containing C=C bonds are also possible. Several are shown below.

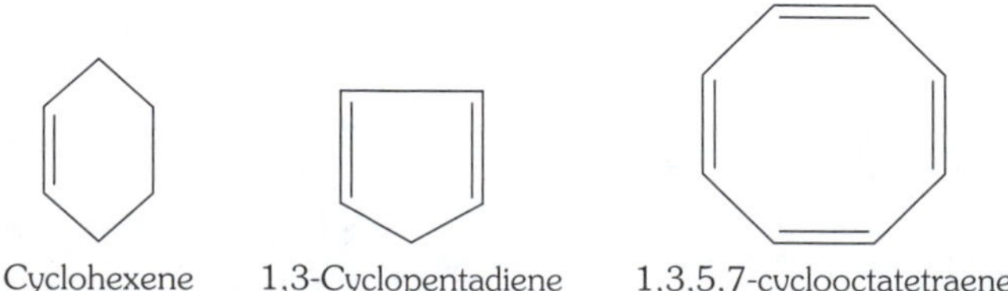

Cyclohexene 1,3-Cyclopentadiene 1,3,5,7-cyclooctatetraene

Unsaturated Hydrocarbons: The Alkynes

Hydrocarbons containing a carbon–carbon triple bond are named **alkynes.** The two carbon atoms forming the triple bond are joined by one sigma and two pi bonds, and are sp hybridized. The alkynes have the general molecular formula C_nH_{2n-2} (if they have one triple bond). Several representatives of this class are shown below. Alkynes are named in the same manner as alkenes, except that the ending *-yne* replaces the ending *-ene*.

$H-C\equiv C-H$ Ethyne (Acetylene)

$CH_3-C\equiv C-H$ Propyne (Methyl acetylene)

$CH_3-C\equiv C-CH_3$ 2-Butyne (Dimethyl acetylene)

Aromatic Hydrocarbons

Benzene, C_6H_6, is the most important member of this important class of hydrocarbons. The molecule contains a ring of six sp^2 hybridized carbon atoms with the unhybridized p orbitals perpendicular to the ring system. The six electrons in the π parts of the bonds are **delocalized** (spread evenly over the six carbon nuclei). Such molecules are often unusually stable. There are two equally valid ways of representing the structure of benzene. These two ways are called resonance forms. The molecule is therefore a **resonance hybrid,** with the "true structure" of benzene lying midway between the two resonance forms, in the same way that a mule is a genetic hybrid descendant of a male donkey and a female horse. The mule does not change back and forth, being a donkey half the time, and a horse the other half. Thus, the properties of a resonance hybrid are fixed.

There are many familiar hydrocarbon species that are based on the benzene structure, which contain various functional groups. A few examples are given below. Note that the delocalized π-electrons are represented by a circle in the ring. This is an alternate representation of the ring, often used by organic chemists.

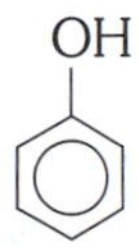

OH
Phenol

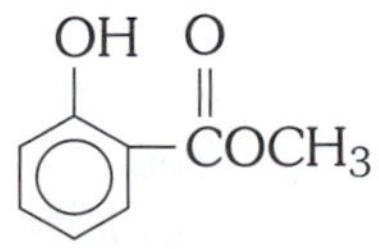

OH O COCH$_3$
Oil of Wintergreen

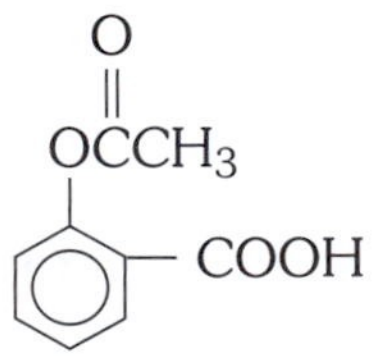

O OCCH$_3$ COOH
Aspirin

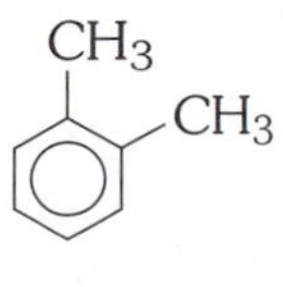

CH$_3$ CH$_3$
ortho-xylene (1,2-dimethylbenzene)

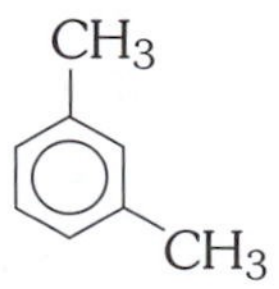

CH$_3$ CH$_3$
meta-xylene (1,3-dimethylbenzene)

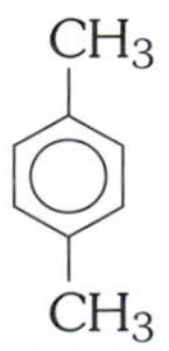

CH$_3$ CH$_3$
para-xylene (1,4-dimethylbenzene)

There are also many important aromatic hydrocarbons that contain more than one fused benzene ring. Several are illustrated here:

Naphthalene

Anthracene

Phenanthrene

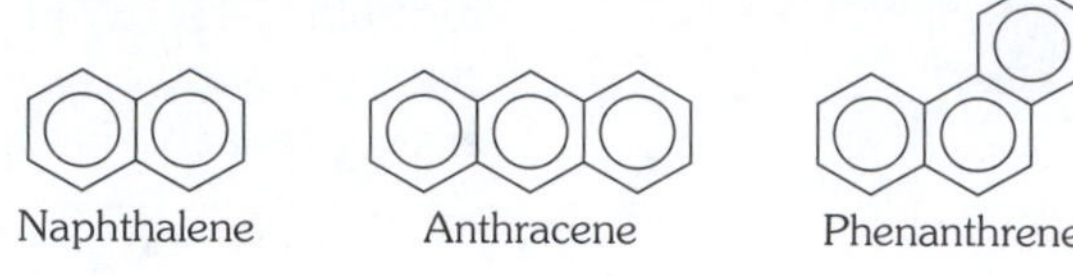

Hydrocarbons Containing Functional Groups

The basic types of hydrocarbon compounds outlined above may have one or more of their hydrogen atoms replaced by a **functional group.** The substituted benzenes above illustrate a number of functionalities (—CH_3, —OH, —COOH) attached to the aromatic ring. Additional examples are shown here:

Functional Group	*Class of Compound*	*Example*	*Name*
—OH	Alcohol	CH_3CH_2—OH	Ethanol (Ethyl Alcohol)
C—O—C	Ether	CH_3—O—CH_3	Dimethyl Ether
—N<	Amine	CH_3—NH_2	Methylamine
O ‖ C—C—H	Aldehyde	O ‖ CH_3—C—H	Ethanal (Acetaldehyde)
O ‖ C—C—C	Ketone	O ‖ CH_3—C—CH_3	Propanone (Acetone)
O ‖ —C—OH	Acid	O ‖ CH_3—C—OH	Ethanoic Acid (Acetic Acid)
O ‖ —C—O—C	Ester	O ‖ CH_3—C—O—CH_3	Methyl Acetate
—X (X═F, Cl, Br, I)	Haloalkane	CH_3CH_2—Cl	Chloroethane (Ethyl Chloride)

EXPERIMENTAL PROCEDURE Prepare models of the molecules indicated on the data sheets, and answer the questions about structure and bonding.

Name: ______________________________

Date: __________ Section: ______________________________

Data Sheet: Experiment 35

1. Prepare models of the straight-chain alkanes methane through hexane. Note that each carbon atom can rotate about its respective carbon–carbon bond.

a. Sketch the methane molecule showing the tetrahedral nature of the carbon atom.

b. Using line structures, sketch the configuration of propane through hexane. The line formula for propane is drawn a guide.

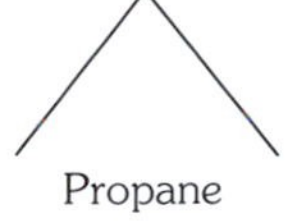

Propane Butane Pentane Hexane

c. Orient the butane model so that the methyl groups are *anti-* to one another (with an angle of 180° between them). Sketch the figure as a sawhorse structure. Rotate the central C—C bond so that the methyl groups are at an angle of 120°. Sketch this figure. Next, rotate the model so that the methyl groups are at a 60° angle and finally so that they eclipse each other (angle of 0°). Also sketch these structures. Explain which configuration is the most stable for butane.

180° 120° 60° 0°

Name: ______________________________

Date: __________ Section: ______________________________

Data Sheet: Experiment 35, Page 2

d. Prepare models of the five structural isomers of the hexane molecule, C_6H_{14}.

Sketch the line (skeleton) formulas for these molecules.

1.

2.

3.

4.

5.

Give a suitable name to each of the numbered hexane isomers. For example, one of them is named 3-methylpentane.

1. ______________________________

2. ______________________________

3. ______________________________

4. ______________________________

5. ______________________________

Name: ______________________________

Date: __________ Section: ______________________________

Data Sheet: Experiment 35, Page 3

2. Prepare models and sketch line (skeleton) formulas for each of the following unsaturated hydrocarbons.

1-Hexene

trans-3-Hexene

cis-3-Hexene

cis-*trans*-2,6-Octadiene

1-Pentyne

2-Pentyne

3. Prepare a model of the aromatic compound benzene, C_6H_6.

a. Comment on the fact that measurement of C—C bond distances in benzene shows that all are the same length (1.397 Å) and that the C—C bond angles are all 120°.

Name: ______________________________

Date: __________ Section: ______________________________

Data Sheet: Experiment 35, Page 4

b. Benzene was at one time used extensively as a solvent. However, since it has been found to be **carcinogenic,** it has largely been replaced by toluene. Toluene contains an aromatic ring substituted with a methyl group and has a molecular formula of C_7H_8.

Sketch a possible structure for toluene. Do you think this material would have a higher or lower boiling point than benzene? Explain.

c. Draw structures of the following aromatic hydrocarbons.

Chlorobenzene

1,2-Dichlorobenzene

3-Nitrotoluene

4-Bromobenzoic acid

Ethylbenzene

4-Nitrophenol

1,3,5-Trimethylbenzene

2,4,6-Trinitrotoluene

Name: ______________________________

Date: __________ Section: ______________________________

Data Sheet: Experiment 35, Page 5

d. Draw two resonance forms for naphthalene using the Kekule type structures.

e. There are three dichlorobenzene isomers. Draw them. Which one would you expect not to have a dipole moment? Explain.

f. Would you expect cyclobutyne to be a stable compound? Explain. (Hint: Make a model.)

Name: ______________________________

Date: __________ Section: ______________________________

Data Sheet: Experiment 35, Page 6

4. a. Identify the functional groups in each of the following compounds.

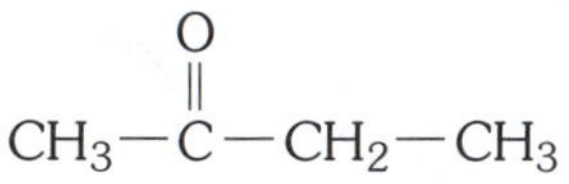

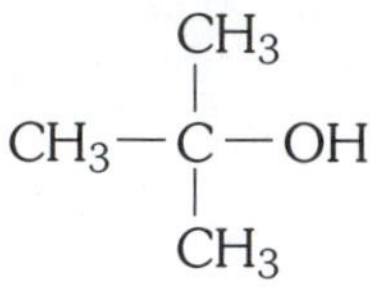

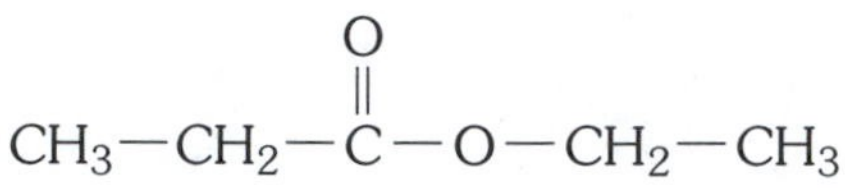

$$CH_3-\overset{\overset{\Large O}{\|}}{C}-CH_2-CH_3 \qquad CH_3-\underset{\underset{\Large CH_3}{|}}{\overset{\overset{\Large CH_3}{|}}{C}}-OH \qquad CH_3-CH_2-\overset{\overset{\Large O}{\|}}{C}-O-CH_2-CH_3$$

$$CH_3-CH_2-CH_2-CH_2-CH_2-CH_2-CH_2-NH_2 \qquad CH_3-\underset{\underset{\Large H}{|}}{\overset{\overset{\Large CH_3}{|}}{C}}-Br$$

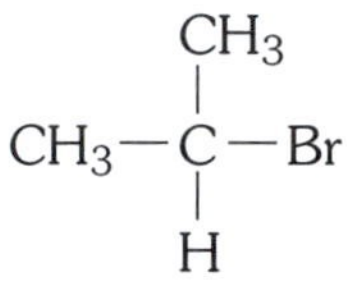

b. Ethyl alcohol is miscible with water in all proportions, whereas diethyl ether is only slightly soluble in water. Explain. (Hint: Think of hydrogen bonding.)

c. Prepare models for acetone and formaldehyde, $H_2C{=}O$. What are the bond angles around the $C{=}O$ carbon atom in each of these compounds?

Would you expect these compounds to be soluble in water? Explain.

Name: ____________________

Date: ________ Section: ____________________

Data Sheet: Experiment 35, Page 7

d. Give structural formulas for

1. An aldehyde

2. An alkyl fluoride

3. An ether

4. A carboxylic acid

5. A ketone

6. An ester

7. An amine

8. An alcohol

Experiment 36 Introduction to Organic Qualitative Analysis

OBJECTIVE

- To utilize qualitative tests to help identify organic compounds.

PRIOR READING Experiment 35

INTRODUCTION

One of the exciting challenges that a chemist faces on a regular basis is the identification of organic compounds. Approximately five million organic compounds have been prepared or isolated to date. This makes the identification of these materials of utmost importance—and it appears to be an insurmountable task! The majority of these substances, however, can be grouped into a comparatively small number of classes. Chemists have at their disposal an enormous database of chemical and spectroscopic information that has been organized over the past years. Forensic chemistry (the identification of drugs and other criminal evidence), environmental chemistry (the identification of toxic pollutants), industrial research and development, to name a few areas, all depend to a large extent on the ability of the chemist to isolate, purify, and identify specific organic chemicals.

A systematic approach has been developed over the years to simplify the identification problem. The task of identification was originally based on the solubility characteristics of the compounds and on certain chemical tests that could be used to detect the presence of various functional groups. Spectroscopic techniques such as infrared (see Experiment 16), nuclear magnetic resonance, ultraviolet, visible (see Experiments 13–15), and mass spectroscopy are now used extensively for this purpose.

In this experiment, the basic chemical and solubility tests that can be used to distinguish compounds that contain the major functional groups (reviewed in Experiment 35) will be explored. These groups include the alkanes, alkenes, alkyl-, and aryl-halides, alcohols, aldehydes, ketones, carboxylic acids, and amines.

In this introductory experiment, the compounds that will be given to you to test will have previously been purified. In advanced work, it is imperative that the purity of the unknown compound be established before test results are performed, otherwise the results are meaningless. Physical measurements are often used to determine purity. For solids, the melting point, and for liquids, the boiling point, refractive index and density are generally measured (see Experiment 2).

In any analysis of unknown compounds, one should follow a systematic approach. One possible sequence is suggested below (all methods are described in the next section):

1. Perform an **ignition test** to determine the general nature of the compound. Does it contain double bonds? Is it aromatic or aliphatic?
2. Determine the **solubility characteristics** of the species. This can often lead to valuable information related to the structural composition of an unknown organic compound.
3. Carry out **chemical tests** to assist in identifying elements other than carbon, hydrogen or oxygen. Nitrogen, phosphorus, sulfur, and the halogens are frequently found in organic compounds.

It is also important to realize that **negative** findings are often as important as positive results in classifying and identifying a given compound. Cultivate the habit of following a systematic pathway or sequence so that no clue or bit of information is lost or overlooked along the way. It is also important to develop the attitude and habit of planning ahead. Outline a logical plan of attack, depending on the nature of the unknown, and follow it. As you gain more and more experience in this type of investigative endeavor, the planning stage will become easier. Record all observations and results of the tests on your data sheet. Review these data as you execute the sequential phases of your plan. This serves to keep you on the path to analytical success.

In this experiment, you will be given a series of known compounds upon which to carry out the tests. In this manner you can observe and record how each species behaves under certain experimental conditions. This exercise will be followed by a series of selected unknown compounds for you to categorize into their respective classes.

EXPERIMENTAL PROCEDURE

A. Ignition Tests

Valuable information can be obtained by carefully noting the manner in which a given compound burns.

- Place 1–2 mg of the sample on a spatula, followed by heating with a microburner. Do not hold the sample directly in the flame—heat the spatula about 1 cm from the flat end and move the sample slowly into the flame.
- Important observations to be made concerning the ignition test are summarized in Table 36.1.
- As the heating of the sample takes place you should make the following observations:

1. Any melting or evidence of sublimation: gives a ballpark idea of the melting point by the temperature necessary to cause melting.
2. The color of the flame as the substance begins to burn (see Table 36.1).
3. The nature of the combustion (flash, quiet, or an explosion). Rapid, almost instantaneous combustion indicates high hydrogen content. Explosion indicates presence of nitrogen or nitrogen-oxygen-containing groups.

    a. If a black residue remains and disappears on further heating at higher temperature, the residue is carbon.
    b. If the residue undergoes swelling during formation, the presence of a carbohydrate (sugar) or similar compound is indicated.

Table 36.1 Ignition Test Observations[a]

Type of Compound	Example	Observation
Aromatic and other unsaturated compounds	Toluene	Yellow, sooty flame
Lower aliphatic compounds	Hexane	Yellow, almost nonsmoky flame
Compounds with oxygen	Ethanol	Clear bluish flame
Polyhalogen compounds	Chloroform	Do not ignite until the flame is applied directly to the substance
Sugars and proteins	Sucrose	Characteristic odor
Acid salts or organometallics	Sodium acetate	Residue

[a] Cheronis, N. D., Entrikin, J. B., *Semimicro Qualitative Analysis*, Interscience: New York, 1947, p. 85.

c. If the residue is initially black, but still remains after heating, an oxide of a heavy metal is indicated.
d. If the residue is white, the presence of an alkali or alkaline earth, carbonate or SiO_2 from a silane or silicone is indicated.

Testing of Known Materials

- To test the nature of basic hydrocarbons (aliphatic and aromatic), ignite small samples of toluene (aromatic) and hexane (aliphatic). Record your observations in the data table.
- To test the nature of alcohols, sugars, and acid salts, ignite small samples of methanol or ethanol, glucose or sucrose, and sodium tartrate or ferrocene (iron residue) or hexamethyl disiloxane (SiO_2 residue). Record your observations in the data table.

The **Beilstein test** is used to detect the presence of halogens (Cl, Br, I). Organic compounds that contain chlorine, bromine, or iodine and hydrogen are decomposed on ignition in the presence of copper oxide to yield the corresponding hydrogen halides. These gases react to form the volatile cupric halides that impart a green or blue-green color to a non-luminous flame. It is a very sensitive test, but some nitrogen compounds and carboxylic acids also give false positive results.

- Pound the end of a copper wire to form a flat surface that can act as a spatula. Push the other end of the wire (~4 in. long) into a cork stopper, which serves as a handle. Heat the flat tip of the wire in a flame until the flame is no longer differently colored.
- Place a drop of the liquid unknown or a few mg of the solid unknown on the cooled flat surface of the spatula. Gently heat the material in the flame. The carbon present in the compound will burn first (the flame will be luminous), but then the characteristic green or blue-green color will be evident. It may be fleeting, so watch carefully.
- To observe the Beilstein test, ignite a small sample of *tert*-butyl chloride or bromobenzene as directed above. Record your observations in the data table.

B. Solubility Characteristics

Determination of the solubility characteristics of an organic compound can often give valuable information as to its structural composition. It is especially useful when correlated with spectral analysis. Several schemes have been proposed that place a substance in a definite group, according to its solubility in various solvents. A simplified version covering those types of compounds is discussed below.

There is no sharp dividing line between soluble and insoluble, and an arbitrary ratio of solute to solvent must be selected. We suggest that a compound be classified as soluble if its solubility is greater than 15 mg/500 μL of solvent. Solubility determinations should be carried out at room temperature in 10 × 75 mm test tubes.

- Place about 15 mg of sample in the test tube, and add a total of 0.5 mL of solvent in three separate portions from a calibrated Pasteur pipet. Between additions, stir the sample vigorously with a glass stirring rod for about 2 minutes.
- If the sample is water soluble, test the solution with litmus paper to assist in classification according to the solubility scheme in Figure E36.1.
- To test with litmus paper, dip the end of a small glass rod into the solution and then gently touch the litmus paper with the rod. **Do not dip the litmus paper into the solution.**

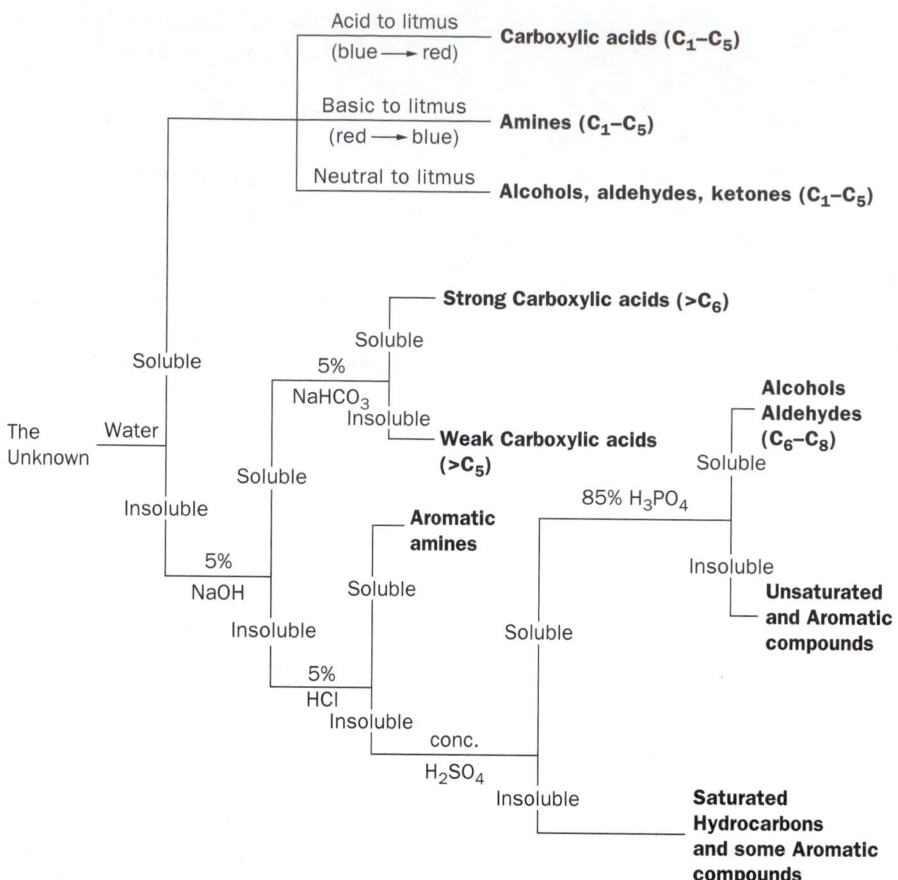

Figure E36.1 *Solubility scheme for organic compounds.*

In performing the solubility tests, follow the scheme in the order given below. **Keep a record of all observations.**

- Test for water solubility. If soluble, test the acidity of the solution with litmus paper.
- If the compound is water soluble, determine the solubility in diethyl ether. This test further classifies water-soluble materials.
- Water-insoluble compounds should now be tested with 5% aqueous NaOH solution. If soluble, determine the solubility in 5% aqueous $NaHCO_3$. The use of the $NaHCO_3$ solution aids in distinguishing between strong (soluble) and weak (insoluble) acids.
- Compounds insoluble in 5% aqueous NaOH are tested with 5% HCl.
- Compounds insoluble in 5% aqueous HCl are tested with concentrated H_2SO_4. If soluble, further differentiation is made using 85% H_3PO_4 as shown in the scheme.

Note that it may not be necessary to test solubility in every solvent to classify a given compound. **Do only those tests that are required to place the compound in one of the solubility groups.** Make all observations with care, and proceed in a logical sequence as you make the tests.

Testing of Known Materials

- To observe solubility classifications, check the solubility of toluene, hexane, isopropyl alcohol, *tert*-butyl chloride, 2-pentene, methyl ethyl ketone, and benzoic acid following the scheme in Figure E36.1. Record all observations in the data table.

C. Chemical Test for Nitrogen

The Soda Lime test is used to detect the presence of nitrogen in an organic compound.

- In a 10 × 75 mm test tube, mix approximately 50 mg of soda lime and 50 mg of MnO_2.
- Add one drop of a liquid unknown or about 10 mg of a solid unknown. Place a moist strip of brilliant yellow paper (moist red litmus paper may be used if necessary) over the mouth of the tube.
- Using a test tube holder, hold the tube at an incline and heat the contents gently at first and then quite strongly. Nitrogen containing compounds will usually evolve ammonia. A positive test for nitrogen is the deep red coloration of the brilliant yellow paper (or blue color of the litmus paper).

Testing of Known Materials

- To observe the Soda Lime test, ignite a small sample of triethylamine or N,N-dimethylaniline as directed above **(HOOD).** Record all observations in the data table.

Classification of an Unknown

- Obtain a sample of unknown from your laboratory instructor, and classify it following the method outlined above.

Name: ______________________________

Date: __________ Section: ______________________________

Data Sheet: Experiment 36

A. Ignition Tests: Observations on known compounds.

1. Aromatic

2. Aliphatic

3. Alcohol

4. Sugar

5. Acid salt

6. Beilstein Test

Name: ______________________________

Date: __________ Section: ______________________

Data Sheet: Experiment 36, Page 2

B. Solubility Classification: Observations on known compounds. Outline a solubility scheme for each of the tested compounds.

1. Toluene

2. Hexane

3. Isopropyl alcohol

4. *tert*-Butyl chloride

5. 2-Pentene

6. Methyl ethyl ketone

7. Benzoic acid

Name: ____________________

Date: ________ Section: ____________________

Data Sheet: Experiment 36, Page 3

C. Chemical Test for Nitrogen: Soda lime test.

1. Triethylamine

2. N,N-Dimethylaniline

D. Classification of Unknown.

Unknown number ________

Name: ______________________________

Date: __________ Section: ______________________________

Questions: Experiment 36

1. An unknown compound is insoluble in water, 5% sodium hydroxide, 5% hydrochloric acid, and concentrated sulfuric acid. Would you classify this material as an alcohol, ketone, saturated hydrocarbon, or an amine? Give reasons for your choice.

2. You have investigated several unknown organic compounds and observed their behavior as given below. In each case, indicate the deductions that may be made as to the nature of the compound.

a. A liquid species burned with a yellow-sooty flame. When heated on a copper spatula, a fleeting green flame was observed.

b. A compound having a fish-like smell was soluble in water and gave a positive soda lime test.

3. An unknown compound was observed to burn with a non-sooty yellow flame. It was insoluble in 5% NaOH, 5% HCl, and concentrated sulfuric acid solution. Indicate precisely what each of the solubility tests tell you about the unknown compound. To what class does the unknown compound belong?

Experiment 37 Synthesis of Aspirin

OBJECTIVES

- To synthesize an organic compound.

PRIOR READING

Experiment 35: Introduction
Section 3.7: Suction Filtration

INTRODUCTION

Organic chemistry is heavily involved in many chemical industries, such as plastics, petroleum, polymers and pharmaceuticals. Many everyday products are organic, including food, cloth, and drugs. Aspirin is the most commonly used of all drugs, with more than 40 million pounds produced each year in the United States alone. Aspirin is also known by its chemical name, acetyl-salicylic acid, since it is usually prepared via the reaction of acetic anhydride, $(CH_3CO)_2O$, hence, the acetyl root in its name) and salicylic acid. Aspirin is an analgesic (it relieves pain) and antipyretic (it reduces fever).

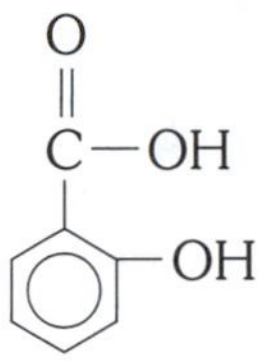

Salicylic Acid	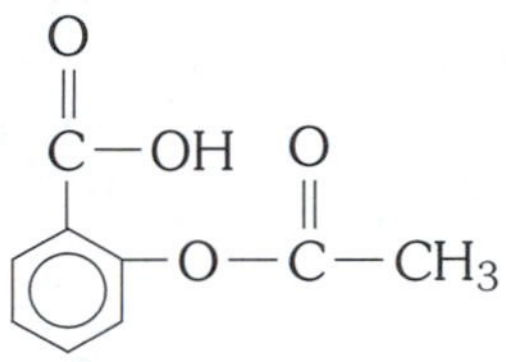 Aspirin (Acetylsalicylic Acid)	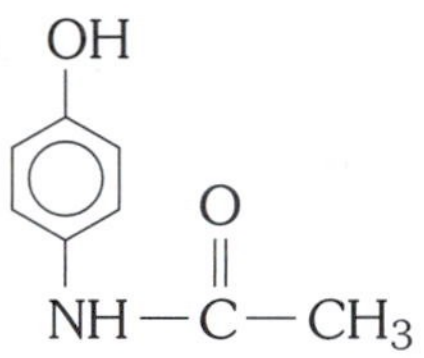 Tylenol (Acetaminophen)	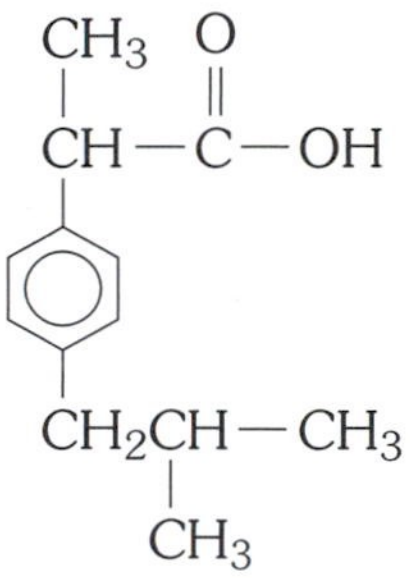 Advil (Ibuprofen)

Polyunsaturated fatty acids (found in corn and sunflower oils) are oxidized in the body to form compounds called prostaglandins. Excessive amounts of prostaglandins can cause various ailments, such as fever, pain, and inflammation. Aspirin interferes with the body's ability to synthesize prostaglandins, thus reducing the fever and inflammation pain associated with them.

Salicylic acid has the same analgesic and antipyretic properties as aspirin. Salicylic acid was prepared (from the bark of the willow tree) prior to the first synthesis of aspirin. It was initially used to cure headaches and fever, but its use was problematic since salicylic acid is very acidic and therefore irritating to the stomach, mouth, and mucous membranes. It was replaced by aspirin, which is somewhat less acidic than its salicylic acid parent, but has the same beneficial effects. Some people even find aspirin to be too acidic, and use the various aspirin substitutes (such as Tylenol and Advil, which are even weaker acids) instead.

When commercial aspirin is swallowed, it travels unchanged to the small intestine, due to an acid-resistant coating. There, the bases present cause it to react to reform the anion of salicylic acid, which is then absorbed into the bloodstream.

As common as aspirin is, there are several serious side effects associated with its use. High doses of aspirin have been known to damage the liver and kidneys over long periods of time. Reye's syndrome has been associated with the taking of aspirin by children with the flu or chicken pox. On the other hand, aspirin has been found to lower the chances of heart attack and stroke. Obviously, care is indicated in the therapeutic use of even the most "innocuous" drug.

EXPERIMENTAL PROCEDURE

- Place about 100 mg of salicylic acid in a dry 10 mL Erlenmeyer flask. If any moisture is present in the flask, the reaction will be extremely slow. Add 250 μL of acetic anhydride to the flask, using a automatic delivery pipet.
- Using a medicine dropper, add three drops of concentrated H_2SO_4 **(Caution: Corrosive!).** Swirl the flask for about 5 minutes.
- Allow the flask to stand at room temperature for an additional 5 minutes. During this time, a precipitate of aspirin may form. While waiting, assemble a suction filtration apparatus using a Hirsch funnel and suction flask (see section 3.7, p. 42).
- Add about 5 mL of cold water (graduated cylinder) to the flask. The remaining aspirin will precipitate, since it is not soluble in water.
- Collect the aspirin using the filtration apparatus assembled above. Wash the product with three 1 mL portions of cold water. Dry the product on a clay tile or a piece of filter paper. Weigh the amount of aspirin product, and determine the percentage yield.

Test of Aspirin Activity

Tests will now be performed to determine if aspirin becomes active in the stomach or the intestine. Remember: it is the salicylic acid that is the active ingredient, not the aspirin itself. Salicylic acid (and all phenols) forms a highly colored complex with iron(III) chloride.

- Place an approximately 10 mg sample of salicylic acid in a small test tube, and add a few drops of water and one drop of 1% iron(III) chloride solution. Observe the results.
- Perform the same test with 10 mg of well-crushed (mortar and pestle) commercial aspirin.
- Take two 10 mg portions of your aspirin product. Place one portion in a clean small test tube, and add five drops of 0.15 M HCl (stomach acid). Add one drop of the iron(III) chloride solution. Has the aspirin become active?
- Place the other portion of aspirin in another small test tube. Add five drops of 0.1 M NaOH (simulating intestinal base). Add one drop of the iron(III) chloride solution. Has the aspirin become active?

Name: ______________________________

Date: __________ Section: ______________________

Pre-Laboratory Questions: Experiment 37

1. What functional groups are present in salicylic acid, aspirin, Tylenol, and Advil?

2. High temperature and inflammation associated with fever are methods the body has to defend itself. In what ways can these be beneficial?

3. Upon standing, especially in a moist atmosphere, aspirin loses its potency and develops a distinct vinegary (acetic acid) smell. Write a balanced equation accounting for the decomposition of aspirin.

Name: ____________________

Date: ________ Section: ____________________

Data Sheet: Experiment 37

1. Mass of salicylic acid used ________ g

Moles of salicylic acid used ________ mol

2. Volume of acetic anhydride used ________ mL

Moles of acetic anhydride used ________ mol
(density = 1.080 g mL^{-1})

3. Mass of aspirin product ________ g

Moles of aspirin product ________ mol

4. Percentage yield: ________ %

5. What was the effect of adding the iron(III) chloride to the salicylic acid?

6. What was the effect of adding the iron(III) chloride solution to the aspirin?

7. From your tests on the aspirin product, does the aspirin become active in the stomach or in the intestine? Explain why.

Name: ______________________

Date: __________ Section: ______________________

Questions: Experiment 37

1. One gram of salicylic acid will dissolve in 460 mL of water, yielding a solution with a pH of 2.4. Determine the K_a value for salicylic acid, and compare it to that of aspirin, 3.27×10^{-4}.

2. Aspirin is prepared in 85% yield (on average) from salicylic acid.

a. How many grams of salicylic acid will be required to form 150 g of aspirin?

b. How many mL of acetic anhydride will be required?

Appendix A
Safety Data for Common Solvents

Acetone: Acetone is an extremely flammable liquid. It is not normally considered dangerous, but normal precautions should be employed. ORL-RAT LD50: 5800 mg/kg.

Chloroform: Chloroform is a potent narcotic agent. It may be fatal if inhaled, swallowed, or absorbed through the skin. It is classified as a carcinogen. ORL-RAT LD50: 908 mg/kg.

Cyclohexane: Cyclohexane is harmful if inhaled or swallowed. It is extremely flammable. ORL-RAT LD50: 1215 mg/kg.

Diethyl ether: Ether is an extremely flammable solvent. On exposure to moisture it tends to form peroxides, which may be explosive. The solvent is a potent narcotic. ORL-RAT LD50: 1215 mg/kg.

Dimethylsulfoxide: Dimethylsulfoxide (DMSO) is harmful if swallowed, inhaled, or absorbed through the skin. Overexposure has been found to have effects on fertility. ORL-RAT LD50: 14,500 mg/kg.

Ethyl alcohol: Ethyl alcohol may be fatal if inhaled, swallowed, or absorbed through the skin in large amounts. It has been shown to have effects on fertility and on embryo development. ORL-RAT LD50: 7060 mg/kg. The vapor may travel considerable distances to the source of ignition and flash back.

Hexane: Hexane is harmful if inhaled, swallowed, or absorbed through the skin. It is a flammable liquid. ORL-RAT LD50: 28,710 mg/kg.

Isopropyl alcohol: Isopropyl alcohol (rubbing alcohol) is not normally considered dangerous, but the usual precautions should be followed. ORL-RAT LD50: 5045 mg/kg.

Methyl alcohol: Methanol may be fatal if swallowed. It is harmful if inhaled or absorbed through the skin. It is a flammable liquid. ORL-RAT LD50: 5628 mg/kg.

Methylene chloride: The compound is harmful if swallowed, inhaled, or absorbed through the skin. ORL-RAT LD50: 600 mg/kg. It is a possible carcinogen.

Toluene: Toluene is a flammable liquid. ORL-RAT LD50: 5000 mg/kg.

All safety data in this table are derived from the Sigma-Aldrich Material Safety Data Sheets on CD-ROM, Aldrich Chemical Co., Inc., Milwaukee, WI.

Appendex B
List of Common Acids and Bases

Acetic acid [CAS 64–19–7]: Glacial acetic acid is available in up to 100% purity. FW = 60.05 g mol^{-1}; mp = 16.2°C; bp = 116–118°C; d = 1.049 g mL^{-1}. Concentration is 17.5 M. The acid is corrosive and toxic and has a pungent odor.

Ammonium hydroxide [CAS 1336–21–6]: Ammonium hydroxide is available as a 28–30% solution. FW = 17 g mol^{-1} (as NH_3); d = 0.900 g mL^{-1}. Concentration is about 15 M. The base is corrosive and toxic and has a pungent odor.

Hydrochloric acid [CAS 7647–37–2]: Hydrochloric acid is available as a 37% solution. FW = 36.46 g mol^{-1}; d = 1.200 g mL^{-1}. Concentration is about 12 M. The acid is extremely corrosive and toxic.

Nitric acid [CAS 7697–37–2]: Nitric acid is available as a 69–71% solution. FW = 63.01 g mol^{-1}; d = 1.400 g mL^{-1}. Concentration is about 15.6 M. The acid is extremely corrosive and toxic. Toxic fumes of NO_2 may be given off. Strong oxidizing agent.

Phosphoric acid [CAS 7664–38–2]: Phosphoric acid is available as an 85% solution. FW = 98.00 g mol^{-1}; d = 1.685 g mL^{-1}. Concentration is about 14.6 M. The acid is corrosive.

Sulfuric acid [CAS 7664–93–9]: Sulfuric acid is available as a 95–98% solution. FW = 98.08 g mol^{-1}; d = 1.840 g mL^{-1}. Concentration is about 18 M. The acid is extremely corrosive. Strong oxidizing and dehydrating agent.

Appendix C
Table of Physical and Chemical Constants

Symbol	Name	Value
[illegible]	Velocity of light	2.988×10^{10} cm sec^{-1}
	Planck's constant	6.626×10^{-27} erg sec
N	Avogadro's number	6.023×10^{23} molecules/mole, atoms/g-atom, formula units/mol
F	Faraday's constant	2.8926×10^{14} esu mol^{-1}
		9.648×10^{4} C mol^{-1}
		2.306×10^{4} cal V^{-1} mol^{-1}
V_o	Molar volume of gas at STP	22.4136 liters mol^{-1}
P^o	Standard pressure	1.000 atm
		760 mm Hg (torr)
		14.696 lb in.$^{-2}$
k	Boltzmann's constant	1.3806×10^{-16} erg K^{-1} molecule^{-1}
e	Charge on the electron	4.803×10^{-10} esu
		1.602×10^{-19} C
m_e	Rest mass of the electron	9.109×10^{-28} g
		5.4859×10^{-4} amu
amu	Atomic mass unit	1.660×10^{-24} g
m_p	Rest mass of the proton	1.6726×10^{-24} g
		1.00727 amu
m_n	Rest mass of the neutron	1.6749×10^{-24} g
		1.00866 amu
R	Gas constant	0.082056 L atm K^{-1} mol^{-1}
		8.3143 JK^{-1} mol^{-1}
		1.9872 cal K^{-1} mol^{-1}
eV	Electron volt	1.602×10^{-12} erg molecule
		2.3061×10^{4} cal mol^{-1}

Appendix D
Table of Common Units and Conversion Factors

Length

SI unit: Meter (m)		
1 kilometer	=	1000 meters
	=	0.62137 mile
1 meter	=	100 centimeters
1 centimeter	=	10 millimeters
1 nanometer	=	1×10^{-9} meter
1 picometer	=	1×10^{-12} meter
1 inch	=	2.54 centimeters (exactly)
1 Ångstrom	=	1×10^{-10} meter

Mass

SI unit: Kilogram (kg)		
1 kilogram	=	1000 grams
1 gram	=	1000 milligrams
1 pound	=	453.6 grams = 16 ounces
1 ton	=	2000 pounds

Volume

SI unit: Cubic meter (m^3)		
1 liter (L)	=	1×10^{-3} m^3
	=	1000 cm^3
	=	1.056710 quarts
1 gallon	=	4 quarts

Pressure

SI unit: Pascal (Pa)		
1 pascal	=	1 N m^{-2}
	=	1 kg m^{-1} s^{-2}
1 atmosphere	=	101.325 kilopascals
	=	760 mm Hg = 760 torr
	=	14.70 lb in.$^{-2}$

Energy

SI unit: Joule (J)		
	=	1 Nm
1 joule	=	1kg m^2 s^{-2}
	=	0.23901 calorie
	=	1 C × 1 V
1 calorie	=	4.184 joules
1 electron volt	=	96.485 kJ mol^{-1}

Appendix E
Acid (K_a) and Base (K_b) Dissociation Constants[a]

Acids	Equation	K_a	pK_a
Acetic	$HC_2H_3O_2 + H_2O = H_3O^+ + C_2H_3O_2^-$	1.8×10^{-5}	4.75
Ammonium ion	$NH_4^+ + H_2O = H_3O^+ + NH_3$	5.6×10^{-10}	9.26
Arsenic*	$H_3AsO_4 + H_2O = H_3O^+ + H_2AsO_4^-$	5.6×10^{-3}	2.25
	$H_2AsO_4^- + H_2O = H_3O^+ + HAsO_4^{2-}$	1.7×10^{-7}	6.77
	$HAsO_4^{2-} + H_2O = H_3O^+ + AsO_4^{3-}$	2.5×10^{-12}	11.60
Boric	$H_3BO_3 + H_2O = H_3O^+ + H_2BO_3^-$	7.3×10^{-10}	9.14
	$H_2BO_3^- + H_2O = H_3O^+ + HBO_3^{2-}$	1.8×10^{-13}	12.74
	$HBO_3^{2-} + H_2O = H_3O^+ + BO_3^{3-}$	1.6×10^{-14}	13.80
Carbonic	$H_2CO_3 + H_2O = H_3O^+ + HCO_3^-$	4.3×10^{-7}	6.37
	$HCO_3^- + H_2O = H_3O^+ + CO_3^{2-}$	5.6×10^{-11}	10.25
Chromic	$H_2CrO_4 + H_2O = H_3O^+ + HCrO_4^-$	1.8×10^{-1}	0.74
	$HCrO_4^- + H_2O = H_3O^+ + CrO_4^{2-}$	3.2×10^{-7}	6.49
Hydrocyanic	$HCN + H_2O = H_3O^+ + CN^-$	4.9×10^{-10}	9.31
Hydrofluoric	$HF + H_2O = H_3O^+ + F^-$	3.5×10^{-4}	3.45
Hydrogen sulfide*	$H_2S + H_2O = H_3O^+ + HS^-$	9.1×10^{-8}	7.64
	$HS^- + H_2O = H_3O^+ + S^{2-}$	1.1×10^{-12}	11.96
Hypochlorous	$HClO + H_2O = H_3O^+ + ClO^-$	2.9×10^{-5}	4.53
Nitrous**	$HNO_2 + H_2O = H_3O^+ + NO_2^-$	4.6×10^{-4}	3.37
Oxalic	$H_2C_2O_4 + H_2O = H_3O^+ + HC_2O_4^-$	5.90×10^{-2}	1.23
	$HC_2O_4^- + H_2O = H_3O^+ + C_2O_4^{2-}$	6.40×10^{-5}	4.19
Phosphorous	$H_3PO_4 + H_2O = H_3O^+ + H_2PO_4^-$	7.1×10^{-3}	2.21
	$H_2PO_4^- + H_2O = H_3O^+ + HPO_4^{2-}$	6.2×10^{-8}	7.21
	$HPO_4^{2-} + H_2O = H_3O^+ + PO_4^{3-}$	2.2×10^{-13}	12.67
Sulfurous*	$H_2SO_3 + H_2O = H_3O^+ + HSO_3^-$	1.5×10^{-2}	1.81
	$HSO_3^- + H_2O = H_3O^+ + SO_3^{2-}$	1.0×10^{-7}	6.91

Bases	Equation	K_b	pK_b
Acetate ion	$C_2H_3O_2^- + H_2O = HC_2H_3O_2 + OH^-$	5.6×10^{-10}	9.26
Ammonia	$NH_3 + H_2O = NH_4^+ + OH^-$	1.8×10^{-5}	4.74
Bicarbonate	$HCO_3^- + H_2O = H_2CO_3 + OH^-$	2.2×10^{-8}	7.66
Carbonate	$CO_3^{2-} + H_2O = HCO_3^- + OH^-$	2.3×10^{-4}	3.64
Cyanide	$CN^- + H_2O = HCN + OH^-$	1.4×10^{-5}	4.86
Flouride	$F^- + H_2O = HF + OH^-$	1.7×10^{-11}	10.78
Hydrazine	$N_2H_4 + H_2O = N_2H_5^+ + OH^-$	3.0×10^{-6}	5.52
Nitrite	$NO_2^- + H_2O = HNO_2 + OH^-$	2.2×10^{-11}	10.65
Oxalate	$C_2O_4^{2-} + H_2O = HC_2O_4^- + OH^-$	1.6×10^{-10}	9.79
Phosphate	$PO_4^{3-} + H_2O = HPO_4^- + OH^-$	2.1×10^{-2}	1.68
Sulfate	$SO_4^{2-} + H_2O = HSO_4^- + OH^-$	8.3×10^{-13}	12.08
Sulfite	$SO_3^{2-} + H_2O = HSO_3^- + OH^-$	1.6×10^{-7}	6.79

[a]All values are given at 25°C except as indicated: *(18°C) and **(12.5°C).

Appendix F
Solubility Product Constants at 25°C[a]

Compound	K_{sp}	Compound	K_{sp}
Carbonates		**Hydroxides**	
Ag_2CO_3	8.45×10^{-12}	$Ca(OH)_2$	4.68×10^{-6}
$BaCO_3$	2.58×10^{-9}	$Co(OH)_2$ (pink)	1.09×10^{-15}
$CaCO_3$	4.96×10^{-9}	$Fe(OH)_2$	4.87×10^{-17}
$FeCO_3$	3.07×10^{-11}	$Fe(OH)_3$	2.64×10^{-39}
$MgCO_3$	6.82×10^{-6}	$Mg(OH)_2$	5.61×10^{-12}
$MnCO_3$	2.24×10^{-11}	$Mn(OH)_2$	2.06×10^{-13}
$NiCO_3$	1.42×10^{-7}	$Pb(OH)_2$	1.42×10^{-20}
$PbCO_3$	1.46×10^{-13}	$Sn(OH)_2$	5.45×10^{-27}
$SrCO_3$	5.60×10^{-10}	$Zn(OH)_2$	6.68×10^{-17}
$ZnCO_3$	1.19×10^{-10}		
Chlorides		**Iodides**	
$AgCl$	1.77×10^{-10}	AgI	8.51×10^{-17}
$CuCl$	1.72×10^{-7}	CuI	1.27×10^{-12}
Hg_2Cl_2	1.45×10^{-18}	Hg_2I_2	5.33×10^{-29}
$PbCl_2$	1.17×10^{-5}	HgI_2	2.82×10^{-29}
		PbI_2	8.49×10^{-9}
Fluorides		**Sulfates**	
BaF_2	1.81×10^{-7}	Ag_2SO_4	1.20×10^{-5}
CaF_2	1.46×10^{-10}	Ba_2SO_4	1.07×10^{-10}
MgF_2	7.42×10^{-11}	$CaSO_4$	7.10×10^{-5}
PbF_2	7.12×10^{-7}	Hg_2SO_4	7.99×10^{-7}
SrF_2	4.33×10^{-9}	$PbSO_4$	1.82×10^{-7}
ZnF_2	3.04×10^{-2}	$SrSO_4$	3.44×10^{-7}
Chromates		**Sulfides**	
Ag_2CrO_4	1.12×10^{-12}	Ag_2S	6.69×10^{-50}
$BaCrO_4$	1.17×10^{-10}	CdS	1.40×10^{-29}
$PbCrO_4$	2.0×10^{-10}	CuS	1.27×10^{-36}
$SrCrO_4$	3.6×10^{-5}	FeS	1.59×10^{-19}
		HgS	6.44×10^{-53}
		MnS	4.65×10^{-14}
		NiS	1.07×10^{-21}
		PbS	9.04×10^{-29}
		SnS	3.25×10^{-28}
		ZnS	2.93×10^{-25}

[a]Weast, R. C., Ed., *Handbook of Chemistry and Physics,* 69th ed., CRC Press; Boca Raton, FL, 1988–1989.

Appendix G
General Solubility Rules for Ionic Compounds in Water

1. All nitrates **(NO_3^-)** are soluble.
2. Salts of the group 1 cations **(Na^+, K^+, Rb^+,** and **Cs^+)** or of ammonium **(NH_4^+)** are soluble.
3. Halides **(Cl^-, Br^-,** and **I^-)** and thiocyanate **(SCN^-)**—are soluble, except for Ag^+, Cu^+, Tl^+, Pb^{2+}, and Hg_2^{2+}. Bromide and iodide are oxidized by some cations.
4. Sulfates **(SO_4^{2-})** are all soluble except for Pb^{2+}, Ba^{2+}, and Sr^{2+}. (Ca^{2+}, Hg_2^{2+}, and Ag^+ are somewhat soluble.)
5. Nitrites **(NO_2^-)** and permanganates **(MnO_4^-)** are all soluble, except for $AgNO_2$. These ions are powerful oxidizing agents, so they are unstable with cations which are easily oxidized.
6. Thiosulfates **($S_2O_3^{2-}$)** are soluble, except for Pb^{2+}, Ba^{2+}, and Ag^+. ($Ag_2S_2O_3$ decomposes in excess thiosulfate with reduction of Ag^+.)
7. Sulfites **(SO_3^{2-})**, carbonates **(CO_3^{2-})**, phosphates **(PO_4^{3-})**, oxalates **($C_2O_4^{2-}$)**, and chromates **(CrO_4^{2-})**, are all insoluble in neutral of basic solution, except for Ca^{2+} (except calcium oxalate) and Mg^{2+} and ions listed in Rule 2. All are soluble in acidic solution. Sulfite and oxalate can form soluble complexes.
8. Fluorides **(F^-)** are insoluble, except for Ag^+, Fe^{3+}, and the ions listed in Rule 2. Some transition metal fluorides are soluble, especially in excess fluoride, due to complex ion formation.
9. Ferrocyanides **($Fe[CN]_6^{4-}$)** are insoluble, except for ions listed in Rule 2.
10. Hydroxides **(OH^-)** are insoluble, except for Sr^{2+}, Ba^{2+}, and the ions listed in Rule 2. Many metals are soluble in excess hydroxide, due to complex ion formation.

Appendix H
pH Ranges of Common Acid–Base Indicators

Indicator	pH Range	Color Change
Methyl violet	0.0–1.6	Yellow to blue
Thymol blue	1.2–2.8	Red to yellow
	8.0–9.6	Yellow to blue
Bromophenol blue	3.0–4.6	Yellow to blue
Methyl orange	3.2–4.4	Red to yellow
Bromocresol green	3.8–5.4	Yellow to blue
Methyl red	4.8–6.0	Red to yellow
Bromocresol purple	5.2–6.8	Yellow to purple
Litmus	6.0–8.0	Red to blue
Bromothymol blue	6.0–7.6	Yellow to blue
Phenol red	6.6–8.0	Yellow to red
Thymol blue	1.2–2.8	Red to yellow
	8.0–9.6	Yellow to blue
Phenolphthalein	8.2–10.0	Colorless to pink
Alazirin yellow	10.1–12.0	Yellow to red

Glossary

absolute temperature The temperature referenced to absolute zero. For the Celsius scale, absolute temperature is given in Kelvins, K = °C + 273.15.

absolute zero The temperature at which molecular motion ceases.

absorbance The amount of incident light absorbed by a particular sample.

absorption spectrum The spectrum obtained when electromagnetic radiation is passed through a substance which absorbs some frequencies of the radiation.

accuracy The extent to which a measured value corresponds to the true value.

acid A compound that donates a proton to another species (Brønsted-Lowry), that increases the concentration of H^+ in solution (Arrhenius) or that accepts an electron pair (Lewis).

acid ionization constant (K_a) The equilibrium constant for the ionization of a weak acid in water.

activation energy The amount of energy necessary for the reactants to form the transition state.

adduct The product of the reaction of a Lewis acid with a Lewis base. For example, H_3O^+ is the adduct that forms upon the reaction of the Lewis acid H^+ with the Lewis base H_2O.

aliquot A subsample, usually a carefully measured fraction of a larger sample.

amorphous A species having no definite crystal structure.

ampere (A) Unit of electric current. An ampere is a coulomb per second.

analyte The material being analyzed.

anion A negatively charged ion.

anode The electrode at which oxidation takes place in an electrochemical cell.

aqueous solution A solution in which the solvent is water.

atmosphere A unit of pressure. 1 atm = 760 torr = 1.013 bar = 101,325 Pa.

atom The smallest particle of an element that retains the physical properties of that element.

atomic mass The average mass of an atom, as compared to the most abundant isotope of carbon.

atomic number The number of protons found in the nucleus of an atom.

Avogadro's number The number of particles in a mole, 6.02×10^{23}.

bald Without hair. *See* Pike, Szafran but not Foster.

barometer An instrument for measuring atmospheric pressure.

base A species that accepts a proton from an acid.

battery A package of one or more galvanic cells. Batteries are used to store or produce electrical energy.

binary compound A compound composed of only two elements.

boiling point The temperature at which the vapor pressure of a liquid equals the external pressure on the liquid.

Boyle's Law The volume of a given quantity of gas is inversely proportional to the pressure applied to it at constant temperature.

buffer A solution consisting of an acid and a salt of its conjugate base (or a base and a salt of its conjugate acid). Buffer solutions resist changes in pH when small amounts of acid or base is added.

buffer capacity The ability of a buffer solution to resist pH changes upon addition of acid or base.

calorimeter constant The amount of heat required to raise the temperature of a calorimeter by 1°C.

calorimetry The process of measuring the amount of heat given off or absorbed by a physical change or chemical reaction.

catalyst A species that speeds up a reaction by lowering the energy of activation.

catalyze To act as a catalyst.

cathode The electrode at which reduction takes place in an electrochemical cell.

cation A positively charged ion.

Celsius Temperature Scale Temperature scale in which the freezing point of water is zero, and the boiling point of water is 100.

Charles' Law The volume of a given quantity of gas is directly proportional to its absolute temperature (K) at constant pressure.

colligative properties Properties of a solution that depend only on the concentration of the solute species.

combustion The act of burning, usually in oxygen or air.

common ion effect A shift in equilibrium due to the addition of an ion present in the equilibrium expression.

Coulomb (C) The SI unit of electric charge.

Dalton's Law of Partial Pressures The total pressure of a gas mixture is equal to the sum of the individual partial pressures.

density The mass of a unit volume of a material.

diffusion The ability of a gas to spontaneously spread out in all directions.

disproportionation A redox reaction in which a single reactant is partially oxidized and partially reduced.

dissociation The process by which a solute breaks apart and its ions move or spread throughout a solution.

double blind test A test where the person giving the drug doesn't know who gets the drug and who gets a placebo, and neither does the recipient.

electrode A conducting substance, through which electrons enter or leave a conducting medium.

electrolysis The use of an electric current to carry out an oxidation reduction reaction.

electronegativity The ability of an element or group of elements to attract bond electrons to themselves.

electron The negatively charged fundamental particle. The electron is found outside the atomic nucleus.

element A pure substance that cannot be further broken down by chemical means.
empirical formula The formula representing the lowest whole-number ratio of elements in a compound.
endothermic A reaction that absorbs energy (heat) from its surroundings.
end point The point during a titration when a sufficient amount of titrant has been added for complete reaction.
enthalpy (ΔH) of reaction The heat lost or absorbed during a chemical or physical change.
entropy (S) The amount of disorder of a system.
enzyme A complex natural polymer that occurs in living systems. Enzymes cause specific reactions to occur more quickly by lowering their energies of activation.
equilibrium The condition at which the forward and reverse rates of reaction are equal.
equilibrium constant The value of the reaction quotient when the system is at equilibrium.
exothermic A reaction that releases energy (heat) to its surroundings.

faraday (F) The quantity of charge on 1 mol of electrons. 1 F = 96,485 C.
filter cake The solid that is collected on the filtration apparatus.
filtrate The liquid that passes through a filter, usually collected in a filter flask.
fluid A material in the vapor or liquid state.
free energy (G) A thermodynamic function that predicts the spontaneity of a chemical reaction.
freezing point The temperature at which the solid and liquid phases of a material are in equilibrium.
functional group A part of a molecule responsible for specific chemical behavior in that molecule.

galvanic cell A device that uses an oxidation–reduction reaction to do work and produce current.
gas The state of matter in which molecules have neither definite macroscopic volume or shape.
gas constant (R) The proportionality constant derived from the Ideal Gas Law, $PV = nRT$.
Graham's Law The Ideal Gas Law dealing with diffusion (effusion).

half-cell The portion of an electrochemical cell in which a half-reaction takes place.
half-reaction The oxidation or reduction part of a redox reaction.
heat capacity The amount of heat necessary to increase the temperature of a given amount (usually a mole or a gram) of material one degree.
Hess's Law The enthalpy change for an overall reaction is the sum of the enthalpies of the individual steps following any path.
hydrocarbon A compound consisting primarily of hydrogen and carbon. An organic compound.

Ideal Gas Law A gas law that is based on the kinetic–molecular model of gases.
indicator A dye that changes color, indicating the endpoint of a reaction.
inert Nonreactive.
ion A charged particle resulting from the loss or gain of electrons.
ionization energy The energy necessary to remove an electron from a gaseous atom.
isotopes Atoms whose nuclei contain the same number of protons but different numbers of neutrons.

joule (J) The SI unit of energy. 1 J = 1 N m.

Kelvin (K) The temperature unit on the Kelvin temperature scale. The magnitude of the Kelvin degree is the same as a Celsius degree.
kinetics The study of reaction rates and mechanisms.

le Chatelier's principle When a stress is applied to a system at equilibrium, the equilibrium shifts to alleviate that stress.
Lewis acid A species that is an electron pair acceptor.
Lewis base A species that is an electron pair donor.
ligand A Lewis base that has donated an electron pair to a metal, which acts as a Lewis acid.
limiting reagent The reagent that is used up first, thus limiting the amount of product obtainable.

mass The quantity of matter contained by an object.
melting point The temperature at which the solid and liquid phases of a material are in equilibrium.
metathesis reaction A reaction in which two or more compounds exchange ions.
micro- Prefix meaning 10^{-6}.
miscible Two liquids are miscible if they mix together intead of forming two layers.
molality (m) The number of moles of solute per kilogram of solvent.
molarity (M) The number of moles of solute in 1 liter of solution.
mole Avogadro's number of particles.
molecular formula The formula representing the elements found in a molecule of a compound.

nanometer A wavelength unit commonly used in spectroscopy, equal to 10^{-9} meters.
neutralization A reaction that occurs when equivalent amounts of acid and base are mixed.
neutron A fundamental, noncharged particle found in the atomic nucleus.
nitrogen fixation The ability to remove molecular nitrogen from the atmosphere and convert it into nitrates (or other fertilizers).
nucleus The positive center of an atom that contains most of the mass.

oxidation The loss of electrons.
oxidation number The charge on an atom that would result if the molecule or ion were entirely ionic.
oxidation–reduction reaction *See* redox reaction.

oxidizing agent The substance that accepts electrons in a redox reaction.

partial pressure The pressure that one component of a gas mixture would exert if it were alone under the same conditions.
periodic law The properties of the elements are periodic functions of their atomic structure.
pH The negative log of the hydrogen ion concentration in a solution.
precipitate An insoluble material that settles out of a solution.
proton The positively charged fundamental particle, found in the atomic nucleus.

qualitative analysis Analysis to determine the presence or absence of the analyte.
quantitative analysis Analysis to determine how much of the analyte is present.
quantized Occurring in discrete steps.

rate equation The equation representing the relationship between the reaction rate and the concentration of reactants or products.
redox reaction A reaction in which one material is oxidized and another reduced.
reduction The gain of electrons.
resonance forms Two or more equivalent Lewis structures having the same arrangements of atoms but different arrangements of electrons.

salt An ionic compound resulting from the reaction of an acid and a base.
saturated A solution containing the maximum possible amount of a given solute.
solubility constant The equilibrium constant corresponding to the dissolution of a slightly soluble salt at a given temperature in a given solvent.
spectrum The compound's response to an electromagnetic field corresponding to a specific compound.
standard solution A solution with an exactly known concentration of solute.
standardization The process of determining the exact concentration of an analyte in solution, usually by titrating with a standard solution.
sublime Changing directly from the solid to the gas phase (without going through the liquid phase). Carbon dioxide (dry ice) is an example of a material that sublimes at room temperature and pressure.
supersaturated solution A solution containing excess solute relative to its solubility constant.

tared When an object is tared, it is set on top of a balance, and the balance reading is adjusted to zero. Many electronic digital balances can be tared at the touch of a button.
titration Method of determination of the concentration of an unknown by adding a solution of reactant of known concentration until a particular endpoint is reached.
torr A unit of pressure equal to 1/760 of an atmosphere. 1 torr = 1 mm Hg.
transition metal A metal with an incomplete d electron shell in at least one of its common oxidation states. Examples of transition metals include Fe, Ni, Cr, Co, Pt, and Au.
transition state The (usually unstable) material that is formed on the reaction pathway between reactants and products.
transmittance The percentage of incident light that is not absorbed (that is transmitted) by a given sample.
triturate To agitate by stirring, usually with a glass rod or spatula.

valence electron An electron in the outermost shell of an atom, a bonding electron.
vapor pressure The partial pressure of a substance in equilibrium with its liquid or solid.

weak acid An acid that is only partially ionized in a given solution.

Zvi The first name of the first author of this textbook. Hebrew for "deer"—in case you were wondering.